D1606718

POWER SYSTEM QUALITY ASSESSMENT

POWER SYSTEM QUALITY ASSESSMENT

J Arrillaga, N R Watson

University of Canterbury, Christchurch, New Zealand

S Chen

Nanyang Technological University, Singapore

JOHN WILEY & SONS

Chichester • New York • Weinheim • Brisbane • Singapore • Toronto

National 01243 779777
International (+44) 1243 779777

e-mail (for orders and customer service enquiries): cs-books@wiley.co.uk

Visit our Home Page on http://www.wiley.co.uk or http://www.wiley.com

Other Wiley Editorial Offices

John Wiley & Sons, Inc., 605 Third Avenue,
New York, NY 10158-0012, USA

Wiley-VCH Verlag GmbH,
Pappelallee 3, D-69469 Weinheim, Germany

Jacaranda Wiley Ltd, 33 Park Road, Milton,
Queensland 4064, Australia

John Wiley & Sons (Asia) Pte Ltd, 2 Clementi Loop #02-01,
Jin Xing Distripark, Singapore 129809

John Wiley & Sons (Canada) Ltd, 22 Worcester Road,
Rexdale, Ontario M9W 1L1, Canada

Library of Congress Cataloguing-in-Publication Data

Arrillaga, J.
Power system quality assessment / J. Arrillaga, N.R. Watson, S. Chen.
p. cm.
Includes bibliographical references (p.)
ISBN 0-471-98865-0 (cased : alk. paper)
1. Electric power system stability. 2. Electric power systems—Quality control. I. Watson, N. R. II. Chen, S. III. Title.
TK1010.A77 2000
621.31—dc21 99-38990
CIP

British Library Cataloguing in Publication Data

A catalogue record for this book is available from the British Library

ISBN 0 471 98865 0

Typeset in 10/12pt Times by Laser Words, Madras, India
Printed and bound in Great Britain by Bookcraft (Bath) Ltd
This book is printed on acid-free paper responsibly manufactured from sustainable forestry in which at least two trees are planted for each one used for paper production

CONTENTS

PREFACE

The dependence of modern life upon the continuous supply of electrical energy makes system reliability and power quality topics of utmost importance in the power systems area.

Although the subject of power quality has not been clearly defined so far, an increasing number of problems are being discussed under its banner. New restrictions are being continually added under the general heading of electromagnetic compatibility to meet the critical requirements of electronic-controlled loads and limit the otherwise ever-increasing content of waveform distortion created by the proliferation of non-linear plant components in modern power systems.

The design and maintenance of high power quality levels relies, first, on the predictive ability of system simulation and later on the use of accurate monitoring and diagnostic techniques. System simulation, monitoring and state estimation are thus the basis of power quality assessment.

To some extent, these topics have been described in previous books and, particularly, in chronological order, *Power System Harmonics* (John Wiley & Sons, 1985), *Electrical Power Quality* (Stars in a Circle Publications, 1991), *Electrical Power Systems Quality* (McGraw-Hill, 1997) and *Power System Harmonic Analysis* (John Wiley & Sons, 1997). The first of these described the distorting causes and effects in power systems, as well as the techniques then available to predict, monitor and eliminate the harmonic content. The next two covered more specifically the topic of power quality. In particular, Dr G.T. Heydt in the book *Electric Power Quality* presented a most valuable contribution to the understanding of power quality issues. The exclusive object of *Power System Harmonic Analysis* was the simulation of the interaction between non-linear plant components and the linear power network.

Power quality has now become an important part of international power conferences and is the specific topic of at least two held biennially, i.e. the International Conference on Harmonics and Quality of Power (ICHQP) and Power Quality Assessment (PQA). The topic is also under continuous development by working groups of the International Electrotechnical Commission (IEC), International Conference of Large Electrical Networks (CIGRE) and the Institute of Electrical and Electronics Engineers (IEEE).

As the name indicates, the sole purpose of the present book is power quality assessment and this is motivated by a growing need to design systems and interconnections with acceptable quality standards, and to monitor the quality of electrical energy exchanges in the new deregulated environment in order to allocate responsibilities for possible infringements.

The authors would like to acknowledge the valuable assistance received directly or indirectly from their present and previous colleagues of the Power Systems Group at the University of Canterbury, New Zealand. In particular, G Bathurst, P B Bodger, M B Dewe, Z P Du, A Miller, B C Smith and A R Wood. Finally, they wish to thank Mrs G M Arrillaga for her active participation in the preparation of the manuscript.

1

INTRODUCTION

1.1 General Purpose

The term *quality* is sometimes used as synonymous with supply reliability [1,2] to indicate the existence of an adequate and secure power supply. A broader definition [3] has described *service quality*, encompassing the three aspects of reliability of supply, quality of power offered and provision of information. Judging by the content of the innumerable contributions to the topic in recent years, *power quality* is generally used to express the quality of the voltage. With the expansion of power electronic control in the transmission and utilisation of electrical energy, there is increasing acceptability of the latter interpretation.

Much of the early work in this area has been concerned with harmonics [4,5]. While harmonics distortion is an increasing quality problem, a wider power quality concept is needed that includes non-periodical and transient deviations from the ideal waveforms. Such deviations are used to assess Electromagnetic Compatibility (EMC), a subject concerned with the satisfactory operation of components and systems without interfering with or being interfered by other system components. As the power system is the conducting vehicle for possible interference between consumers, an important aspect of power system quality is the system ability to transmit and deliver electrical energy to the consumers within the limits specified by the EMC standards (those are discussed in Chapter 2).

To set the scene, this chapter introduces the *ultimate* purpose of the book, i.e. the maintenance of power system quality, with a brief description of the main deviations from the ideal waveforms and their effect on system operation. This is followed by a brief introduction to the monitoring and estimation techniques used in power system quality assessment, the specific object of the remaining chapters.

1.2 Disturbances

In the context of power quality, a disturbance is a temporary deviation from the steady-state waveform caused by faults of brief duration or by sudden changes in the power system. The disturbances considered by the International Electrotechnical Commission include voltage dips, brief interruptions, voltage increases, and impulsive and oscillatory transients.

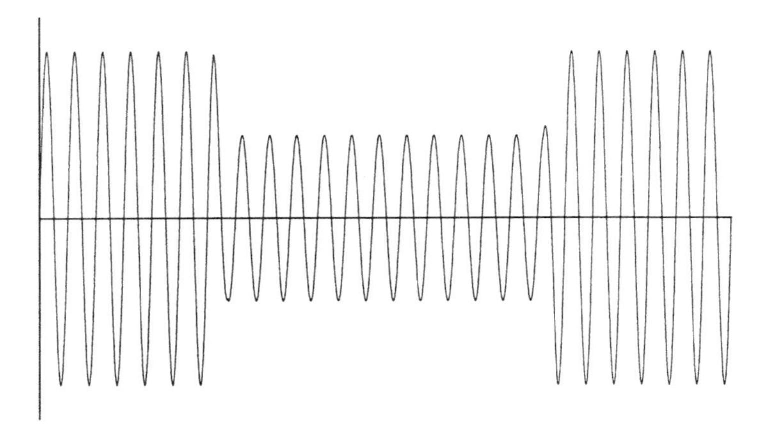

Figure 1.1 Voltage sag

1.2.1 Voltage Dips (sags)

A voltage dip is a sudden reduction (between 10% and 90%) of the voltage, at a point in the electrical system, such as shown in Figure 1.1, and lasting for 0.5 cycle to several seconds. Dips with durations of less than half a cycle are regarded as transients.

A voltage dip may be caused by switching operations associated with a temporary disconnection of supply, the flow of heavy current associated with the start of large motor loads or the flow of fault currents. These events may emanate from customers' systems or from the public supply network. The main cause of momentary voltage dips is probably the lightning strike.

In terms of duration, dips tend to cluster around three values: 4 cycles (the typical clearing time for faults) 30 cycles (the instantaneous reclosing time for breakers), and 120 cycles (the delayed reclosing time of breakers). The effect of a voltage dip on equipment depends on both its magnitude and its duration; in about 40% of the cases observed to date, they are severe enough to exceed the tolerance standard adopted by computer manufactures.

Possible effects are: extinction of discharge lamps; incorrect operation of control devices; speed variation or stopping of motors; tripping of contactors; computer system crash; or commutation failure in line commutated inverters.

Possible solutions against voltage sags are the use of uninterrupted power supplies or power conditioners.

1.2.2 Brief Interruptions

Brief interruptions can be considered as voltage sags with 100% amplitude (see Figure 1.2). The cause may be a blown fuse or breaker opening and the effect an expensive shutdown. For instance, supply interruptions of a few cycles (in the case of a glass factory) or a few seconds (at a major computer centre) may cost hundreds of thousands of dollars. The main protection of the customer against such events is the installation of uninterruptible power supplies.

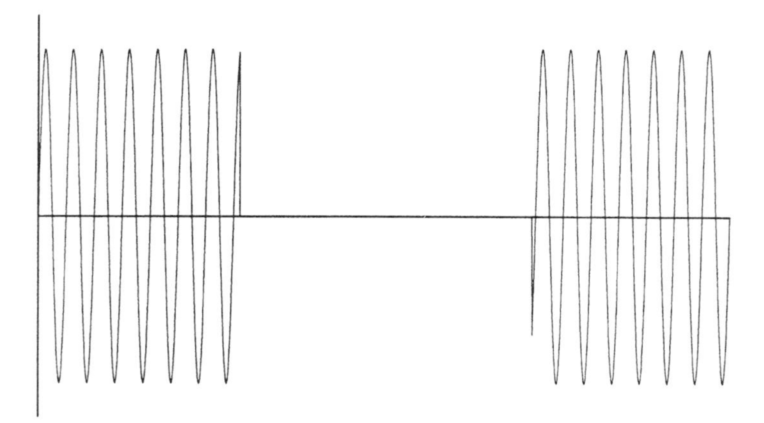

Figure 1.2 Voltage interruption

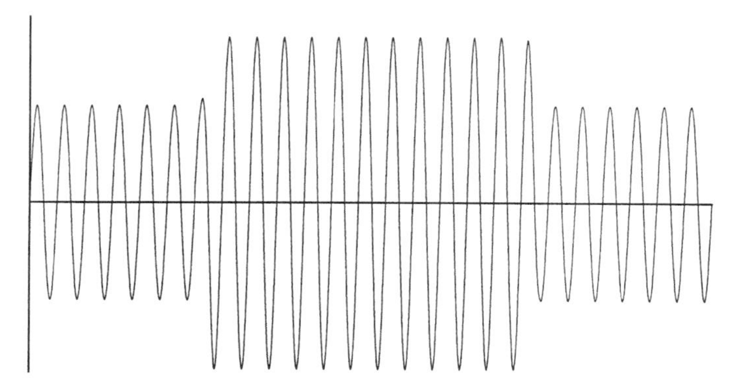

Figure 1.3 Voltage swell

1.2.3 Brief Voltage Increases (Swells)

Voltage swells, shown in Figure 1.3, are brief increases in r.m.s. voltage that sometimes accompany voltage sags. They appear on the unfaulted phases of a three-phase circuit that has developed a single-phase short circuit. They also occur following load rejection.

Swells can upset electric controls and electric motor drives, particularly common adjustable-speed drives, which can trip because of their built-in protective circuitry. Swells may also stress delicate computer components and shorten their life.

Possible solutions to limit this problem are, as in the case of sags, the use of uninterruptible power supplies and conditioners.

1.2.4 Transients

Voltage disturbances shorter than sags or swells are classified as transients and are caused by sudden changes in the power system [6].

According to their duration, transient overvoltages can be divided into switching surge (duration in the range of milliseconds), and impulse spike (duration in the range of microseconds).

Surges are high-energy pulses arising from power system switching disturbances, either directly or as a result of resonating circuits associated with switching devices. They also occur during step load changes.

In particular, capacitor switching can cause resonant oscillations leading to an overvoltage some three to four times the nominal rating, causing tripping or even damaging protective devices and equipment. Electronically based controls for industrial motors are particularly susceptible to these transients.

Figure 1.4 shows the development of a switching oscillation from the initial clearing instant and trapping of the charge on the capacitor to a subsequent restrike.

Impulses, shown in Figure 1.5, result from direct or indirect lightning strokes, arcing, insulation breakdown, etc.

Protection against surges and impulses is normally achieved by surge-diverters and arc-gaps at high voltages and avalanche diodes at low voltages.

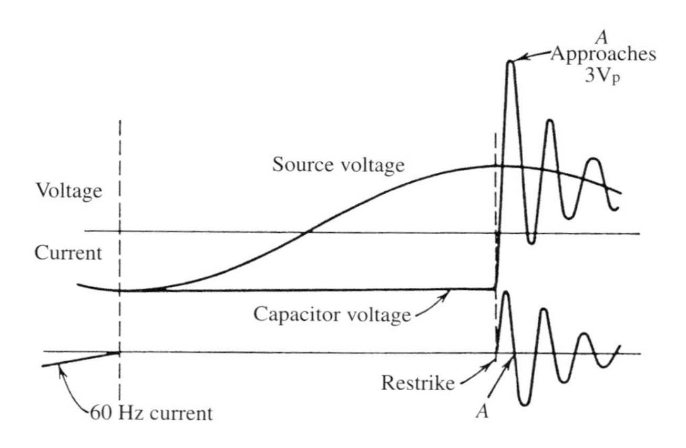

Figure 1.4 Capacitance switching with a restrike at peak voltage [6]

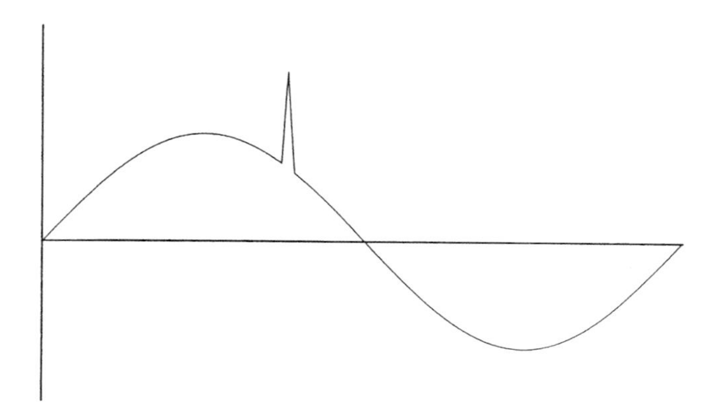

Figure 1.5 Impulse

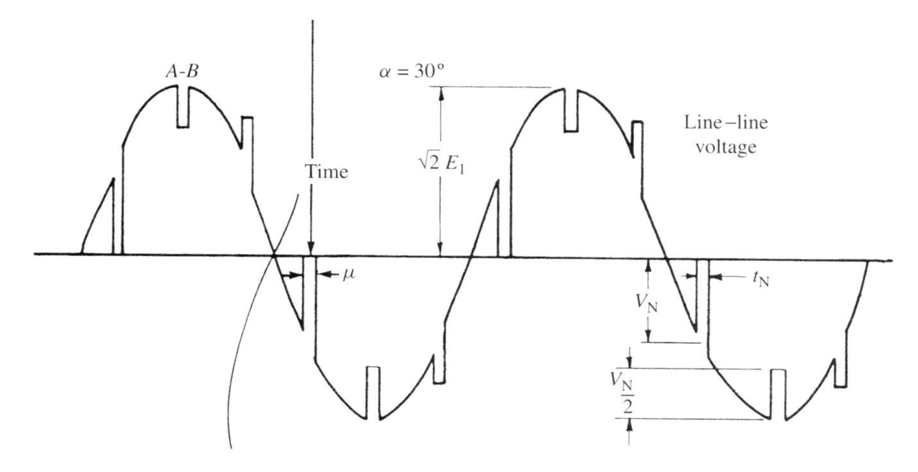

Figure 1.6 Voltage notches

Faster transients in nanoseconds due to electrostatic discharges, an important category of EMC, are not normally discussed under power quality.

1.2.5 Voltage Notches

Notching is a periodic transient occurring within each cycle as a result of the phase-to-phase short circuits caused by the commutation process in a.c.–d.c. converters. Being periodic, this disturbance is already characterised by the harmonic spectrum of the voltage waveform. However, the sharp edges created by the switching instants also contain high frequency oscillations that affect the insulation co-ordination of the plant and can give rise to radiated interference. This effect can be reduced by the provision of damper circuits (snubbers) across the switching devices.

An example of the line voltage waveform (with perfect snubbing) on the secondary side of a six-pulse converter transformer is shown in Figure 1.6, where μ is the notch width corresponding to the commutation overlap.

Notches can upset electronic equipment and damage inductive components by their high level of voltage rise and the additional zero crossing of the mains voltage. However, most of the high frequency content of notches not under the utilities' control is filtered by the power transformer at the service entrance, which explains why it is not propagated on MV lines. Most problems caused by notches are, therefore, confined to the customer's own installation.

1.3 Unbalance

Unbalance describes a situation, as shown in Figure 1.7, in which either the voltages of a three-phase voltage source are not identical in magnitude, or the phase differences between them are not 120 electrical degrees, or both.

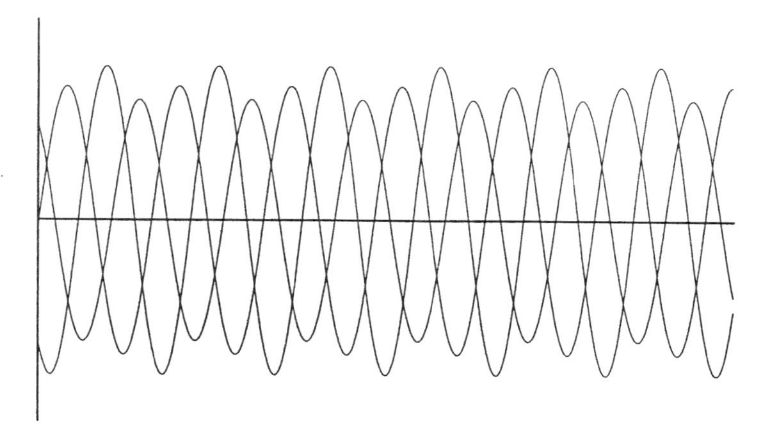

Figure 1.7 Voltage unbalance

The degree of unbalance is usually defined by the proportion of negative and zero *sequence components.*

The main causes of unbalance are single-phase loads (such as electric railways) and untransposed overhead transmission lines.

A machine operating on an unbalanced supply will draw a current with a degree of unbalance several times that of the supply voltage. As a result, the three-phase currents may differ considerably and a temperature rise will take place in the machine. Motors and generators, particularly the large and more expensive ones, may be fitted with protection to detect extreme unbalance. If the supply unbalance is sufficient, the *single-phasing* protection may respond to the unbalanced currents and trip the machine.

Polyphase converters, in which the individual input phase voltages contribute in turn to the d.c. output, are also affected by an unbalanced supply, which causes an undesirable ripple component on the d.c. side, and non-characteristic harmonics on the a.c. side.

1.4 Distortion

Waveform distortion is generally discussed in terms of harmonics, which are sinusoidal voltages or currents having frequencies that are whole multiples of the frequency at which the supply system is designed to operate (e.g. 50 Hz or 60 Hz). An illustration of fifth harmonic distortion is shown in Figure 1.8. When the frequencies of these voltages and currents are not an integer of the fundamental they are termed interharmonics.

Both harmonic and interharmonic distortion is generally caused by equipment with non-linear voltage/current characteristics.

In general, distorting equipment produces harmonic currents which in turn cause harmonic voltage drops across the impedances of the network. Harmonic currents of the same frequency from different sources add vectorially.

In the UK, the fifth harmonic has been identified as the harmonic order exhibiting the highest peak levels of high voltage systems, with values between 2.5% and 3.0% at some locations. The fifth also most frequently presents the highest mean harmonic

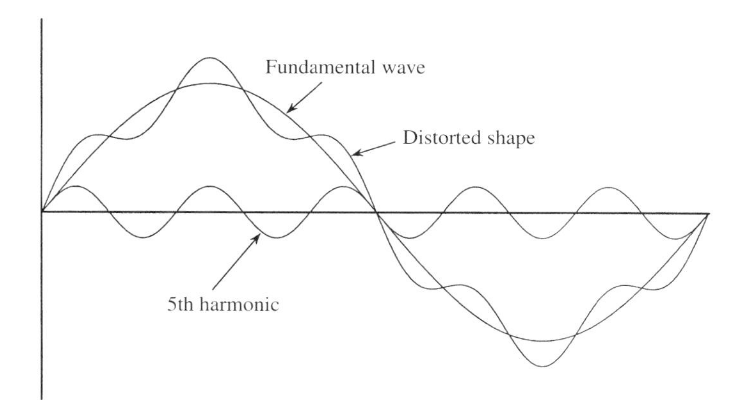

Figure 1.8 Example of a distorted sine wave

levels, a characteristic which has been found to be consistent both geographically and with time. This feature illustrates the potential impact on future transmission equipment immunity levels of emissions which originate in the low voltage network.

Whilst seasonal variation in harmonic magnitudes has been apparent at individual sites, the pattern is not consistent. Neither was evidence discovered that a particular geographical area is prone to high harmonic voltages. It is believed that, in general, harmonic levels tend to be influenced primarily by local and immediately adjacent conditions rather than wider zonal effects.

The main detrimental effects of harmonics are [4].

- maloperation of control devices, mains signalling systems and protective relays,
- extra losses in capacitors, transformers and rotating machines,
- additional noise from motors and other apparatus,
- telephone interference,
- the presence of power factor correction capacitors and cable capacitance. These can cause shunt and series resonances in the network producing voltage amplification even at a remote point from the distorting load.

As well as the above, interharmonics can perturb *ripple control signals* and at subharmonic levels can cause flicker.

To keep the harmonic voltage content within the recommended levels the main solutions in current use are:

- the use of high pulse rectification (e.g. smelters and *high-voltage d.c.* (HVd.c.) converters),
- passive filters, either tuned to individual frequencies or of the band-pass type,
- active filters and conditioners.

The harmonic sources can be grouped into three categories according to their origin, size and predictability, i.e. small and predictable (domestic and residential), large and random (arc furnaces) and large and predictable (static converters).

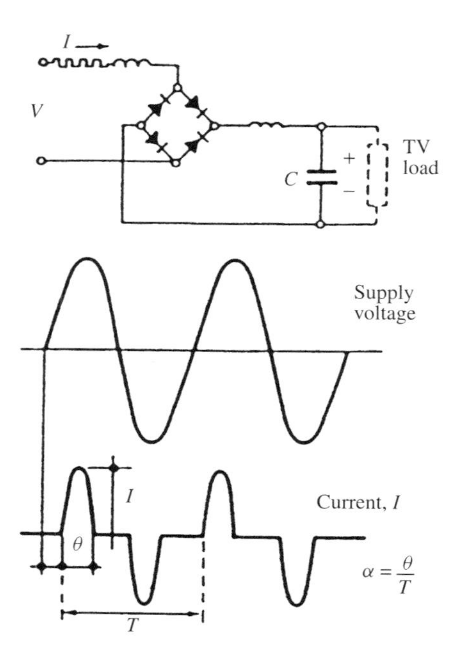

Figure 1.9 Supply voltage and current waveforms of a television set

1.4.1 Small Sources

The residential and commercial power system contains large numbers of single-phase converter-fed power supplies with capacitor output smoothing, such as TVs and PCs, requiring the current waveform shown in Figure 1.9. Although their individual rating is not significant, there is little diversity in their operation and their combined effect produces considerable odd-harmonic distortion. The gas discharge lamps add to that effect since they produce the same harmonic components.

Figure 1.10 illustrates the harmonic current spectrum of a typical high-efficiency lamp. The total harmonic distortion of such lamps can be between 50% and 150% of the fundamental current.

The phase displacements of the third harmonic currents with respect to the fundamental voltage waveform are the same in the threephases of the transformer, i.e. they are of zero sequence. This fact must be taken into account when rating transformer neutrals and calculating telephone interference.

1.4.2 Large and Random Sources

The most common and damaging load of this type is the arc furnace. Arc furnaces produce random variations of harmonic and interharmonic content which is uneconomical to eliminate by conventional filters.

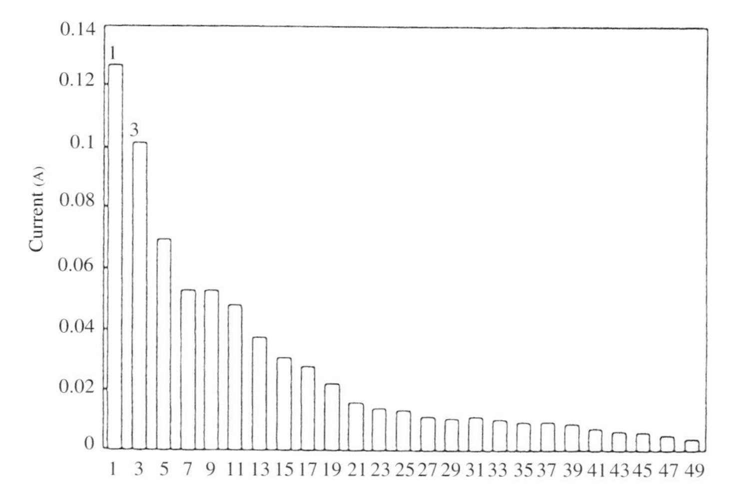

Figure 1.10 Harmonic spectrum of a high-efficiency lamp

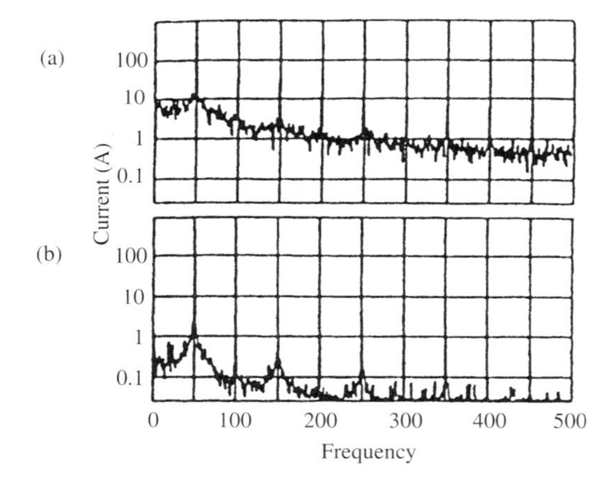

Figure 1.11 Typical frequency spectra of arc furnace operation: (a) during fusion; (b) during refiring

Figure 1.11 shows a snapshot of the frequency spectra produced by an arc furnace during the melting and refining processes, respectively. These are greatly in excess of the recommended levels.

These loads also produce voltage fluctuations and flicker. Connection to the highest possible voltage level and the use of series reactances are among the measures currently taken to reduce their impact on power quality.

1.4.3 Static Converters

Large power converters, such as those found in smelters and HVd.c. transmission are the main producers of harmonic current and considerable thought is given to their local elimination in their design.

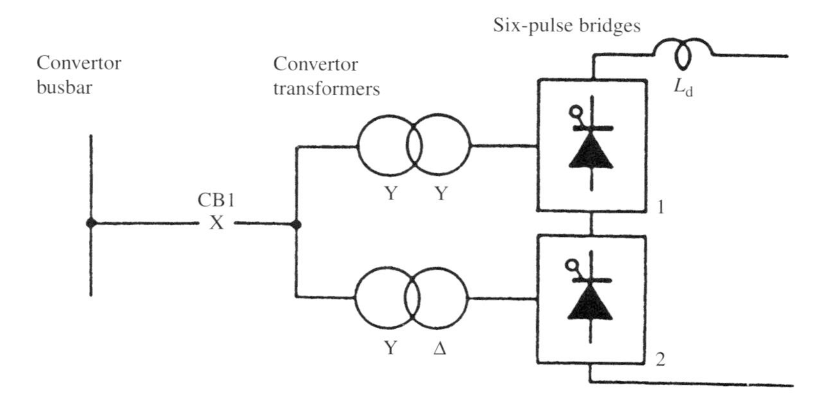

Figure 1.12 The 12-pulse converter

The standard configuration for industrial and HVd.c. applications is the 12 pulse converter, shown in Figure 1.12. The *characteristic* harmonic currents for the configuration are of orders $12k + 1$ (of positive sequence) and $12k - 1$ (of negative sequence) and their amplitudes are inversely proportional to the harmonic order, as shown by the spectrum of Figure 1.13(b) which corresponds to the time waveform of Figure 1.13(a). These are, of course, maximum levels for ideal system conditions, i.e. with an infinite (zero impedance) a.c. system and a perfectly flat direct current (i.e. infinite smoothing reactance).

When the a.c. system is weak and the operation not perfectly symmetrical *uncharacteristic harmonics* appear.

While the characteristic harmonics of the large power converter are reduced by filters, it is not economical to reduce the uncharacteristic harmonics in that way and, therefore, even a small injection of these harmonic currents can, via parallel resonant conditions, produce very large voltage distortion levels.

An example of uncharacteristic converter behaviour is the presence of a fundamental frequency on the d.c. side of the converter, often induced from a.c. transmission lines in the proximity of the d.c. line, which produces second harmonic and direct current on the a.c. side.

Even harmonics, particularly the second, are very disruptive to power electronic devices and are, therefore, heavily penalised in the regulations.

The flow of d.c. current in the a.c. system is even more distorting, the most immediate effect being asymmetrical saturation of the converter's or other transformers with a considerable increase in even harmonics which, under certain conditions, can lead to *harmonic instabilities* [7].

Another common example is the appearance of triplen harmonics. Asymmetrical voltages, when using a common firing angle control for all the valves, result in current pulse width differences between the three phases which produces triplen harmonics. To prevent this effect, modern large power converters use the Equidistant Firing concept instead [8]. However, this controller cannot eliminate second harmonic amplitude modulation of the d.c. current which, via the converter modulation process, returns third harmonic current of positive sequence. This current can flow through the converter

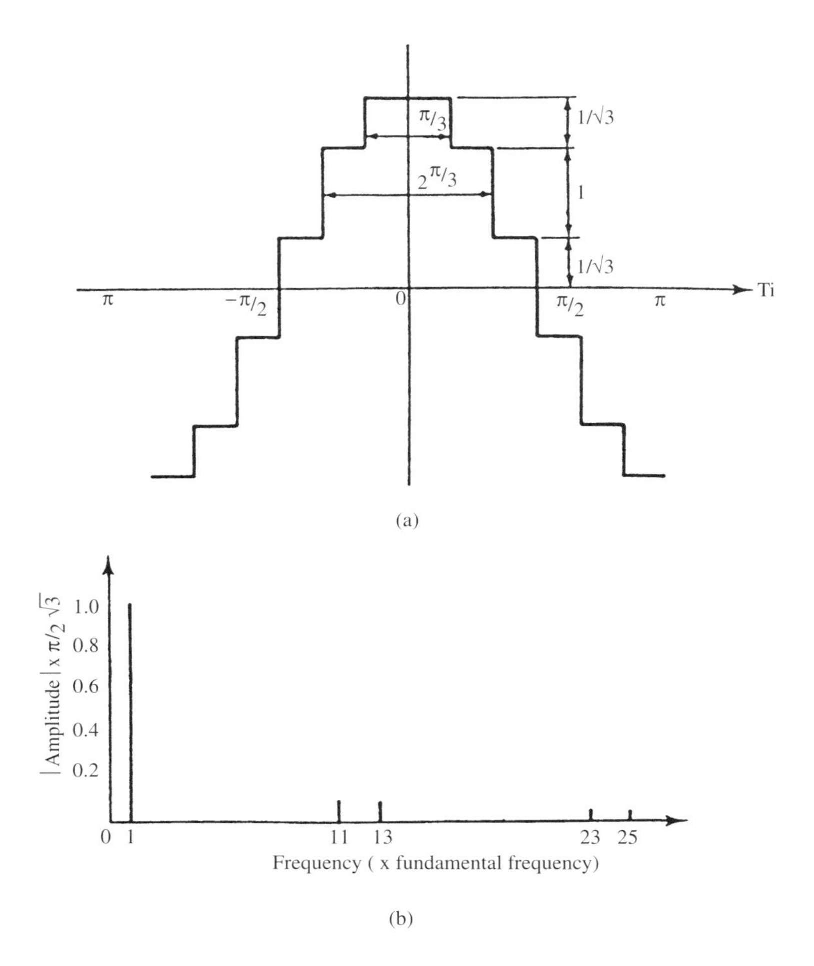

Figure 1.13 The 12-pulse converter: (a) current waveform; (b) spectrum

transformer regardless of its connection and penetrate far into the a.c. system. Again, the presence of triplen harmonics is discouraged by stricter limits in the regulations.

The frequency-conversion process used in earlier railway systems is an important source of quasi-stationary interharmonics. Most of these systems operate at $16\frac{2}{3}$ Hz and this interharmonic frequency, as well as its harmonic components, especially $83\frac{1}{3}$ and $116\frac{2}{3}$, propagate into the 50/60 Hz supply system.

1.5 Voltage Fluctuations

Voltage fluctuations are divided in two broad categories:

1 Step voltage changes, regular or irregular in time, such as those produced by welding machines, rolling mills, mine winders, (Figures 1.14(a) (b)).

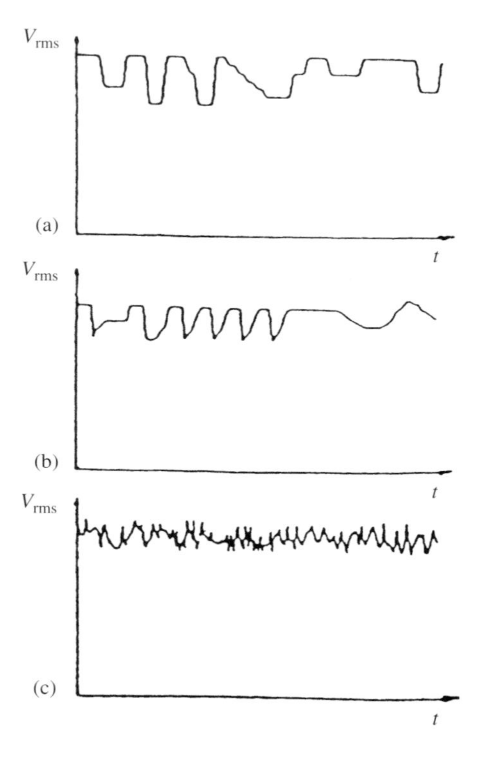

Figure 1.14 Voltage fluctuations

2 Cyclic or random voltage changes produced by corresponding variations in the load impedance, the most typical case being the arc furnace load (Figure 1.14(c)).

Possible effects of voltage fluctuations include the degradation of performance in equipment using capacitors, control system disturbances and instabilities of internal voltage and currents in electronic equipment.

Generally, since voltage fluctuations have an amplitude not exceeding $\pm 10\%$, most equipment is not affected by this type of disturbance. The main problem is their effect on flicker, as discussed in the following section.

1.6 Flicker

Flicker has been described [9,10] as '*the impression of fluctuating luminance or colour occurring when the frequency of the variation of the light stimulus lies between a few hertz and the fusion frequency of images*'. This is a very loose definition considering that '*the fusion frequency of images*' varies from person to person and depends on many factors.

Fluctuations in the system voltage (more specifically in its r.m.s. value) can cause perceptible (low frequency) light flicker depending on the magnitude and frequency of the variation. Power system engineers hence call this type of disturbance *voltage flicker* but often it is just shortened to *flicker*. Figure 1.15 represents a simple case of voltage flicker, where the a.c. voltage is amplitude modulated (AM) by a sine wave seen as the envelope of the voltage waveform. The voltage can be expressed as

$$v(t) = V(1 + m\cos\omega_{\mathrm{m}}t)\cos\omega_0 t, \tag{1.1}$$

where ω_0 is the a.c. system fundamental frequency, ω_{m} the frequency of the modulating sine wave, V the nominal amplitude of the a.c. voltage and $m(= \Delta V/2V)$ the modulation factor.

The above equation can be expanded into

$$v(t) = V\left[\cos\omega_0 t + \frac{m}{2}\cos(\omega_0 + \omega_{\mathrm{m}})t + \frac{m}{2}\cos(\omega_0 - \omega_{\mathrm{m}})t\right]. \tag{1.2}$$

Equation (1.2) shows that the AM signal consists of three spectral components which, by analogy with telecommunication theory, are the carrier (ω_0) and two sideband components ($\omega_0 + \omega_{\mathrm{m}}$ and $\omega_0 - \omega_{\mathrm{m}}$). It is important to understand the terminology used. The voltage flicker frequency in Figure 1.15 is 7 Hz, which manifests itself as a 43 Hz and a 57 Hz component in the spectrum of the voltage, i.e. *flicker frequency* refers to the spectral components of the enveloping waveform and not to the frequency components of the voltage.

The range of modulation frequency that causes noticeable flicker is in the 0–30 Hz band.

Non-periodic events can also cause perceptible light flicker. Any potentially perceptible change in brightness should therefore be termed *(light) flicker*, broadening the given definitions further by extending the lower frequency limit to non-periodic disturbances.

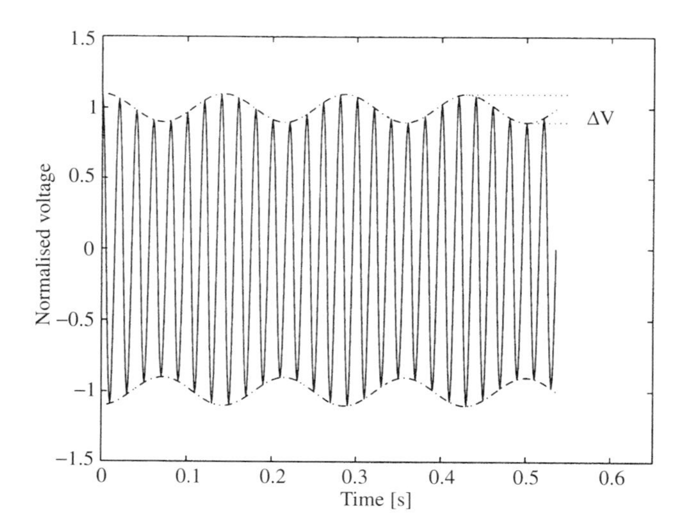

Figure 1.15 Sinusoidal voltage flicker. The 50 Hz a.c. voltage (solid) is amplitude modulated by a 7 Hz sinusoidal waveform (dash dotted). The modulation factor is $m = \Delta V/2V$, where V is the amplitude of the unmodulated voltage

Similarly, the meaning of *voltage flicker* should not be restricted to disturbances like the described amplitude modulation. Switching events, interharmonics and modulated harmonics can also lead to light flicker. The term *voltage flicker* is also used when effects on devices other than electric lamps are considered where light flicker is of no concern.

1.6.1 Causes of Flicker

The main causes of flicker are loads drawing large and highly variable currents. Due to the impedance of the power system (generators, transformers and transmission lines) these changes produce amplitude modulation of the voltage at the load bus and even at remote buses. As an example, Figure 1.16 represents the voltage recorded at the transformer secondary supplying an arc furnace installation and its corresponding spectrum.

Another common source of flicker is the starting of electric motors. To produce a sufficient starting torque they may draw a current many times their full load running current. Similarly, their operation in applications that require an irregular torque is problematic. Critical motor applications range from household appliances (e.g. washing machines, drills, mixers) up to more powerful equipment such as heat pumps or steel rolling mills. The flicker waveshapes are mainly triangular and may be periodic as for reciprocating compressors (e.g. refrigerators) or non-periodic in the case of infrequent starts of large motors.

These flicker sources lead to electric lamp luminance fluctuations by means of amplitude modulation of the supply voltage. Interharmonics present in the voltage spectrum can, however, also produce low frequent light flicker [11,12]. This occurs by beating with either the a.c. system fundamental, other harmonics or interharmonics that might be present. For example, a 100 Hz harmonic together with a 90 Hz interharmonic will lead to a 10 Hz light flicker.

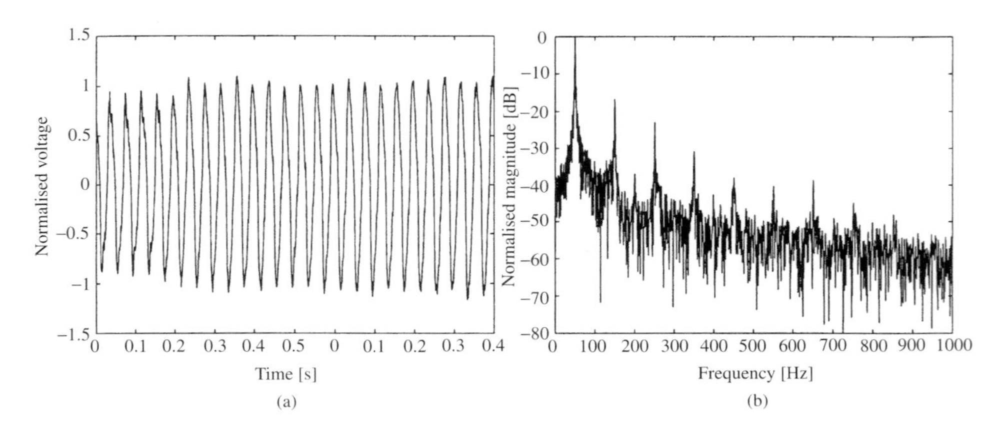

Figure 1.16 Typical arc furnace voltage flicker measured at the supply transformer secondary: (a) time series displaying fluctuations in the envelope; (b) spectrum showing odd harmonics and interharmonics

1.6.2 Effects of Flicker

The flickering of electric lights causes annoyance to human observers. Its detrimental effect on human health and well being is usually more critical than possible effects on other devices. This is due to the high sensitivity of the human eye to changes in brightness. Incandescent light flicker is perceptible for modulation factors (defined in section 1.6) as small as 0.15%.

Even though the gas discharge light (UV) responds to fluctuations in voltage instantaneously, fluorescent lamps are considered to be less sensitive to voltage flicker because of the effect of the phosphor coating and their operation with ballast circuits.

Modern compact fluorescent lamps (CFLs) operate at high frequencies using solid state ballasts. Consequently, the flicker due to a.c. operation is not perceptible at any stage of the human eye–brain system and complaints are fewer [13].

The high-frequency light flicker from fluorescent tubes has, however, been associated with headaches and eyestrain showing that it is perceived to a higher degree than generally recognised [14]. This gives flicker a somewhat broader meaning.

Other reported effects of voltage flicker include: reduced life of electronic, incandescent, fluorescent and cathode-ray tube (CRT) devices; malfunction of phase-locked loops (PLLs) and loss of synchronism in uninterruptible power supplies (UPSs); maloperation of electronic controllers and protection devices.

Sensitive electronic devices can be (and usually are) protected from the adverse effects of voltage flicker at a small cost. This approach, however, is not viable for electric lights, since they exist in such large numbers and are relatively cheap to produce.

1.7 Quality Assessment

In an existing power system the waveforms can be obtained from measurements at points of common coupling (PCC) and their frequency components are then derived using signal processing. This process is at the core of quality assessment, the purpose of this book, and is discussed in Chapter 5. However, progress in quality assessment has been relatively slow due to the cost and limitations of existing transducer and monitoring equipment.

The transducer's analogue signals, normally converted to digital form, are subjected to either time or frequency domain processing, depending on the type of information required and its subsequent use. For active power conditioning, essential to some industrial processes, the information is derived directly from the waveforms in real time. The same applies to disturbance detections to obtain the magnitude and duration of the event.

On the other hand, the processing of steady or quasi-steady state waveform distortion requires some form of signal analysis in the frequency domain. Processing speed is less critical in this case since the results are normally used to provide statistical information. Until recently, continuous real time processing had been beyond the capability of commercial instrumentation, and discrete intervals of time domain information had to be stored on tape for subsequent off-line processing. It will be shown in Chapter 5 that the use of dedicated Digital Signal Processors (DSPs) and parallel processing can provide sufficient speed for continuous signal processing in real time [15].

The addition of non-linear loads and general reinforcements of the power system require extensive computer simulation to predict their effect on power quality. This is a well-documented field [15] that plays an important role in modern power system design.

In general, the immediate effect of the load non-linearity is current waveform distortion, as discussed in Section 1.4.3. For system planning and design purposes the current waveforms of distorting loads are derived from knowledge of their non-linear characteristics and their corresponding frequency components. With the non-linearities represented as current sources, models of the network components in the frequency domain are used to derive information of voltage and current distortion throughout the network.

Waveform analysis and system modelling are two important components in the assessment of power quality and are considered in Chapters 4 and 6, respectively.

1.7.1 Power Quality State Estimation

Generally, power quality assessment is not carried out regularly and systematically; instead, *ad hoc* measurement procedures are used to ensure that the PCC voltage waveform meets the specified distortion levels. A local solution, however, affects the rest of the system to some degree and in a positive or negative way. It is therefore far from ideal, both in terms of overall cost and performance. The eventual combination of local monitoring and global assessment should provide cost-effective and better technical solutions.

The global assessment also has practical limitations, owing to the restricted number of monitoring points and to insufficient knowledge of the system configuration and characteristics. An optimal location of the given number of instruments and monitoring channels, combined with computer system simulation, can be used to derive acceptable information and locate the problem areas. By association with power system state estimation, this type of assessment is referred to as Power Quality State Estimation.

Of course, global assessment requires adequate synchronisation of the signals derived at the individual locations; this can easily be achieved by GPS time stamping. In the context of power quality, the state estimation concept has only been discussed with reference to harmonic distortion and lacks general acceptability by the industry. However, considering its potential applications, it is expected to play an important role in future power quality assessment. Therefore, the book would not be complete without reference to harmonic state estimation (HSE) [16], a subject discussed in Chapter 7.

1.8 References

1. Glossary of terms and definitions concerning electric power transmission system access and wheeling, *IEEE Power Engineering Society*, 96, TP 110-0.
2. Billington, R and Allan, R N, (1988). *Reliability Assessment of Large Electric Power Systems*, Kluwer Academic Publishers.

3. Meeuwsen, J and Kling, W, (1997). The influence of different network structures on power supply reliability, *Quality of Power Supply*, ETG Conference, November.
4. Arrillaga, J, Bradley, D and Bodger, P S, (1985). *Power System Harmonics*, John Wiley and Sons, London.
5. Arrillaga, J, Smith, B C, Watson, N R and Wood, A R, (1997). *Power System Harmonic Analysis*, John Wiley and Sons, London.
6. Greenwood, A, (1971). *Electrical Transients in Power Systems*, Wiley Interscience, USA.
7. Chen, S, Wood, A R and Arrillaga, J, (1996). HVdc converter transformer core saturation instability: A frequency domain analysis, *IEE on Proceedings of the Generation Transmission and Distribution* **143**(1), pp. 75–81.
8. Ainsworth, J, (1968). The phase-locked oscillator. A new control system for controlled static converters, *IEEE Transactions*, **PAS-87**, pp. 859–865.
9. IEEE, (1984). Standard Dictionary of Electrical And Electronics Terms, Std. 100, IEEE.
10. IEC, (1983). Multilingual Dictionary of Electricity, IEC.
11. Mombawer, W, (1990). Flicker caused by interharmonics, *etz Archiv*, **12**(12), pp. 391–396.
12. Peretto, L and Emanuel, A E, (1997). A theoretical study of the incandescent filament lamp performance under voltage flicker, *IEEE Transactions on Power Delivery*, **12**(1), pp. 279–285.
13. Wilkins, A J, Nimmo-Smith, I, Slater A I and Bedocs, L, (1989). Fluorescent lighting, headaches and eyestrain, *Lighting Research and Technology*, **21**(1), pp. 11–18.
14. Eysel, U T and Burandt, U, (1984). Fluorescent tube light evokes flicker responses in visual neurons, *Vision Research*, **24**(9), pp. 943–948.
15. Miller, A J and Dewe, M B, (1992). Multichannel continuous harmonic analysis in real time, *IEEE Transactions on Power Delivery*, **7**(4), pp. 1913–1919.
16. Du, Z P, Arrillaga, J and Watson, N R, (1996). Continuous harmonic state estimation of power systems, *IEE on Proceedings of the Generation, Transmission and Distribution*, **143**(4), pp. 329–336.

2

POWER QUALITY INDICES AND STANDARDS

2.1 Introduction

The development of standards and guidelines is centred around the following objectives [1]:

- description and characterisation of the phenomena.
- major sources of power quality problems.
- impact on other equipment and on the power system.
- mathematical description of the phenomena using indices or statistical analysis to provide a quantitative assessment of its significance.
- measurement techniques and guidelines.
- emission limits for different types and classes of equipment.
- immunity or tolerance level of different types of equipment.
- testing methods and procedures for compliance with the limits
- mitigation guidelines.

The international body widely recognised as the curator of electric power quality standards is the International Electrotechnical Commission or Commission Electrotechnique Internationale IEC which is based in Geneva. IEC has defined a series of standards, called Electromagnetic Compatibility (EMC) Standards, to deal with power quality issues. This series is published in separate parts according to the following structures.

1 *General (IEC 61000-1-x)*: the general section introduces and provides fundamental principles on EMC issues and describes the various definitions and terminologies used in the standards.

2 *Environment (IEC 61000-2-x)*: this part describes and classifies the characteristics of the environment or surroundings where equipment will be used. It also provides guidelines on compatibility levels for various disturbances.

3 *Limits (IEC 61000-3-x)*: this section defines the maximum levels of disturbances caused by equipment or appliances that can be tolerated within the power system. It also defines the immunity limits for equipment sensitive to EMC disturbances.

4 *Testing and Measurement Techniques (IEC 61000-4-x)*: these provide guidelines on the design of equipment for measuring and monitoring power quality disturbances. They also outline the equipment testing procedures to ensure compliance with other parts of the standards.

5 *Installation and Mitigation Guidelines (IEC 61000-5-x)*: this section provides guidelines on the installation techniques to minimise emission as well as to strengthen immunity against EMC disturbances. It also describes the use of various devices for solving power quality problems.

6 *Generic Standards (IEC 61000-6-x)*: these include the standards specific to a certain category of equipment or for certain environments. They contain both emission limits and immunity levels standards.

Although the above structure is fairly new, the majority of the contents have been widely known to the industry in one form or another. Much of the material contained in this series was adopted from standards and guidelines developed by individual countries. Some of the other organisations who have developed their own standards are CENELEC, UIE, IEEE, ANSI, NEMA. These standards are usually very much application-based or specific to a certain environment. For example, the IEEE 141 [2] describes the effect of voltage disturbances on equipment within an industrial area. Some standards may also be based on the capability or state of technology of the monitoring equipment. The majority of the standards concentrate on the emission limits and susceptibility of a particular type or class of equipment or appliance under certain environmental conditions.

2.2 Classification of Power Quality Phenomena

Despite numerous disagreements on what constitutes electric power quality, there is a concerted call to establish a common description of the power quality phenomena and the environment in which they occur. The two most widely referenced standards or guidelines are the IEC EMC series as outlined in section 2.1 and the IEEE 1159 [3].

IEC 61000-2-5: 1995 [4] The IEC has devised a classification system to identify a limited set of parameters and associated values, which may be used to specify performance requirements. The electromagnetic environment in which modern electronic systems are expected to operate without interference is very complex. Three categories of environmental phenomena have been defined to describe all disturbances:

1 low-frequency phenomena (<9 kHz),
2 high-frequency phenomena (>9 kHz),
3 electrostatic discharge (ESD) phenomena.

Each phenomenon is then divided into radiated and conducted disturbances depending on the medium within which they occur. Radiated disturbances occur in

Table 2.1 Principal phenomena causing electromagnetic disturbances

Conducted low-frequency phenomena
• Harmonics, inter-harmonics • Signalling voltages • Voltage fluctuations • Voltage dips and interruptions • Voltage unbalance • Power frequency variations • Induced low frequency voltages • d.c. in a.c. networks
Radiated low-frequency phenomena
• Magnetic fields • Electric fields
Conducted high-frequency phenomena
• Induced CW (continuous wave) voltages or currents • Unidirectional transients • Oscillatory transients
Radiated high-frequency phenomena
• Magnetic fields • Electric fields • Electromagnetic fields — Continuous waves — Transients
Electrostatic discharge phenomena (ESD) Nuclear electromagnetic pulse (NEMP)

the medium surrounding the equipment, while conducted disturbances occur in various metallic media. Table 2.1 lists the types of phenomena according to IEC classifications.

Although it is arguable that all of these phenomena can be considered as power quality issues depending on the application or environment considered, it is generally accepted by the power industry that only the two conducted categories constitute power quality. The boundary between low frequency and high frequency is assumed by the report as being 9 kHz. However, when addressing a type of disturbance prevailing in one frequency range with a small overlap into the other range, the boundary might be slightly shifted to keep the phenomenon within one descriptive range.

IEC 61000-2-1: 1990 [5] This technical report is concerned with the conducted disturbances in the frequency range up to 10 kHz with an extension for mains signalling systems. It describes the environment for those conducted low-frequency phenomena listed in Table 2.1. It provides simple descriptions of the phenomena, their common source or causes and the detrimental effects of these disturbances. This report acts as the reference for specifying compatibility levels for low voltage a.c. distribution systems with a nominal voltage up to 240 V, single-phase or 415 V, three-phase and a nominal frequency of 50 Hz or 60 Hz. These levels are specified in IEC 61000-2-2: 1990 [6].

IEEE 1159: 1995 [3] The IEEE 1159 contains several additional terms related to the IEC terminology. The term *sag* is used as a synonym to the IEC term *dip*. The category *short duration variations* is used to refer to the voltage dips and short interruptions as defined by the IEC. The term *swell* is introduced as an inverse to *sag (dip)*. The category *long duration variations* has been added to deal with ANSI C84.1:1989 limits. The category *noise* has been added to deal with broad-band conducted phenomena. The category *waveform distortion* is used as a container category for the IEC *harmonics, interharmonics*, and *d.c. in a.c. networks* phenomena as well as an additional phenomenon from IEEE 519 [13] called *notching*. Table 2.2 shows the IEEE categorisation of electromagnetic phenomena used for the power quality community.

These phenomena can be described by listing appropriate attributes. For steady-state phenomena, the following attributes can be used:

- Amplitude
- Frequency
- Spectrum
- Modulation
- Source impedance
- Notch depth
- Notch area

For non-steady-state phenomena, other attributes may be required:

- Rate of rise
- Duration
- Rate of occurrence
- Energy potential
- Source impedance

Table 2.2 provides information regarding typical spectral content, duration and magnitude for each category of electromagnetic phenomena. This categorisation, when used with the attributes mentioned above, provide a means clearly to describe an electromagnetic disturbance. The categories and their descriptions are important to be able to classify measurement results and to describe electromagnetic phenomena that can cause power quality problems.

Although the majority of the standards and guidelines are meant for testing certain types or classes of equipment or appliance under certain electromagnetic environments, they are still applicable for analysing the quality of the power system as a whole. This is because the quality of a system is generally not decided on its aggregate, but rather from a single point of common coupling that is of interest or concern to the customer or utility. However, one has to exercise judgement on which standards or indices are appropriate for evaluating the system quality. Generally, the tolerance limits specified for various types or classes of equipment, particularly sensitive equipment such as computer and communication devices, can be used as the basis for determining the quality of a power system.

Table 2.2. Categories and typical characteristics of power system electromagnetic phenomena defined in IEEE 1159:1995

Categories	Typical spectral content	Typical duration	Typical voltage magnitude
1. Transients			
1.1. Impulsive			
1.1.1. Nanosecond	5 ns rise	<50 ns	
1.1.2. Microsecond	1 μs rise	50 ns–1 ms	
1.1.3. Millisecond	0.1 ms rise	>1 ms	
1.2. Oscillatory			
1.2.1. Low frequency	<5 kHz	0.3–50 ms	0–4 p.u.
1.2.2. Medium frequency	5–500 kHz	20 μs	0–8 p.u.
1.2.3. High frequency	0.5–5 MHz	5 μs	0–4 p.u.
2. Short-duration variations			
2.1. Instantaneous			
2.1.1. Sag		0.5–30 cycles	0.1–0.9 p.u.
2.1.2. Swell		0.5–30 cycles	1.1–1.8 p.u.
2.2. Momentary			
2.2.1. Interruption		0.5 cycles–3 s	<0.1 p.u.
2.2.2. Sag		30 cycles–3 s	0.1–0.9 p.u.
2.2.3. Swell		30 cycles–3 s	1.1–1.4 p.u.
2.3. Temporary			
2.3.1. Interruption		3 s–1 min	<0.1 p.u.
2.3.2. Sag		3 s–1 min	0.1–0.9 p.u.
2.3.3. Swell		3 s–1 min	1.1–1.2 p.u.
3. Long-duration variations			
3.1. Interruption, sustained		>1 min	0.0 p.u.
3.2. Undervoltages		>1 min	0.8–0.9 p.u.
3.3. Overvoltages		>1 min	1.1–1.2 p.u.
4. Voltage imbalance		Steady state	0.5%–2%
5. Waveform distortion			
5.1. d.c. offset		Steady state	0%–0.1%
5.2. Harmonics	0–100th H	Steady state	0%–20%
5.3. Interharmonics	0–6 kHz	Steady state	0%–2%
5.4. Notching		Steady state	
5.5. Noise	Broad-band	Steady state	0%–1%
6. Voltage fluctuation	<25 Hz	Intermittent	0.1%–7%
7. Power frequency variations		<10 s	

The electromagnetic phenomena considered in this chapter are limited to those listed in Table 2.2 and their corresponding terms in the IEC standards. Their simplified descriptions, major sources and main effects have been introduced in Chapter 1. This chapter looks into the mathematical descriptions of the phenomena and concentrates on the following items:

- The standards and guidelines that are pertinent to the disturbances,

- Commonly used indices for the characterisation of the disturbances,
- Compatibility levels: equipment emission limits and minimum immunity capabilities.

2.3 Disturbances

Disturbance has been introduced as a temporary deviation from the steady-state waveform. This definition can be used to describe just about anything unfavourable in a power system, but it is used in this chapter to refer to one-off (i.e. non-repetitive) change in the amplitude of the system voltage at the fundamental frequency for a short period of time. This deviation can be a high-frequency phenomenon or a low-frequency phenomenon. High-frequency phenomena are termed transients, while the low-frequency phenomena comprise voltage dips (sags), interruptions and swells.

2.3.1 High-Frequency Disturbances: Transients

The term transient is commonly used to refer to abrupt changes to the voltage or current waveform. The word surge is sometimes used to describe transients caused by lightning strikes. IEEE 100:1992 defines the transient as 'that part of the change in a variable that disappears during transition from one steady-state operating condition to another'. Unfortunately, this definition can be used to describe just about any disturbances that occur in the power system.

IEC 61000-2-5: 1990 The IEC classification divides transients into two groups of unidirectional and oscillatory in accordance with the sources that are responsible for the occurrence of these disturbances. Unidirectional transients are unidirectional in polarity, while the instantaneous value of oscillatory transients changes polarity rapidly. *Oscillatory surges* range from less than 1 kHz caused primarily by capacitor switching, to several megahertz due mainly to local oscillations and disconnect switching. Transients of high-frequency range have limited energy deposition capability, but can have high peak voltages. Those of lower-frequency range can have higher energy deposition capability but lower peak voltages.

High-energy surges are associated with nearby direct lightning discharges or fuse operation. They can be lightning surges originating on overhead distribution systems, lightning surges originating on overhead lines and travelling in cables, or surges generated by fuse operation involving trapped energy in the power system inductance. *Very fast surges* occur as single events, such as electrostatic discharges, or as bursts associated with local load switching. Both involve very little energy but are capable of producing serious interference or upset. Dielectric breakdown is also a source of high-frequency disturbances. *Coupled disturbances* are a type of conducted disturbance originating from radiated waves; the radiated transients can couple into the wiring systems and propagate further into the system.

Surges can sometimes be described by their energy in order to help in the selection of a surge-protection device. However, the energy distribution among the circuit elements depends not only on the impedance of the source (a.c. mains), but also on the impedance of the surge-protection device called upon to divert the surge. The energy delivered to the end equipment is the significant factor, but it depends on the distribution between the source and the load (equipment or surge-diverting protective device or both).

IEEE c62.41: (1991) [7] Transients have a wide variety of waveforms, which depend upon the mechanism of generation. In IEEE c62.41 (1991), they are classified according to their origin, the type of effect on the system or equipment, or according to their electromagnetic characteristics. The two major sources of transients in a power system are:

- environment-produced transients that are the direct or indirect effect of lightning strikes.
- appliance-produced transients resulting from the operation of a mechanical or semiconductor switch; inrush currents during transformer energisation and the effects of transformer core saturation; or faults within a piece of equipment or within the power system.

IEEE c62.41 (1991) describes the different mechanisms by which lightning produces surge voltages on power conductors. Lightning strike produces electromagnetic fields that can induce voltages on nearby conductors. The flow of lightning ground current can couple to the common ground impedance paths of the grounding network, causing voltage differences across its length and breath. The operation of gap-type arrestors causes a rapid voltage drop and, when coupled through the capacitance of a transformer, creates surge voltages at its secondary. Direct lightning strikes can produce very high currents; the resulting voltages can exceed the withstand capability of equipment and conventional surge-protective devices in low-voltage (LV) circuits.

Switching transients can be due to normal or abnormal conditions. Turning appliances on or off near the point of interest causes minor transients. The commutation process in electronic power converters causes periodic transients, also known as notching, due to a momentary phase-to-phase short circuit in the 100 μs range. Multiple reignitions or restrikes can escalate to surge voltage of complex waveforms and of high amplitudes. Switching of major power equipment such as capacitor banks causes transients with longer duration lasting several hundred milliseconds, compared with typical durations in the order of one microsecond to tens of microseconds for other minor switching transients and lightning-induced transients. System short circuits and arcing faults can also result in high transients. The operation of fast-acting (less than 2 μs) overcurrent protective devices leaves inductive energy trapped in the circuit upstream; upon collapse of the field, high voltages are generated.

Transients can be described according to their many characteristic components, which include amplitude, duration, rise time, frequency of ringing, polarity, energy delivery capability, amplitude spectral density, position with respect to the mains waveform, and frequency of occurrence. However, it is very difficult to specify which of these attributes is the more important or useful. The peak amplitude and frequency of occurrence attributes have found favour in some past surveys. Duration, ringing frequency and rate of voltage change are also commonly used but primarily for switching transients only. Lightning transients can be characterised by the amount of energy contained in the surge that can be deposited in a surge-absorbing device. However, it is necessary to measure the surge current as well as the voltage in order to calculate the energy level.

Without a commonly recognised attribute, it is difficult to define the compatibility levels for transients. No prescribed tolerance limits are imposed on equipment, other than the expected capability of each class of equipment to withstand a certain range of transients. The IEEE c62.41:1991 outlines the waveshapes of representative voltage transients that can be used for verifying the sustaining capability of various types of equipment.

IEEE 1159: 1995 The IEEE 1159:1995 standard defines unidirectional transients as impulsive (see Figure 2.1) and these are characterised by their high rise time implicating high-frequencies and high peak values. They are usually caused by lightning strikes and their high-frequency contents are damped quickly by the system's resistive elements. They may excite the system's natural frequency precipitating to oscillatory transients. Oscillatory transients (see Figure 2.1) are usually described by their spectral content and in particular their predominant frequency, magnitude and duration.

The common types of oscillatory transient phenomena are classified into three subclasses of high-frequency, medium-frequency and low-frequency transients, coinciding with the common types of switching in power systems. High-frequency oscillatory transients are almost always due to some form of switching event and are often the result of a local system response to an impulsive transient. A typical example is the oscillatory response of the RLC snubber circuit to the commutation process of power electronic devices. Its primary frequency component is of the order of 500 kHz with a typical duration measured in microseconds.

Medium-frequency transients have a primary frequency between 5 kHz and 500 kHz with a duration measured in tens of microseconds. Back-to-back capacitor energisation results in oscillatory transients of tens of Kilohertz. When a capacitor is energised in close proximity to one already in service, the energised bank sees the de-energised bank as low impedance, resulting in current transients oscillating between them. These oscillations can last up to tens of microseconds depending on the size of the capacitors and the damping (usually resistive losses) between them. Cable switching may also result in oscillatory voltage transients in the medium-frequency range.

Low-frequency oscillatory transients have primary frequency components of less than 5 kHz, and a duration from 0.3 ms to 50 ms. They are frequently encountered in the sub-transmission and distribution systems and are normally caused by capacitor bank or transformer energisation. The presence of series capacitors may induce resonances magnifying the low-frequency components in the transformer inrush current.

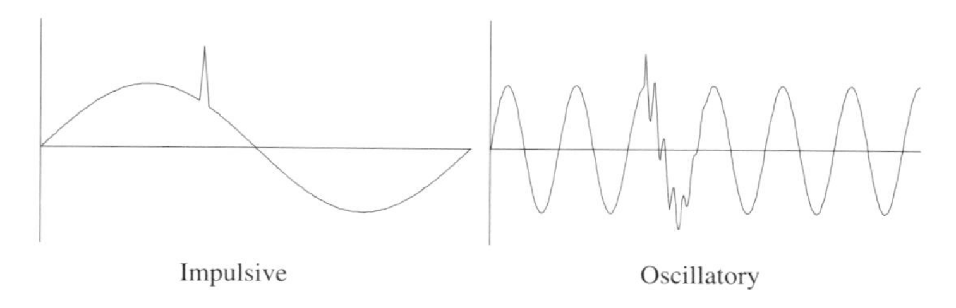

Figure 2.1 Voltage transients as defined in IEEE 1159:1995

Such oscillatory transients can also occur when unusual conditions result in ferro-resonances.

Many attributes can be used to describe the characteristics of a transient waveform. However, the random nature of occurrence and the wide variations of transient wave-shapes make it difficult to decide which attributes should be measured. The selection of attributes and the significance of each of the selected attributes depend on the susceptibility characteristics of the equipment under consideration and the environment under study. Some transients occur without causing any problems, some can cause maloperation of certain types of equipment, while others may cause damage. The presence of other circuit components, which react to the transient, may affect the measurement results. The shape of a transient can be changed quickly by the action of a protective surge arrester and may contain different characteristics when viewed from another part of the power system network. Similarly, the characteristics of similar transients (i.e. those that originated from the same source at the same point of a power system) may change that when the load or system configuration is altered.

IEC 816: 1984 [8] Because of the complex and variable nature of transients, equipment susceptible to transients is divided into three categories in order to determine which of the parameters are to be measured:

1 those which are susceptible to a restricted band of frequencies, such as radio or carrier frequency receivers,

2 those which are susceptible to a broad band of low radio frequencies, where the peak voltage is usually the critical parameter, but the energy can also be an important parameter,

3 those which are susceptible to a broad band of frequencies in the higher frequency bands, where digital equipment is often susceptible to high rate of rise of transient pulses and may even be damaged.

Measurement of transients can be undertaken in the time domain or the frequency domain. Time-domain measurements have limits on viewing time, while frequency-domain measurements have corresponding limits on bandwidth. It is stipulated that both time-domain and frequency-domain measurements are to be performed when measuring transients of unknown characteristics, in order to obtain maximum information about the transients.

The peak amplitudes and the rate of occurrence are normally considered important under most situations. In general, the peak amplitude versus rate of occurrence curves can provide an overview of the system quality. Detailed analysis of other attributes would depend on the nature of the equipment subjected to the transients. The rise time characterises the transient in its amplitude–frequency relation. The shorter the rise time, the more extensive will be the disturbing action in the frequency spectrum. Amplitude is especially significant for long transients (>1 μs) and it is usually the main quantity influencing performance degradation or device destruction. Component destruction also depends on the energy content of the transient. This energy level is dependent on the amplitude of the transient and the internal impedance of the disturbance source. Transient duration can be important but is dependent on the time constant of the susceptible equipment in question. Similarly, repetition frequency may be useful for estimating the disturbing effect of the transients, but its importance is also dependent on the time constant of the susceptible equipment. Obviously, the chances that the logic systems will malfunction will increase if the transient duration is longer or if the repetition frequency is high.

2.3.2 Low-Frequency Disturbances: Dips, Interruptions and Swells

In the context of this chapter, low-frequency disturbances are regarded as momentary deviations from the sinusoidal supply voltage waveform lasting from half a cycle to less than one minute. Disturbances of duration of less than half a cycle are regarded as transients and these have been described in Section 2.3.1. Figure 2.2 shows the three types of disturbances considered in this chapter: voltage dips (sags), interruptions (outages) and swells. Voltage dips and short interruptions are included in the IEC EMC standards but voltage swell is only introduced in IEEE 1159:1995.

IEC 61000-2-1: A voltage dip is a sudden reduction of the voltage at a point in the electrical system, followed by voltage recovery after a short period of time, from half a cycle to a few seconds. A short interruption is a complete loss of supply voltage for a period not exceeding 1 min and is sometimes regarded as a 100% voltage dip. Voltage dips are regarded as having a voltage drop Δ V ranging from 10% to 99% of the system nominal voltage, and short interruptions are regarded as 100% dips, with durations ranging from one half-cycle to several seconds. Interruptions longer than 1 min are no longer considered a low-frequency EMC issue, but power supply interruption. These voltage dips and interruptions can be caused by:

- short circuits in LV networks cleared by fuse operation lasting up to several milliseconds,
- faults on MV and HV lines or other equipment followed possibly by automatic reclosing of 100 ms to 600 ms,
- switching of large loads, especially motors and capacitor banks.

Voltage dips can be characterised by their amplitude and duration. A simple dip whose amplitude is constant during its duration can be described by a single set of these two values, while those with complex shapes may be characterised by two or more pairs of the values. Voltage changes that do not reduce the system voltage to less than 90% of the nominal voltage are not considered to be voltage dips, since this is the range of slow voltage variations, due to gradual load changes, and voltage fluctuations due to rapid and repetitive voltage changes.

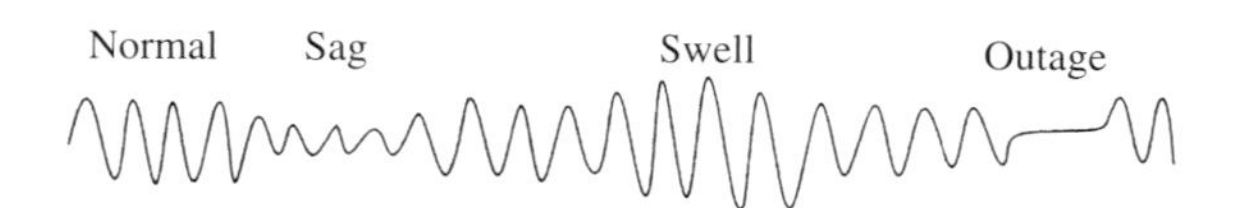

Figure 2.2 Voltage disturbances

IEEE 1159: In this standard, the low-frequency phenomena are divided according to their duration; short-duration variations ranging from one-half cycle to one minute and long-duration variations lasting more than one minute. The short-duration disturbances are further divided into instantaneous, momentary and temporary disturbances; in each case, the durations are chosen to correlate with typical operation times of protective devices. This duration division is also recommended for differentiating the disturbances during monitoring. The amplitude of these variations is defined to be greater than the ±10% (of system nominal) commonly used as the voltage regulation limit. Voltage drops and rises with duration lasting more than one minute are considered as long-duration undervoltages and overvoltages.

Short-duration variations are almost always caused by fault conditions, the energisation of large loads that require high currents, or intermittent loose connections in power wiring. A voltage sag may precede interruptions when the interruptions are due to a fault in the source system. The duration of the interruption depends on the reclosing capability of the protective devices. A fault on a parallel feeder circuit will result in a voltage drop at the substation bus that affects all the other feeders until the fault is cleared. An induction motor will draw six to eight times its full load current during starting. If this current is large relative to the system fault current, the resultant voltage sag can be significant. Voltage swells are much less common than sags and are usually associated with system fault conditions. Swells can occur on the unfaulted phases when a single line-to-ground fault occurs on another phase. It can also be caused by switching off a large load or switching on a large capacitor bank.

Long-duration variations are generally not the result of system faults. They are caused by the constant changes in system loads and the system switching operations. Poor voltage regulation or control or incorrect transformer tap settings can result in system overvoltages or undervoltages. Insufficient voltage regulation capabilities and overloaded circuits can result in undervoltages, which may lead to sustained interruptions if the system can no longer cope with the demand.

The main attributes for characterising these brief disturbances are the change in the amplitude and the duration of the occurrence. A voltage dip or swell whose amplitude is constant during its duration may be characterised by the amplitude change and the duration values. However, a complex dip or swell whose voltage amplitude changes in stages may be described by two or more pairs of amplitude change and duration values. It can also be simplified into a maximum amplitude change value and the overall duration value. The rate of occurrence should also be recorded alongside the change in the amplitude and the length of the duration. The amplitude changes may not be similar on the three phases and, hence, all three phases are to be treated as independent although the disturbances on one phase may originate from another phase. Root-mean-square (r.m.s.) voltage values can also be used but are generally reserved for long-duration disturbances.

UIE-DWG-2-92-D [9]: This guide was put together by the Disturbance Working Group of Union Internationale Electrothermie (UIE) for recommending measurements in order to obtain information on the number and character of voltage dips emanating from industrial installations. The measurements are aimed at pointing out and characterising the types of loads causing disturbances, and the equipment being disturbed. The results are to be collated to serve as a guide for the choice of immunity levels for susceptible equipment installed inside, as well as outside, the plant. This information will be of great value for equipment manufacturers and industrial users to design their installations.

Measurements are co-ordinated for industries with various types of loads and different conditions so as to achieve the widest possible database. The guide also provides several specifications for the measurement equipment.

- Simultaneous measurements are to be conducted on all phases.
- The instrument shall have an accuracy of 1% so that it is possible to identify and quantify dips and interruptions of between 99% and 100% of nominal voltage.
- Equipment shall be able to be triggered so as to provide information only when a voltage dip occurs.
- The measurement threshold should be able to be set at 10% below nominal voltage.
- The minimum voltage (minimum half-cycle value) reached during a voltage dip should be recorded.
- The equipment shall record the duration of the dip as the time during which the voltage is 10% or more below the nominal value.
- The minimum duration of voltage dips to be recorded is half a cycle implying a sampling rate of at least 100/120 Hz.
- The maximum duration of voltage dips is 60 s.
- The equipment shall be immune to interruptions of supply in terms of the integrity of stored data as well as pre-set measurement parameters.

The monitoring shall continue long enough to allow a statistical analysis to be completed. A minimum recording period of three months is required for all normal types of operation of the plant. Loads causing disturbances and equipment suffering much from dips are also to be analysed.

2.3.3 CBEMA and ITI Curves

The well-known '(CBEMA) curve' [10] Association Computer Business Equipment Manufacturer, shown in Figure 2.3, can be used to evaluate the voltage quality of a power system with respect to voltage interruptions, dips or undervoltages and swells or overvoltages. This curve was originally produced as a guideline to help CBEMA members in the design of the power supply for their computer and electronic equipment. By noting the changes of power supply voltage on the curve, it is possible to assess whether the supply is reliable for operating electronic equipment, which is generally the most susceptive equipment in the power system.

The curve shows the magnitude and duration of voltage variations on the power system. The region between the two sides of the curve is the tolerance envelope within which electronic equipment is expected to operate reliably. Rather than noting a point on the plot for every measured disturbance, the plot can be divided into small regions with certain ranges of magnitude and duration. The number of occurrences within each

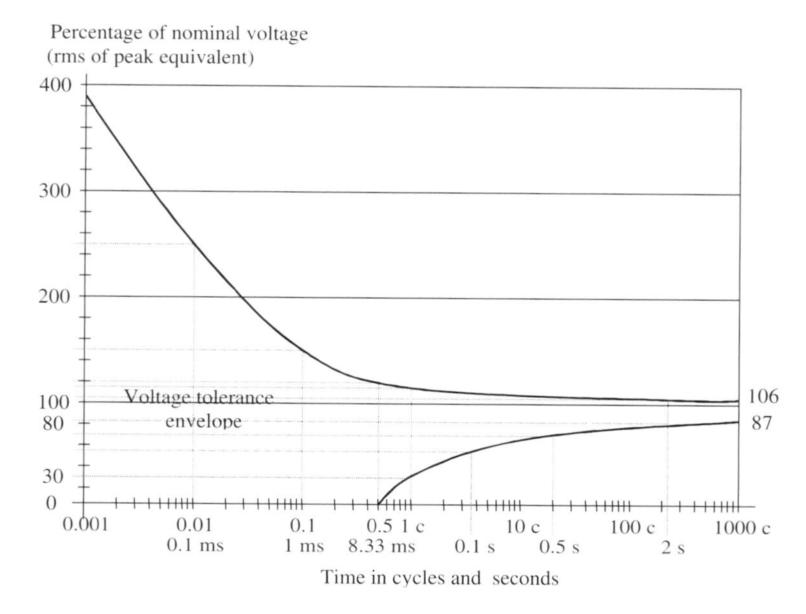

Figure 2.3 CBEMA curve

small region can be recorded to provide a reasonable indication of the quality of the system.

CBEMA has been renamed as the Information Technology Industry Council (ITI) and a new curve [11], as shown in Figure 2.4, has been developed to replace CBEMA's. However, due to the prominence of the CBEMA curve among the computer and electronic industries, the ITI curve is being regarded as the new CBEMA within the high-technology circle. The main difference between them is that the ITI version is piecewise, and hence easier to digitise than the continuous CBEMA curve. The tolerance limits at different durations are very similar in both cases. Although currently only the CBEMA curve has been officially endorsed in IEEE Standard 446, it is anticipated that the ITI curve will also be endorsed by various standard bodies in the near future.

The boundary of the ITI curve is defined by seven possible disturbance events.

1 *Steady-state tolerances*: this range describes an r.m.s. variation between ±10 % from the nominal voltage, which is either very slowly varying or is constant. Any voltages in this range may be present for an indefinite period, and are function of normal loadings and losses in the system.

2 *Line voltage swell*: this region describes a r.m.s. voltage rise of up to 120% of the r.m.s. nominal voltage, with a duration of up to 0.5 s. This transient may occur when large loads are removed from the system.

3 *Low-frequency decaying ringwave*: this region describes a decaying ringwave transient, which typically results from the connection of power-factor-correction capacitors to a distribution system. The transient may have a frequency ranging from 200 Hz to 5 kHz, depending on the resonant frequency of the system. It is assumed to have completely decayed by the end of the half-cycle in which it occurs, and it

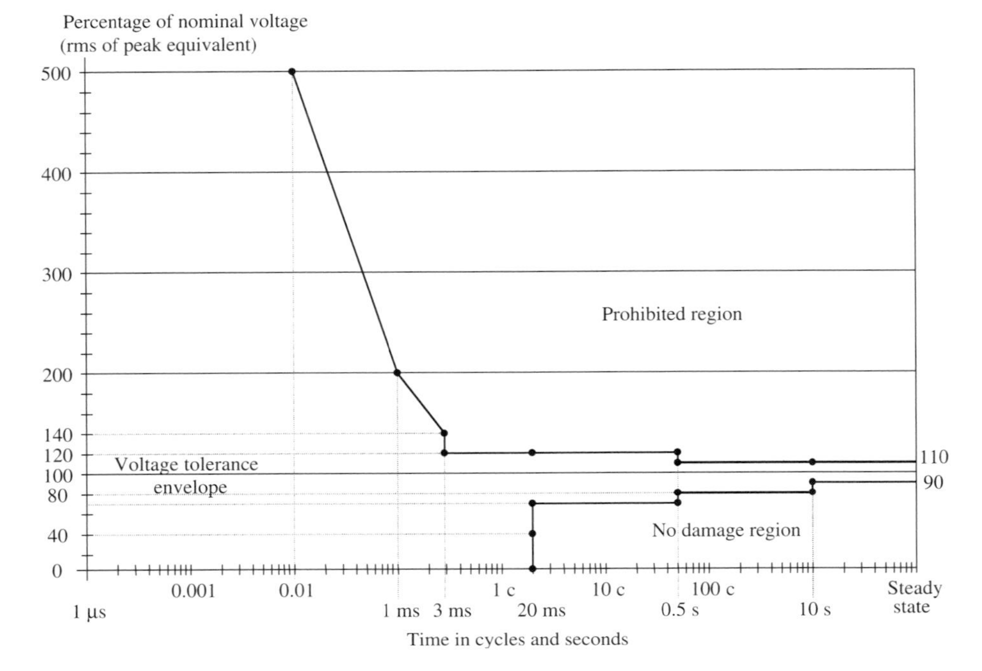

Figure 2.4 ITI curve

occurs near the peak of the nominal voltage waveform. Its amplitude varies from 140% for 200 Hz ringwaves to 200% for 5 kHz ringwaves, with a linear increase in amplitude with frequency.

4. *High-frequency impulse and ringwave*: this region describes the transients which typically occur as a result of lightning strikes, and is characterised by both amplitude and duration (energy) rather than r.m.s. amplitude.

5. *Voltage sags to 80% of nominal*: these sags are the result of the application of heavy loads, as well as fault conditions, at various points in the distribution system. They have a typical duration of up to 10 s.

6. *Voltage sags to 70% of nominal*: these also result from heavy loads switching and system faults. Their typical duration is up to 0.5 s.

7. *Dropout*: this transient is typically the result of the occurrence and subsequent clearing of faults in the distribution system. It includes both severe r.m.s. voltage sags and complete interruptions, followed by immediate re-application of the nominal voltage; the total interruption may last up to 20 ms.

Outside this bounded tolerance region, two other unfavourable regions are defined. The No-damage region includes sags and dropouts that are more severe than those described above. The voltages applied continuously are less than the lower limit of the steady-state tolerance range. Information technology equipment is not expected to function normally in this region, but no damage to the equipment should result. In the other region, called the prohibited region, the voltage swells exceed the upper limit of the curve boundary; damage to the equipment is expected if it is subjected to voltages with these characteristics.

Both the CBEMA and ITI curves were specifically derived for use in the 60 Hz 120 V distribution voltage system. To our knowledge, no study has been carried out to find whether they are suitable for use in the 50 Hz 240 V distribution voltage system. The guideline expects the user to exercise their own judgement when applying those curves on equipment operating under different voltage levels from those specified.

2.3.4 R.M.S. Variation Indices

The basic idea behind r.m.s. variation indices is similar to that of sustained interruption indices such as *SAIFI* and *CAIDI*, which are in wide use by electricity distribution companies. They are generally recognised as reliability indices and are useful in assessing the reliability of service with respect to sustained interruptions. However, they would neglect voltage dips and swells, even though these disturbances may result in similar shutdowns or maloperations to those caused by the interruptions. Hence, several new indices [12] have recently being proposed to take into consideration the voltage dips and swells in determining the reliability of the system to supply quality power to sensitive loads. The proposal divides the r.m.s. voltage variations into three levels, each of which is identified as a type of event.

1 *Component events*: in these events, each phase of r.m.s. variation is characterised by a magnitude–duration pair. The duration is designated as the period of time that the r.m.s. voltage exceeds a specified threshold voltage level used to characterise the disturbance; e.g. $T_{80\%}$ is the characterised duration of the event for an assessment of voltage dips having a magnitude $\leq 80\%$ of the nominal voltage. The measurement is single phase and the three phases are reported separately.

2 *Measurement events*: these events consider the effect of the disturbances on all three phases of the distribution system. A fault may affect one, two or all three phases and the magnitude and duration of the resulting r.m.s. variation on each phase may differ substantially. The proposed three-phase method characterises each r.m.s. variation measurement event by designating the magnitude and duration as those of the phase with the greatest voltage deviation from nominal voltage.

3 *Aggregate events*: an aggregate event is a collection of all measurement events associated with a single power system occurrence into a single set of event characteristics. The purpose of aggregating events is to provide a truer assessment of service quality than is given by the measurement events. If a device trips or maloperates on the initial r.m.s. variation, subsequent r.m.s. variations will have no effect on the process. Hence, it would be more accurate if subsequent variations were not characterised as separate events, since that would present an unnecessarily poorer impression of the service quality. Many characteristics can be used as the basis for aggregation but the proposed method chooses a time period of one minute. Once the first measurement event is identified, subsequent measurements recorded over a period of one minute will be considered as parts of the same aggregate event. The magnitude and duration of an aggregate event are designated as those of the worst measurement event that exhibits the greatest voltage deviation within the aggregate period.

There are many properties of r.m.s. variations that would be useful to quantify, such as the frequency of occurrence, the duration of disturbances and the number of phases involved. This is because the susceptibility of various devices and processes to r.m.s. variations differs. Many devices are susceptible to varying magnitude only, while others are susceptible to the combination of magnitude and duration. The proposed indices as depicted in Table 2.3 assess both the r.m.s. variation magnitude and the combination of magnitude and duration.

$SARFI_x$ represents the average number of specified r.m.s. variation measurement events that occurred over the assessment period per customer served. The specified disturbances are those with a magnitude less than x for dips or a magnitude greater than x for swells. The other three indices of $SIARFI_x$, $SMARFI_x$ and $STARFI_x$ are subsets of $SARFI_x$ and are for characterising short-duration variations as depicted in Table 2.2 according to the IEEE 1159 standard. The r.m.s. variations are instantaneous, momentary and temporary variations. Other than these indices, similar indices are defined for long-duration variations of undervoltages and overvoltages.

2.4 Waveform Distortions

Waveform distortions were traditionally regarded as harmonic distortions, since harmonics are the major distorting component in the mains voltage and load current waveform. However, increased interharmonic content in the power system has prompted a need to give them greater attention. The growing use of three-phase converters has also raised the awareness of notchings caused by the commutation process.

Table 2.3 R.M.S. variation indices

Type of r.m.s. frequency index	Indices
System average	$SARFI_x = \frac{\sum N_i}{N_T}$
System instantaneous average	$SIARFI_x = \frac{\sum NI_i}{N_T}$
System momentary average	$SMARFI_x = \frac{\sum NM_i}{N_T}$
System temporary average	$STARFI_x = \frac{\sum NT_i}{N_T}$

Note: x = r.m.s. voltage threshold, possible values - 140, 120, 110, 90, 87, 70, 50 and 10; N_i = number of customers experiencing short-duration voltage deviations with magnitudes above $X\%$ for $X > 100$ or below $X\%$ for $X < 100$ due to measurement event I; NI_i = number of customers experiencing instantaneous voltage deviations with magnitudes above $X\%$ for $X > 100$ or below $X\%$ for $X < 100$ due to measurement event I; NM_i = number of customers experiencing momentary voltage deviations with magnitudes above $X\%$ for $X > 100$ or below $X\%$ for $X < 100$ due to measurement event I; NT_i = number of customers experiencing temporary voltage deviations with magnitudes above $X\%$ for $X > 100$ or below $X\%$ for $X < 100$ due to measurement event I; N_T = number of customers served from the section of the system to be assessed

Different countries have developed their own harmonic standards and set their own emission limits according to their individual conditions and requirements. However, with the growth of global trade, the need for appliances manufactured in one country to comply with standards in another country has prompted concerted effort in formulating international standards on harmonics and interharmonics. The IEC 61000 series of the EMC standards includes harmonics and interharmonics as two of the conducted low-frequency electromagnetic phenomena. However, IEEE 519:1992 provides guidelines on harmonics only.

2.4.1 Harmonics

IEC 61000-2-1: 1990 IEC 61000-2-1 defines harmonics as sinusoidal voltages and currents having frequencies that are whole multiples of the frequency at which the supply system is designed to operate (50 Hz or 60 Hz). It also outlines the major sources of harmonics from the three categories of equipment, which are power system equipment, industrial loads and residential loads.

Harmonic currents are generated to a small extent and at low distortion level by generation, transmission and distribution equipment, and to a larger extent by the industrial and domestic loads. High-power types of equipment using phase-angle control and uncontrolled rectification, especially those with capacitive smoothing, are major harmonic contributors. Traditionally, saturation of transformer core during energisation is the main source of harmonics from power utility equipment, but the increasing use of Flexible AC Transmission systems (FACTS) will add further contributions to the harmonic distortion.

The most polluting industrial load is the static power converter, which uses some form of power electronic switching to supply power to virtually every piece of industrial equipment. Induction furnaces and arc furnaces also contribute significantly to the distortions. Although of low power rating, the shear large volume of residential appliances being used simultaneously, and for long periods, cumulatively cause significant distortion in the power system. The main harmonics producing residential loads are those powered by rectifiers with high smoothing capacitors, such as PCs and TV receivers.

Past standards such as IEC 555-3 concentrate on specifying emission limits for various types of equipment connected to the power system. IEC 61000-3-2 (1998) and IEC 61000-3-4 (1998) outline the emission limits from equipment with input current less than 16 A per phase and more than 16 A per phase, respectively.

IEEE 519:1992 [13] IEEE 519:1992 also identifies the major sources of harmonics in power systems. The harmonic sources described in this standard include power converters, arc furnaces, static VAR compensators, inverters of dispersed generation, electronic phase control of power, cycloconverters, switch mode power supplies and pulse wide modulated drives. It illustrates the typical distorted waveshapes, the harmonic order numbers and the level of each harmonic component in the distortion caused by these devices. It also describes how the system may respond to the presence of harmonic pollution. The discussed responses comprise parallel resonance and series resonance, and the effect of system loading on the magnitude of these resonances. On the basis of typical characteristics of low-voltage distribution systems, industrial systems and transmission systems, it discusses the general response of these systems to harmonic distortion.

The effects of harmonic distortion on the operation of various devices or loads are also included in the standard. These devices comprise motors and generators, transformers, power cables, capacitors, electronic equipment, metering equipment, switchgear, relays and static power converters. Interference to the telephone networks as a result of harmonic distortion in the power systems is also discussed and the discussion is largely based on the C-message weighting system created jointly by Bell Telephone Systems (BTS) and Edison Electric Institute EEI. This weighting system is further described in section 2.4.2. The standard also outlines several possible methods of reducing the amount of telephone interference caused by harmonic distortion in the power system.

This standard also describes the analysis methods and measurement requirements for assessing the levels of harmonic distortion in the power system. It summarises the methods for calculating harmonic currents, for calculating system frequency responses and for modelling various power system components for the analysis of harmonic propagation. The section on measurements highlights their importance and lists various harmonic monitors that are currently available. It describes the accuracy and selectivity (the ability to distinguish one harmonic component from others) requirements on these monitors; it also describes the averaging or snap-shot techniques that can be used to 'smooth-out' the rapidly fluctuating harmonic components and thus reduce the overall data bandwidth and storage requirements.

The standard describes methods for designing reactive compensation for systems with harmonic distortion. Various types of reactive compensation schemes are discussed indicating that some of the equipment, such as Thyristor Controlled Recordors, are themselves sources of harmonic distortion. It also outlines the various techniques for reducing the amount of harmonic current penetrating into the a.c. systems. Recommended practices are suggested to both the individual consumers and the utilities for controlling the harmonic distortion to tolerable levels. This standard concludes with recommendations for evaluating new harmonic sources by measurements and detailed modelling and simulation studies. It provides several examples to illustrate how these recommendations can be implemented effectively in several practical systems.

IEC 61000-4-7 [14]: 1993 The main IEC standard on harmonics, IEC 61000-4-7, describes the techniques for measuring harmonic distortion in the power system. For the purpose of formulating the requirements for measurement instruments, it divides the harmonics broadly into three categories:

1 Quasi-stationary (slowly varying)
2 Fluctuating
3 Rapidly changing (or very short bursts of harmonics).

The IEC 61000-4-7 standard covers both time-domain and frequency-domain-based instruments, but this section considers only the frequency-domain techniques which are based on the Fast-Fourier Transform (FFT). Chapter 4 illustrates other frequency extracting techniques, which may be appropriate under certain conditions. The differing characteristics of the three categories of harmonics place different requirements on the design of the measuring instrument. For measuring quasi-stationary harmonics, there can be gaps between the rectangular observation window of 0.1 s to 0.5 s wide. On the other hand, in order to access fluctuating harmonics, the rectangular window width has to be decreased to 0.32 s or a Hanning's window of 0.4 s to 0.5 s width has to be used. Moreover, there should not be any gap between successive rectangular windows and there should be a half-by-half overlapping of the successive Hanning's windows. Lastly, rapidly changing harmonics have to be measured with a 0.08 s to 0.16 s wide rectangular window without any gap between successive windows.

Instruments designed for measuring quasi-stationary harmonics are only appropriate to survey the long-term (such as thermal) effects of harmonics, or for the measurement of constant harmonic currents, such as those produced by television receivers. The measurement of fluctuating harmonic currents, such as those produced by motor reversal or speed change in household appliances with phase control and regulation, has to be made continuously without any gaps between successive observation intervals. Continuous real-time measurement capability is absolutely necessary for assessing the instantaneous effects of the rapidly changing harmonics, or short bursts of harmonics, on sensitive equipment such as electronic controls or ripple control receivers.

Measurements of up to the 50th harmonic order are commonly recommended but there are discussions on increasing it up to the 100th in certain cases. With such a large amount of data to be recorded, statistical evaluation over different observation intervals can be used to compress the data. Five time intervals are recommended in this standard:

1 Very short interval ($T_{\nu s}$): 3 s
2 Short interval (T_{sh}): 10 min
3 Long interval (T_L): 1 h
4 One day interval (T_D): 24 h
5 One week interval (T_W): 7 d.

If instantaneous effects are considered important, the maximum value of each harmonic should be recorded and the cumulative probability (at least 95% and 99%) of these maxima should be calculated. On the other hand, if long-term thermal effects are considered, the maximum of the r.m.s. value at each harmonic and its cumulative probabilities (at 1%, 10%, 50%, 90%, 95% and 99%) are to be calculated and recorded:

$$C_{n,\mathrm{rms}} = \sqrt{\frac{\left(\sum_{k=1}^{M} C_{n,k}^2\right)}{M}}. \tag{2.1}$$

Besides statistical evaluation of the system's harmonic content, several distortion factors have been proposed for characterising the overall harmonic distortion level in the system voltage and system current. They are depicted in Table 2.4 and are described in Section 2.4.2.

Besides outlining the measurement of the magnitude of individual harmonic components, it also describes several special cases of measurements. These cases are the measurements of the phase angle of each harmonic component, the measurements of symmetrical components at harmonic frequencies and interharmonic measurements. These special cases are dealt with in subsequent subsections in this chapter.

Table 2.4 General harmonic indices

Total Harmonic Distortion (*THD*)	$\dfrac{\sqrt{\left(\sum_{n=2}^{50} U_n^2\right)}}{U_1}$
Total Harmonic Distortion adapted to inductance (THD_{ind})	$\dfrac{\sqrt{\sum_{n=2}^{50}\left(\dfrac{U_n^2}{n^\alpha}\right)}}{U_1}$, where $\alpha = 1\ldots2$
Total Harmonic Distortion adapted to capacitance (THD_{cap})	$\dfrac{\sqrt{\sum_{n=2}^{50}(n \times U_n^2)}}{U_1}$
Total Demand Distortion (*TDD*)	$\dfrac{\sqrt{\left(\sum_{n=2}^{50} I_n^2\right)}}{I_{\text{rated}}}$

2.4.2 Harmonic Indices

The most common harmonic index is Total Harmonic Distortion (*THD*). *THD* is defined as the r.m.s. of the harmonics expressed as a percentage of the fundamental component. For most applications, it is sufficient to consider the harmonic range from 2nd to 25th, but most standards specify up to the 50th. The *THD* can be weighted to indicate the stress on various system devices. The weighted distortion factor adapted to inductance is an approximate measure for the additional thermal stress of inductances such as coils and induction motors. For most applications it is sufficient to consider the harmonic range from 2 to 20. On the other hand, the weighted *THD* adapted to capacitors is an approximate measure for the additional thermal stress of capacitors directly connected to the system without series inductance. The weighted distortion factors are only used for the assessment of the quality of system voltage.

Current distortion levels can also be characterised by a *THD* value but it can be misleading when the fundamental load current is low. A high *THD* value for input current may not be of significant concern if the load is light, since the magnitude of the harmonic current is low, even though its relative distortion to the fundamental frequency is high. To avoid such ambiguity, IEEE Standard 519:1992 defined a new factor called the Total Demand Distortion (*TDD*) factor. This term is similar to *THD* except that the distortion is expressed as a percentage of some rated or maximum load current magnitude, rather than as a percentage of the fundamental current. Since electrical power supply systems are designed to withstand the rated or maximum load current, the impact of current distortion on the system will be more realistic if the assessment is based on the designed values, rather than on a reference that fluctuates with the load levels.

Other than these general factors, there are several indices specific to the type of equipment being affected by harmonics. For interference on telecommunication systems, two weighting systems are in wide use by the industry, i.e.

1 the psophometric weighting system [15] proposed by the International Consultation Commission on Telephone and Telegraph Systems (CCITT), used in Europe;

2 the C-message weighting system [16] proposed jointly by BTS and EEI, used in the United States and Canada.

They are used to provide a reasonable indication of the interference from each harmonic, taking into account the response of the telephone equipment and the sensitivity of the human ear. These systems acknowledge that the effect of harmonic interference is not uniform over the audiofrequency spectrum. This non-uniformity is accounted for by having different weights at different audiofrequencies. Figure 2.5 shows the weighting curves of both systems and it can be deduced from the curves that a human ear in combination with a telephone set has a sensitivity to audiofrequencies that peaks at about 1 kHz.

In both systems, the weights are used to calculate the indices to describe the severity of the interference. Table 2.5 provides a comparison of the indices used in both systems.

With the CCITT system, the level of interference is described in terms of a Telephone Form Factor (TFF), which is a dimensionless value that ignores the geometrical configuration of the coupling. It represents an equivalent disturbing voltage at 800 Hz which, if applied to the power line, would cause the same interfering effect to be experienced in a nearby telephone line as does the measured voltage on the power line and its harmonics. The C-message weighting system uses a Telephone Influence Factor (TIF) which is also a dimensionless value. Moreover, the BTS–EEI system also has $I \cdot T$ and $kV \cdot T$ products, which are weighted currents and voltages in the power system. The analogous quantities in the CCITT system to these products are

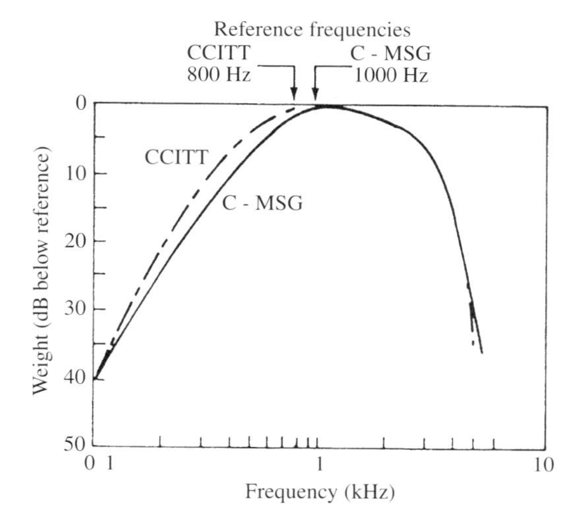

Figure 2.5 Comparison of BTS C-message and CCITT weights

Table 2.5 Harmonic indices for measuring telecommunication interference

Psophometric weighting	C-message weighting
$TFF = \frac{1}{U}\sqrt{\sum_{n=1}^{\infty}(K_f p_f U_f)^2}$	$TIF = \frac{1}{U}\sqrt{\sum_{n=1}^{\infty}(K_f p_f U_f)^2}$
$K_f = f/800$ is a coupling factor p_f is the weight divided by 1000.	$K_f = 5f$ is the coupling coefficient, p_f is the weight of the harmonic
$I.T = \sqrt{\sum_{n=1}^{\infty}(K_f p_f I_f)^2}$	$EDI = \sqrt{\sum_{n=2}^{50}(n p_n I_n)^2}$
$kV.T = \sqrt{\sum_{n=1}^{\infty}(K_f p_f V_f)^2}$	$EDV = \sqrt{\sum_{n=2}^{50}(n p_n V_n)^2}$

Note : U = r.m.s. voltage of the transmission line; U_f = harmonic voltage of frequency f.

the Equivalent Distorting Current (EDI) and Equivalent Distorting Voltage (EDV). These indices are only guideline measurements, since they do not consider coupling and exposure issues. Hence, it is not satisfactory to use them as the sole measures of interference to a communication line.

Other than the above mentioned standards and guidelines, there are several standards that deal with specific equipment under the influence of harmonic distortions, the main ones being:

- ANSI/IEEE C57.110: 1986 [17] on transformers
- IEEE Standard 18: 1980 [18] on shunt power capacitors.

They provide guidelines on how to select the appropriate rating for the equipment operating in an environment with harmonic distortions.

2.4.3 Interharmonics

IEC 61000-2-1 also defines interharmonics, because between the harmonics of the power frequency voltage and current, further frequencies can be observed which are not integer multiples of the fundamental. The document outlines the major sources of interharmonics and explains that they can appear as discrete frequencies or as a wide-band spectrum. Most standards to date do not contain a description of this phenomenon or guideline on its measurement techniques and limits. Continuing efforts are being made to include interharmonics in various harmonic standards or guidelines such as the IEC 61000-4-7 and the IEEE 519.

The IEC 61000-4-7 contains a small subsection on interharmonics as a broad extension of harmonic phenomena. However, it leaves several important issues unresolved, and recommends that issues such as the range of frequencies to be considered, and the centre frequency, should be selected in accordance with the studied phenomenon, e.g. their influence on ripple control receivers or on flicker.

A study report prepared by a joint IEEE/CIGRE/CIRED working group on interharmonics [19] identifies the main problem associated with the measurement of interharmonics as being that a waveform consisting of two or more non-harmonically related frequencies may not be periodic. Hence, most power system monitoring equipment, which is based on the FFT, will encounter errors due to the end-effect. This effect can be minimised by the signal processing techniques commonly used in the communication and broadcast industries, whereby the sampling of the signal need not be synchronised to the power frequency. The use of proper windowing functions and the application of zero padding before performing the FFT can improve the frequency resolution of the measured interharmonic magnitudes significantly. For instance, the Hanning window with four-fold of zero padding technique [19] should also be suitable for measuring interharmonics in power systems.

The type of interharmonic measurement to be used also depends on the purpose of the assessment. Purposes include the diagnosis of a specific problem, a general survey of an electromagnetic environment, compatibility testing and compliance monitoring. The IEC proposes to fix the sampling interval of the waveform to 10 and 12 cycles for a 50 Hz and 60 Hz system, resulting in a fixed set of spectra with 5 Hz resolution, for harmonic and interharmonic evaluation. The sampling will be phase-locked to the mains frequency, thereby minimising the contamination of the harmonic components by the interharmonic components. However, recent indications are that the IEC opt to simplify the assessment process by summing the components between harmonics into one single interharmonic group, and reserving the original method of showing all interharmonic components at 5 Hz steps for specific cases. The frequency bins directly adjacent to the harmonic bins are omitted.

$$X_{\mathrm{IH}}^2 = \sum_{i=2}^{8} X_{10n+i}^2 \quad \text{(50 Hz system)}, \tag{2.2}$$

$$X_{\mathrm{IH}}^2 = \sum_{i=2}^{10} X_{12n+i}^2 \quad \text{(60 Hz system)}, \tag{2.3}$$

where n is the interharmonic group of interest and i is the interharmonic bin being summed.

Similarly, equivalent distortion indices can be defined for interharmonics as those for harmonics. The corresponding Total Interharmonic Distortion Factor ($TIHD$) is

$$TIHD = \frac{\sqrt{\sum_{i=1}^{n} V_i^2}}{V_1}, \tag{2.4}$$

where i is the total number of interharmonics considered and n is the total number of frequency bins present including subharmonics (i.e. interharmonic frequencies that are less than the fundamental frequency). If the subharmonics are important, they can be analysed separately as another index called appropriately the Total Subharmonic

Distortion ($TSHD$).

$$TSHD = \frac{\sqrt{\sum_{s=1}^{S} V_s^2}}{V_1}, \tag{2.5}$$

where S is the total number of frequency bins present below the fundamental frequency. Other distortion factors and statistical evaluation of harmonics as described in Section 2.4.1 can also be applied for the assessment of interharmonics in power systems.

2.4.4 Harmonic Phase-Angle Measurement

The measurement of phase angles between harmonic voltages and currents, together with their amplitudes, is required for the following purposes:

1 to evaluate harmonic flows throughout the system,

2 to identify harmonic sources and harmonic sinks,

3 to assess summation factors of harmonic currents from different disturbing loads if they are connected to the same node,

4 to establish system-equivalent circuits for calculating the impact of new disturbing loads, or the effectiveness of the countermeasures such as filters.

The direction of the active power flow at the harmonic order of interest can help to identify the source of the disturbance. To find the direction of the active power flow, the phase angle between the harmonic voltage at the point of common coupling and the plant feeder current has to be measured. If the active power flows into the public system, the plant is a harmonic source, otherwise it is a sink of harmonic currents from the system. The phase lag of harmonic voltage and current in relation to the fundamental (absolute phase angle) need not be known in this case. Such an absolute phase angle is only needed for evaluating the coupling between frequencies of non-linear loads. However, the measurement of absolute phase angles provides the following additional advantages.

1 Measurements at different nodes of similar or different systems can be compared.

2 It becomes possible to deduce whether the connection or rearrangement of different systems, or locally spread disturbing loads, will increase or decrease the harmonic level in the system. Harmonic distortions with similar phase angles will superimpose raising the harmonic level, while those with opposite phase angles will compensate each other, thereby lowering the harmonic level.

3 Phase angles of disturbing loads, especially from rectifier circuits without firing-control, can be detected in order to evaluate their overall disturbing effect or to find countermeasures.

Extra care is needed in operating the measuring instrument and in interpreting the results, when precise synchronisation across multiple channels is required to measure absolute phase angles. Chapter 5 further illustrates the necessary requirements for implementing such precise synchronisation and measurement.

2.4.5 Harmonic Symmetrical Components

If the loads and transmission and distribution systems are balanced, the three voltages and currents have identical wave shapes and are separated by exactly $\pm\frac{1}{3}$ of the fundamental period. In such a case, only characteristic harmonics exist: these are of zero sequence for orders $n = 3m$ $(m = 1, 2, 3, \ldots)$, of positive sequence for the $n = 3m-2$ orders, and of negative sequence for the $n = 3m-1$ orders. However, asymmetries always exist, causing non-characteristic harmonics in the system. These asymmetries can be evaluated by monitoring the symmetrical components of the harmonics.

Positive-sequence (or negative-sequence) impedances differ from zero-sequence impedances for nearly all loads and network equipment, including transmission lines, cables and transformers. Therefore, a separate treatment of the system is necessary for assessing the harmonic voltages caused by the injected currents. Secondly, the effect of each sequence component differs for most loads and network equipment. Zero-sequence voltages do not affect delta-connected loads such as motors and capacitor banks. Only the non-characteristic components (positive sequence or negative sequence) of the third harmonic voltages cause additional losses in delta-connected motors. Moreover, commonly used transformers with delta-star or star-zigzag winding connections do not transfer zero-sequence currents and voltages.

2.4.6 Notching

Notching is the distortion caused on the line voltage waveform by the commutation process of one valve to another in some power electronic devices. It is repetitive and can be characterised by its frequency spectra. However, notches present several specific concerns to certain types of power equipment such as the extra stress on the insulation of the transformers and generators. Therefore, notching is sometimes characterised by the following properties:

- notch depth — average depth of the line voltage notch from the sinusoidal waveform at the fundamental frequency,
- notch width — the duration of the commutation process,
- notch area — the product of notch depth and width,
- position on the sinusoidal waveform where the notches occur.

Notching is briefly covered in IEEE 1159 and is described in detail in IEEE 519. This document analyses the converter commutation phenomena in detail and describes the notch depth and duration with respect to the system impedance and load current. It also outlines the limits in terms of the notch depth, the *THD* of the supply voltage and the notch area for different supply systems.

2.5 Voltage Unbalance

The simplest method of expressing voltage unbalance is to measure the voltage deviation at each of the three phases, and compare it to the average phase voltage:

$$\text{phase-voltage unbalance} = \frac{\text{maximum deviation from average phase voltage}}{\text{average phase voltage}}.$$

However, the degree of voltage unbalance of the three-phase network is better expressed in terms of symmetrical components as the ratio of the negative-sequence or zero-sequence voltage components to the positive sequence component:

$$\text{voltage-unbalance factor} = \frac{\text{negative-sequence voltage}}{\text{positive-sequence voltage}},$$

$$\text{voltage-unbalance factor} = \frac{\text{zero-sequence voltage}}{\text{positive-sequence voltage}}.$$

There is no consensus on the voltage unbalance limits at which motor and generator can still operate reliably. Since the main effect of voltage unbalance is the heating of machine windings, higher short-term levels of unbalance may be acceptable for a few seconds or even minutes.

2.6 Voltage Fluctuation and Flicker

Voltage fluctuation is described as a cyclical variation of the voltage envelope or a series of random voltage changes, the magnitude of which does not exceed the range of permissible operational voltage changes mentioned in IEC 38 (i.e. up to $\pm$ 10%). The rate of occurrence ranges from 25 per second to one per minute and it should be differentiated from the normal slow variations within the same $\pm$10% limit due to gradual load changes in the networks. Generally, such a phenomenon is characterised by the amplitude of the voltage changes and the rate of repetition.

Many past flicker standards and guidelines prescribe limits for voltage fluctuation in order to limit annoying light flicker. Many voltage flicker curves give the percentage magnitude of the voltage pulsation and the pulsation frequency that leads to perceptible flicker. These curves were developed through considerable testing and actual operating experience. Different curves may provide different acceptable or tolerable pulsation magnitudes and frequencies but they all follow a similar trend as depicted in Figure 2.6. The troughs of the curves show the most sensitive frequency to be around 8 Hz.

IEC 868: 1986 [20]; IEC 61000-4-15: 1997 [22] A common method of analysing the severity of a flicker disturbance is to measure the fluctuation of light luminosity of an incandescent lamp. The IEC 61000-4-15 standard outlines the functional and design specification for flicker measuring apparatus, which has evolved from a proposal by the Union Internationale d'Electrothermie [24]. This assessment of flicker can be broadly divided into two parts: measurement of instantaneous flicker sensation as perceived by human eyes, and statistical evaluation of its severity level in the short and long term. These parts are described in greater detail in Chapter 5.

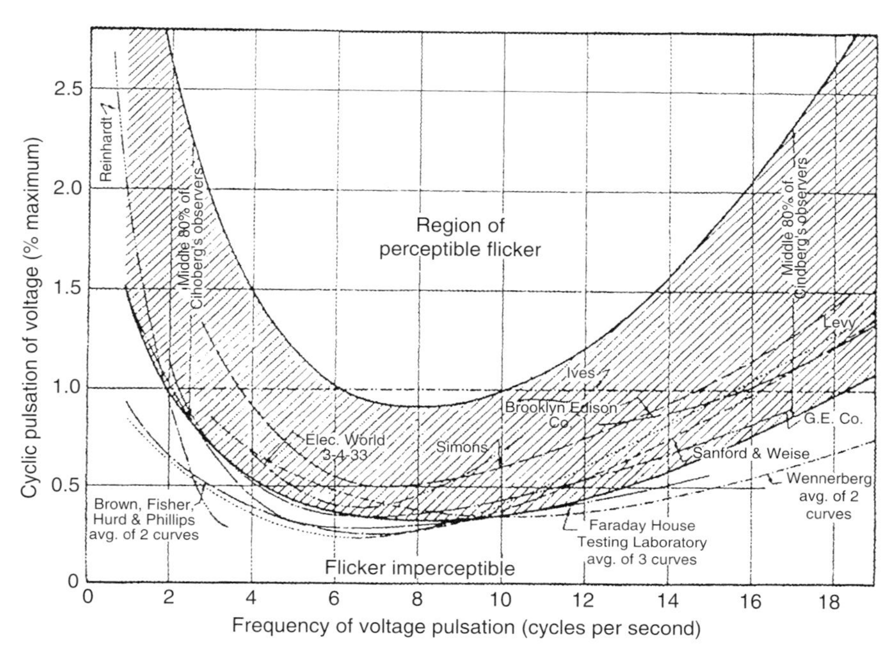

Figure 2.6 Comparison of voltage flicker curves from a number of sources [23] (reproduced from [23] by permission from IEEE)

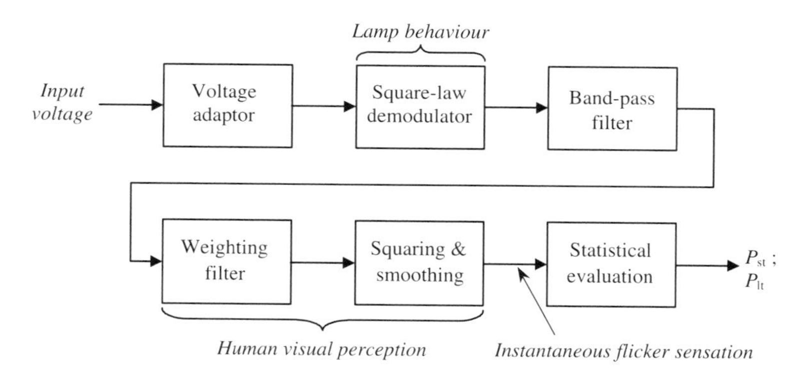

Figure 2.7 Block diagram of IEC flickermeter

Figure 2.7 shows the function block diagram of the flickermeter simplified from IEC 61000-4-15. The input voltage adaptor scales the input voltage dynamically to an average r.m.s. value. This enables the flicker measurement to be made independently of the actual input carrier voltage level. The scaling has a low-pass characteristic with a 60 s time constant tracking slow changes due to voltage regulation. The response of the lamp to voltage fluctuation is simulated with a square law demodulator. Squaring of the voltage yields instantaneous power supplied to the lamp assuming a constant filament resistance. The band-pass filter eliminates the d.c. and doubles the mains frequency ripple of the demodulator output.

The weighting filter reflects the band-pass characteristics of the human eye [25]. The squaring accounts for the fact that positive and negative brightness changes lead to a similar sensation. The smoothing with a low-pass filter realises integration over the waveform so that the flicker sensation is independent of the sequence of the luminance variations. The instantaneous flicker sensation levels are then noted for the statistical evaluation of flicker severity.

IEC 868-0: 1991 [21] This standard is based on a statistical procedure proposed by the UIE and then adopted by the IEC to evaluate the severity of the voltage fluctuation on the light flicker. The amplitude of the flicker sensation is subdivided into a suitable number of classes. Every time that the appropriate value occurs, the counter of the corresponding class is incremented by one. In this way, the frequency distribution function of the flicker sensation is obtained. By choosing a sampling frequency of at least twice the maximum flicker frequency, the result at the end of the measuring interval represents the distribution of flicker level duration in each class. Adding the contents of the counters of all classes, and expressing the count of each class relative to the total, gives the probability density function of the flicker levels.

This function is used to derive the cumulative probability function where significant statistical values such as the mean, standard deviation and the percentage time when the flicker level exceeds certain limits, are obtained.

The cumulative probability, $p(l)$, that the IFL exceeds l is defined as

$$p(l) = \frac{t_l}{T}, \tag{2.6}$$

where t_l = the duration of time during which the signal remains above l, and T = total observation time.

To be practical, only a limited number of $p(l)$ curve points can be computed. Calculating N curve points is equivalent to dividing the full signal range into $N-1$ classes and this method has, therefore, been termed *time at level classification*. Figure 2.8 illustrates this approach using only a small number of evenly spaced classes for clarity. Alternatively, classes can be spaced logarithmically, which allows covering the full dynamic signal range without range switching. The resulting discretised Cumulative Probability Function (CPF) for the observation time is shown in Figure 2.9. Clearly, an increased number of classes will lead to a more accurate distribution but at the expense of more computation.

In order to limit computation and increase the accuracy of the classification methods (linear and logarithmic), several interpolation techniques have been proposed by the IEC [21].

Two severity indices have been proposed for flicker evaluation; namely, short-term flicker severity (P_{st}) and long-term flicker severity (P_{lt}). Short-term severity, P_{st} is generally computed over 10 min and is derived from the time-at-level statistics obtained from the cumulative probability function:

$$P_{st} = \sqrt{0.0314P_{0.1} + 0.0525P_{1s} + 0.0657P_{3s} + 0.28P_{10s} + 0.08P_{50s}}, \tag{2.7}$$

where the percentiles $P_{0.1}$, P_1, P_3, P_{10} and P_{50} are the flicker sensation levels exceeded for 0.1%; 1%; 3%; 10% and 50% of the time during the observation period. The suffix

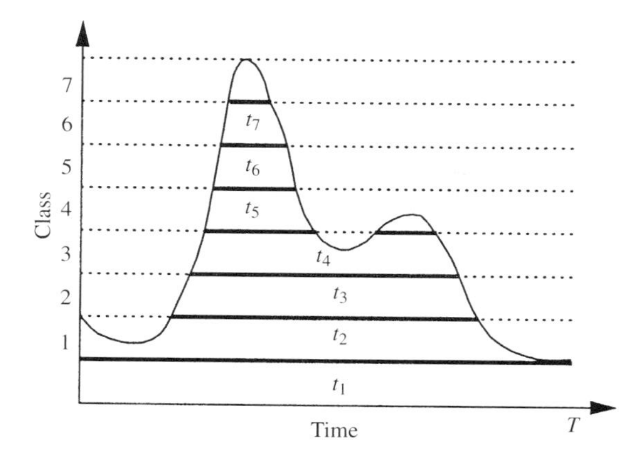

Figure 2.8 Linear classification of time series. Each $t_k (k = 1, \ldots, 7)$ indicates the duration the signal exceeds the lower limit of the corresponding class k

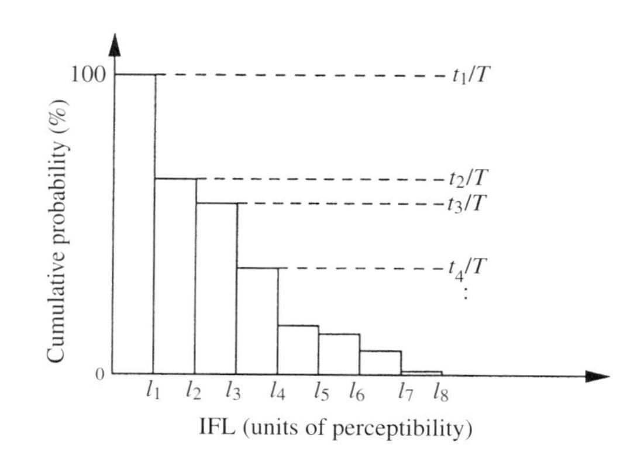

Figure 2.9 Cumulative Probability Function (CPF) of the signal in Figure 2.8 for the observation time T

s in the formula indicates that the smoothed value should be used.

$$\begin{aligned}
P_{50\mathrm{s}} &= (P_{30} + P_{50} + P_{80})/3, \\
P_{10\mathrm{s}} &= (P_6 + P_8 + P_{10} + P_{13} + P_{17})/5, \\
P_{3\mathrm{s}} &= (P_{2.2} + P_3 + P_4)/3, \\
P_{1\mathrm{s}} &= (P_{0.7} + P_1 + P_{1.5})/3.
\end{aligned}$$

The 0.3 s memory time constant in the flickermeter ensures that $P_{0.1}$ cannot change abruptly and no smoothing is needed for this percentile.

The 10 min period on which the short-term flicker severity evaluation is based is suitable for assessing disturbances caused by individual sources with a short duty cycle.

For flicker sources with long and variable duty cycles such as arc furnaces, the long-term flicker severity P_{lt} is derived from the short-term flicker severity levels. The long-term interval is usually several hours long in multiples of the short-term intervals, and is related to the duty cycle of the load or a period over which an observer may react to flicker:

$$P_{\mathrm{lt}} = \sqrt[3]{\frac{\sum_{i=1}^{N} P_{\mathrm{sti}}^{3}}{N}}, \tag{2.9}$$

where $P_{\mathrm{sti}}(i = 1, 2, 3 \ldots)$ are consecutive readings of the short-term severity P_{st}.

2.7 Summary

Most standards concentrate on the effect of power imperfection on equipment and the impact on the system caused by a piece of equipment. The measurement and monitoring techniques and guidelines focus on how to test the immunity and emission levels of equipment. There is no standard or guideline detailing how to perform quality assessment on a system as a whole. In practice, it is up to the individual to decide which parameters are to be analysed under specific conditions, and apply the corresponding indices judiciously. Likewise, the measurement results have to be interpreted according to the issues of concern to the customer or utility. It is impossible to define every probable combination of certain types of equipment and environment, and to outline the necessary procedures or steps for assessing the quality of electric power in a system.

Standards are specific to a certain disturbance and generally assume that no other disturbances occur during the observation interval. However, a power system experiences a multitude of disturbances concurrently and this presents a challenge to the instrument designer as to how to assess the system quality amidst different standards and indices. Therefore, the assessment of power quality in the presence of all variants of disturbances remains quite a challenge for the designers of equipment.

The variety of power quality problems also makes it difficult to search for the most appropriate standard to be used. The IEEE 1159 standard has remedied this problem to a certain extent by bringing all disturbances into one single document. Although the IEC provides a comprehensive structure for cataloguing the EMC series of standards, it still requires a large amount of effort to locate the appropriate or relevant standard.

Although not explicitly used in the chapter, several recent documents have influenced its content and are, therefore, added to the list of references [26]–[34]. A frequently updated list of available standards, categorised according to the phenomena, will simplify the search for appropriate information.

2.8 References

1. Arrillaga, J, Bradley D A and Bodger, P S, (1985) *Power System Harmonics*, John Wiley & Sons.
2. IEEE 141:1986, Recommended Practice for Electric Power Distribution for Industrial Plants.
3. IEEE 1159: 1995, IEEE Recommended Practice on Monitoring Electric Power Quality.
4. IEC 61000-2-5: 1995, Electromagnetic Compatibility (EMC), Part 2: Environment, Section 5: Classifications of Electromagnetic Environments.

5. IEC 61000-2-1: 1990, Electromagnetic Compatibility (EMC), Part 2: Environment, Section 1: Description of the Environment - Electromagnetic Environment for Low-Frequency Conducted Disturbances and Signalling in Public Power Supply Systems.
6. IEC 61000-2-2: 1990, Electromagnetic Compatibility (EMC), Part 2: Environment, Section 2: Compatibility Levels for Low-Frequency Conducted Disturbances and Signalling in Public Power Supply Systems.
7. IEEE c62.41: 1991, IEEE Recommended Practice on Surge Voltages in Low-Voltage AC Power Circuits.
8. IEC 816: 1984, Guide on Methods of Measurement of Short Duration Transients on Low Voltage Power and Signal Lines.
9. UIE-DWG-2-92-D, UIE Guide to Measurements of Voltage Dips and Short Interruptions Occurring in Industrial Installations.
10. National Technical Information Service, (1983), Federal Information Processing Standards Publication 94: Guideline on Electrical Power for ADP Installations,
11. ITI (Information Technology Industry Council, formerly known as the Computer & Business Equipment Manufacturer's Association), ITI Curve Application Note, available at http://www.itic.org/iss_pol/techdocs/curve.Pdf.
12. Brooks, D L, Dugan, R C, Waclawiak, M and Sundaram, S, (1997). Indices for assessing utility distribution system r.m.s. variation performance, *IEEE Transactions on Power Delivery*, PE-920-PWRD-1-04.
13. IEEE 519: 1992, IEEE Recommended Practices and Requirements for Harmonic Control in Electric Power Systems (ANSI).
14. IEC 61000-4-7, 1991, Electromagnetic Compatibility (EMC), Part 4: Limits, Section 7: General Guide on Harmonics and Inter-harmonics Measurements and Instrumentation, for Power Supply Systems and Equipment Connected Thereto.
15. Directives concerning the Protection of Telecommunication Lines against Harmful Effects from Electricity Lines, International Telegraph and Telephone Consultative Committee (CCITT) published by the International Communications Union, Geneva, 1963.
16. Engineering Reports of the Joint Subcommittee on Development and Research of the Edison Electric Institute and the Bell Telephone System, New York, 5 volumes, July 1926 to January 1943.
17. IEEE/ANSI C57.110-1986, Recommended Practice for Establishing Transformer Capability When Supplying Non-sinusoidal Load Currents.
18. IEEE/ANSI Std 18-1980 (Reaff 1991), IEEE Standard for Shunt Power Capacitors.
19. IEEE Interharmonic Task Force, CIGRE 36.05/CIRED 2 CC02 Voltage Quality Working Group, Interharmonics in Power Systems, January 1997.
20. IEC 868: 1986, Flickermeter - Functional and Design Specifications.
21. IEC868-0: 1991, Flickermeter - Evaluation of Flicker Severity.
22. IEC 61000-4-15 Ed. 1, 1997, Electromagnetic Compatibility (EMC), Part 4: Limits, Section 15: Flickermeter - Functional and Design Specifications.
23. Walker, M K, (1979). Electric utility flicker limitations, *IEEE Transactions on Industrial Applications,*. **IA-15**, (6), pp. 644–655.
24. Union Internationale d' Electrothermie, (1991) *WG Disturbances, Flicker Measurement and Evaluation*, 2nd revised edition, Union Internationale d'Electrothermie.
25. de Lange, H,(1954). Relationship between critical flicker-frequency and a set of low-frequency characteristics of the eye, *Journal of the Optical Society of America*, **11**(5), pp. 380–389.
26. IEC 38: 1983, IEC Standard Voltages.
27. ANSI C84.1: 1982, American National Standard For Electric Power Systems and Equipment - Voltage Ratings (60 Hz).

28. UIE-DWG-3-92-G, Guide to Quality of Electrical Supply for Industrial Installations - Part 1: General Introduction to Electromagnetic Compatibility (EMC), Types of Disturbances and Relevant Standards.
29. IEEE 100:1992, IEEE Standard Dictionary of Electrical and Electronics Terms.
30. EN 50160: 1994, Voltage Characteristics of Electricity Supplied by Public Distribution System, CENELEC.
31. IEC 61000-3-2: 1994, Electromagnetic Compatibility (EMC), Part 3: Limits, Section 2: Limits for Harmonic Current Emissions (Equipment Input Current$\leq$ 16 A per Phase).
32. IEC 61000-3-4: 1994, Electromagnetic Compatibility (EMC), Part 3: Limits, Section 4: Limits for Harmonic Current Emissions (Equipment Input Current $\geq$ 16 A per Phase).
33. IEC 61000-3-3: 1994, Electromagnetic Compatibility (EMC), Part 3: Limits, Section 3: Limitation of Voltage Fluctuations and Flicker in Low-Voltage Supply System for Equipment with Rated Current $\leq$ 16 A.
34. IEC 61000-3-5: 1994, Electromagnetic Compatibility (EMC), Part 3: Limits, Section 5: Limitation of Voltage Fluctuations and Flicker in Low-Voltage Supply System for Equipment with Rated Current Greater than 16 A.

3

POWER ASSESSMENT UNDER WAVEFORM DISTORTION

3.1 Introduction

The presence of waveform distortion requires detailed consideration of:

- the definitions of power and energy which, under perfect sinewaves, involve concepts and formulations universally accepted, whereas there are considerable discrepancies among authors in the case of distorted waveforms.
- the techniques used for reactive power compensation, conventionally involving linear capacitors, while their use with distorted waveforms does not achieve high power factors and excites resonant conditions.
- the use of tariffs, often based on the measurement of cos Φ *via* instruments of the Ferraris type, while with non-sinusoidal waveforms the power factor should be derived using the apparent power.

The classical power and energy meters will gradually be replaced by more versatile designs incorporating all the relevant parameters and capable of updating by software the definitions internationally accepted. This will permit establishing new concepts of tariffs penalising not only the demand of reactive power (the present practice) but also the contribution to waveform distortion.

The chapter presents a concise historical background on the development of power definitions and compares their effectiveness in the presence of modern non-linear circuits.

3.2 Single-Phase Definitions

Early attempts to include waveform distortion into the power definitions were made by Budeanu [1] and Fryze [2].

Budeanu divided the apparent power into three orthogonal components, i.e.

$$S^2 = P^2 + Q_B^2 + D^2, \tag{3.1}$$

and defined the terms reactive power,

$$Q_{\mathrm{B}} = \sum_{l=1}^{n} V_l I_l \sin \varphi_l, \tag{3.2}$$

accepted by the IEC and IEEE [3] and complementary power (which he called fictitious)

$$P_{\mathrm{C}} = \sqrt{S^2 - P^2}. \tag{3.3}$$

Fryze separated the current into two orthogonal components, i_{a} (active) and i_{b} (reactive)

$$i = i_{\mathrm{a}} + i_{\mathrm{b}}, \tag{3.4}$$

and proposed for the reactive power the following definition:

$$Q_{\mathrm{F}} = V I_{\mathrm{b}} = \sqrt{S^2 - P^2}. \tag{3.5}$$

To add some physical measuring to the matter, Shepherd [4] proposed the following decomposition for the apparent power:

$$S^2 = S_{\mathrm{R}}^2 + S_{\mathrm{X}}^2 + S_{\mathrm{D}}^2 \tag{3.6}$$

being

$$S_{\mathrm{R}}^2 = \sum_1^n V_n^2 \sum_1^n I_n^2 \cos{}^2 \varphi_n, \tag{3.7}$$

$$S_{\mathrm{X}}^2 = \sum_1^n V_n^2 \sum_1^n I_n^2 \sin^2 \varphi_n, \tag{3.8}$$

$$S_{\mathrm{D}}^2 = \sum_1^n V_n^2 \sum_1^p I_p^2 + \sum_1^m V_m^2 \left(\sum_1^n I_n^2 + \sum_1^p I_p^2 \right). \tag{3.9}$$

In these expressions, S_{R} is said to be active apparent power, S_{X} reactive apparent power and S_{D} distortion apparent power.

The main advantage of this decomposition is that the minimisation of S_{X} immediately leads to the optimisation of the power factor via the addition of a passive linear element, a property not available when applying Budeanu's definition; however, there is no justification for a power decomposition with an active component that differs from the mean value of the instantaneous power over a period (i.e. the active power).

An alternative model also involving three components was proposed by Sharon [5], i.e.

$$S^2 = P^2 + S_{\mathrm{Q}}^2 + S_{\mathrm{C}}^2, \tag{3.10}$$

where

$$P \sim \text{ active power },$$

$$S_{\mathrm{Q}} = V \sqrt{\sum_1^n I^2{}_n \sin^2 \varphi_n} \sim \text{a reactive power in quadrature}, \tag{3.11}$$

and

$$S_C \sim \text{a complementary reactive power.}$$

Similarly to Shepherd's, the minimisation of S_Q in this model results in a maximum power factor via the connection of linear passive elements. Moreover, Sharon replaces Shepherd's S_R questionable term by the more acceptable active power P.

Taking into account that, in general, the main contribution to reactive power comes from the fundamental component of the voltage, Emanuel [6] proposed the following definitions:

$$Q_1 = V_1 I_1 \sin \varphi_1 \tag{3.12}$$

and a complementary power

$$P_C^2 = S^2 - P^2 - Q_1^2. \tag{3.13}$$

Based on Fryze's theory, Kusters and Moore [7] proposed a power definition in the time domain with the current divided into three components:

i_p = an active component with a waveform identical to that consumed in an ideal resistance,
i_{ql}/i_{qc} = a reactive component, corresponding to either a coil or a capacitor,
i_{qlr}/i_{qcr} = a residual reactive component, the remaining current after removing the active and reactive, i.e.

$$i_{qr} = i - i_p - i_q. \tag{3.14}$$

Furthermore, Kusters and Moore suggested the following decomposition for the apparent power:

$$S = P^2 + Q_l^2 + Q_{lr}^2 = P^2 + Q_c^2 + Q_{cr}^2 \tag{3.15}$$

where $P = VI_p$ is the active power, $Q_l = VI_{ql}$ the inductive reactive power, $Q_c = UI_{qc}$ the capacitive reactive power, and Q_{lr}/Q_{cl} the remaining reactive powers obtained from Equation (3.14).

Also based on Fryze's definition, Emanuel [8] proposed two alternative decompositions distinguishing between the power components of the fundamental frequency (P_1, Q_1) and harmonics (P_H, Q_H), i.e.

$$S^2 = (P_1 + P_H)^2 + Q^2{}_F, \tag{3.16}$$

where

$$Q_F^2 = Q_B^2 + D^2, \tag{3.17}$$

and

$$S^2 = (P_1 + P_H)^2 + Q_1^2 + Q_H^2, \tag{3.18}$$

where

$$Q_H^2 = Q_F^2 - Q_1^2. \tag{3.19}$$

Further articles elaborating on the decomposition of power under non-sinusoidal conditions have been published by Czarnecki [9] and Slonin [10].

In a recent contribution, Emanuel [11] makes the following statements.

1 All forms of non-active powers stem from energy manifestations that have a common mark: energy oscillations between different sources, sources and loads or loads and loads. The net energy transfer linked with all the non-active powers is nil.

2 Due to the unique significance of the fundamental powers, S_1, P_1 and Q_1, that comes from the fact that electric energy is a product expected to be generated and delivered and bought in the form of a 60 Hz or 50 Hz electromagnetic field, it is useful to separate the apparent power S in fundamental S_1 and the non-fundamental S_N apparent powers:

$$S^2 = S_1^2 + S_N^2. \tag{3.20}$$

3 As well as all the non-active powers, the term S_N contains also a minute amount of harmonic active power, P_H. The harmonic active power rarely exceeds $0.005P_1$. Thus, in a first approximation, an industrial non-linear load can be evaluated from the measurements of P_1, Q_1 and S_N.

4 The further subdivision of S_N into other components provides information on the required dynamic compensator or static filter capacity and level of current and voltage distortion.

3.2.1 Illustrative Examples [12]

To illustrate the practical consequences of using each of the above definitions, a comparative test study is shown with four single-phase networks with identical r.m.s. values of voltage (113.65 V) and current (16.25 A). In case A, the applied voltage is sinusoidal and the load non-linear; case B contains voltage and current of the same, non-sinusoidal, waveform and in phase with each other; in case C, the two identical waveforms are out of phase and, finally, in D, the voltage and current have harmonics of different orders.

The numerical information corresponding to these four cases is listed in Table 3.1 and the time domain variation of the voltages and currents is shown in Figure 3.1.

Table 3.1 Voltage and current phasors $V = 113.65$ V, $I = 16.25$ A

CASE		$V_1\angle\alpha_1$	$V_3\angle\alpha_3$	$V_5\angle\alpha_5$	$V_7\angle\alpha_7$
A	V_A	$113.65\angle 0°$			
	I_A	$15\angle -30°$	$5.8\angle 0°$	$2\angle 0°$	$1\angle 0°$
B	V_B	$105\angle 0°$	$35\angle 0°$	$21\angle 0°$	$15\angle 0°$
	I_B	$15\angle 0°$	$5\angle 0°$	$3\angle 0°$	$(15/7)\angle 0°$
C	V_C	$105\angle 0°$	$35\angle 0°$	$21\angle 0°$	$15\angle 0°$
	I_C	$15\angle -30°$	$5\angle -90°$	$3\angle -150°$	$(15/7)\angle 150°$
D	V_D	$105\angle 0°$	$40.82\angle 180°$		$15\angle 0°$
	I_D	$15\angle -30°$		$5.44\angle -60°$	$3\angle -30°$

Table 3.2 Alternative powers for four circuits with the same r.m.s. voltage and current but with different waveforms

Magnitude		Case A	Case B	Case C	Case D
Active	P(W)	1476	1845	1282	1403
Budeanu	Q_B(Var)	852	0	978	810
Shepherd	S_X(VA)	852	0	1046	811
Sharon	Q_S(Var)	852	0	1046	870
Emanuel	Q_I(Var)	852	0	788	788
Fryze	Q_F(Var)	1107	0	1327	1198
Budeanu	D(VA)	707	0	897	883
Apparent	S(VA)/PF	1845/0.80	1845/1	1845/0.695	1845/0.76

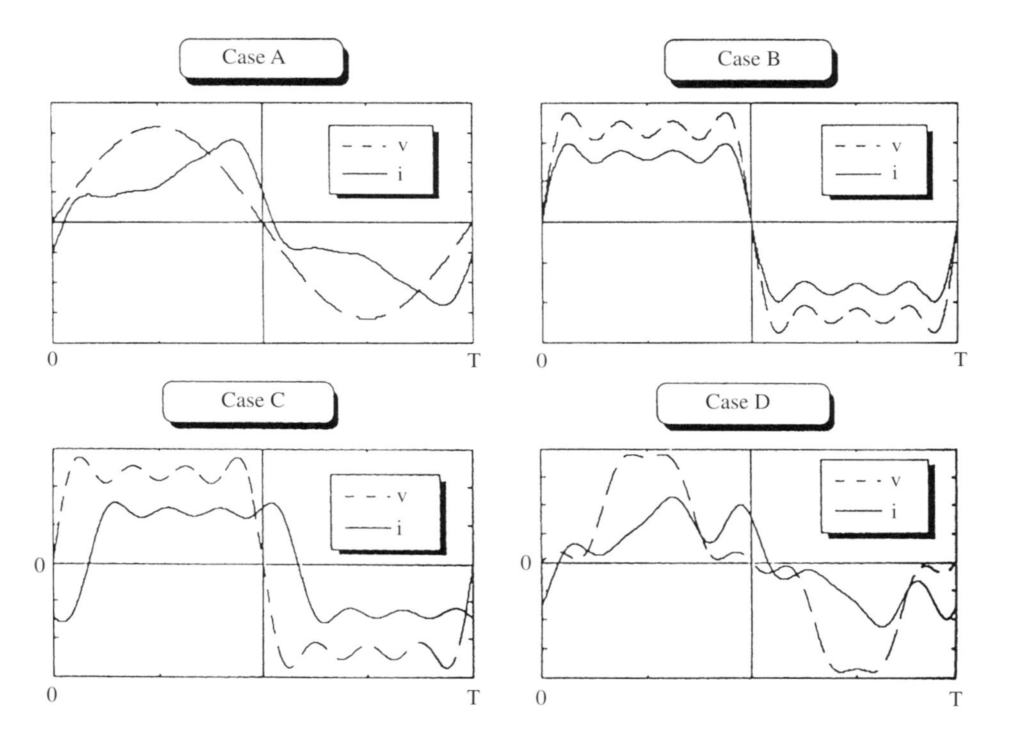

Figure 3.1 Voltage and current waveforms

Table 3.2 lists, for each case, the values of the powers previously defined, i.e. active P, reactive Q_B, apparent reactive S_X, reactive in quadrature Q_S, reactive of fundamental component Q_1, Fryze's reactive Q_F, distortion D, apparent S and also the power factor, PF.

The rest of this section describes the results of three different comparisons made between cases A, C and D.

The first method, type 1 in Table 3.3, relates to conventional capacitive compensation and the table indicates the value of the capacitor (Co) that optimises the power factor $PF_{\max 1}$ and the required apparent power $S_{\min 1}$.

Table 3.3 S_{min} and PF_{max} values for different compensation techniques

	Compensation	Case A	Case B	Case C
Type 1	$S_{min\,1}$(VA)/Co(μF)/$PF_{max\,1}$	1636/210/0.902	1693/98/0.757	1759/81/0.798
Type 2	$S_{min\,2}$(VA)/$PF_{max\,2}$	1636/0.92	1636/0.785	1636/0.858
Type 3	$S_{min\,3}$(VA)/$PF_{max\,3}$	1476/1	1476/0.869	1476/0.950

The second method, type 2, consists of a sinusoidal current injection of fundamental frequency that cancels the reactive current i_{r1} in the load; the corresponding apparent power and power factor are S_{min2} and PF_{max2}, respectively.

Finally, in type 3 compensation, the current injection not only cancels the reactive component i_{r1}, but also the harmonic currents; S_{min3} and PF_{max3} are the apparent power and power factor, respectively.

The more relevant comments resulting from the comparison are as follows.

(1) When the voltage is sinusoidal and the current non-sinusoidal (case A), the load is non-linear; in this case, all the reactive powers obtained are identical, except Fryze's, which includes the distortion component. With reference to the optimisation process, $PF_{max\,1} = PF_{max\,2}$, as could be expected considering the sinusoidal nature of the voltage.

(2) If the voltage and current are in phase and of the same waveform (case B) the instantaneous power never reaches negative values; therefore $P = S$, the reactive and distortion powers are zero and the power factor unity, the load equivalent impedance being a linear resistance.

(3) When the voltage and current have the same waveform but are out of phase with each other (case C) the reactive powers calculated according to the various definitions proposed are different. In this case, the passive circuit does not distort the current waveform but alters its phase such that the power factor increases from an initial value of 0.695 to an optimal value $PF_{max\,3}$ of 0.869.

(4) Case D, due to the load non-linearity, contains different harmonics; the third harmonic is only present in the voltage and the fifth in the current waveforms, respectively. This is the most general case; all the reactive powers calculated are different and the maximum power factor ($PF_{max\,3}$) is 0.95.

(5) The reactive powers calculated according to the various definitions are all different; the smallest is Emanuel's, since he only considers the fundamental component, and the largest Fryze's, which also includes distortion. In general

$$Q_1 \leq Q_B \leq S_X \leq Q_S \leq Q_F \tag{3.21}$$

3.3 Three-Phase Definitions

Apparent power in unbalanced three-phase systems is currently calculated using several definitions that lead to different power factor levels. Consequently, the power bills will also differ due to the reactive power tariffs and, in some countries, to the direct registration of the maximum apparent power demand.

Four different expressions have been proposed for the apparent reactive power, two of them based in Budeanu's and the other two on Fryze's definitions, [2,3,13].

(i)
$$S_{\nu} = \sqrt{\left(\sum_k P_k\right)^2 + \left(\sum_k Q_{\mathrm{b}k}\right)^2 + \left(\sum_k D_k\right)^2} \tag{3.22}$$

or vector apparent power:

(ii)
$$S_{\mathrm{a}} = \sum_k \sqrt{P_k^2 + Q_{\mathrm{b}k}^2 + D_k^2} \tag{3.23}$$

or arithmetic apparent power.

In the above expressions P_k represent the active power, $Q_{\mathrm{b}k}$ Budeanu's reactive power and D_k the distortion power in phase k; these terms are generally accepted by the main international organisations, such as the IEEE and IEC.

(iii)
$$S_{\mathrm{e}} = \sum_k \sqrt{P_k^2 + Q_{\mathrm{f}k}^2} = \sum_k V_k I_k \tag{3.24}$$

an apparent r.m.s. power, that considers independently the power consumed in each phase.

(iv)
$$S_{\mathrm{s}} = \sqrt{P^2 + Q_{\mathrm{f}}^2} = \sqrt{\sum_k V_k^2}\sqrt{\sum_k I_k^2} \tag{3.25}$$

a system apparent power, that considers the three-phase network as a unit.

The last two expressions use the reactive powers as defined by Fryze, Q_{f} and $Q_{\mathrm{f}k}$, which include not only the reactive but also the distortive effects.

In the vector apparent power the phase reactive powers compensate each other, but not in the other expressions; the system apparent power calculates the voltage and current of each phase individually and therefore yields the highest apparent power. In general the following applies:

$$S_{\nu} \leq S_{\mathrm{a}} \leq S_{\mathrm{e}} \leq S_{\mathrm{s}}. \tag{3.26}$$

The power factor of a load or system is generally accepted as a measure of the power transfer efficiency and is defined as the ratio between the electric power transformed into some other form of energy and the apparent power, i.e.

$$PF = P/S. \tag{3.27}$$

Correspondingly, the relative magnitudes of the power factors calculated from the different definitions are

$$PF_{\nu} \geq PF_{\mathrm{a}} \geq PF_{\mathrm{e}} \geq PF_{\mathrm{s}}, \tag{3.28}$$

where PF_{ν}, PF_{a}, PF_{e}, PF_{s} are the power factors corresponding to the vector, arithmetic, r.m.s. and system apparent powers, respectively.

For the power factor to reflect the system efficiency in three-phase networks with a neutral wire, the neutral (zero sequence) currents must be included in the calculation of the equivalent current, i.e. in Equation (3.25):

$$I_k^2 = I_a^2 + I_b^2 + I_c^2 + I_n^2 \tag{3.29}$$

and

$$V_k^2 = V_a^2 + V_b^2 + V_c^2, \tag{3.30}$$

where a, b, and c indicate the individual phase values and n the neutral.

3.3.1 Illustrative Examples [14,15]

The simple test system shown in Figure 3.2 is used to illustrate the different performance of the proposed three-phase power definitions. Asymmetrical three-phase networks, even with sinusoidal voltage excitation and linear resistive loading, interchange reactive energy between the generator phases, despite the absence of energy-storing elements. To show this effect, the circuit of Figure 3.2 is further simplified by making the line resistances equal to zero and assuming that the load is purely resistive, i.e. $Z_a = R_a$, $Z_b = R_b$, $Z_c = R_c$. Moreover, the three-phase source is assumed to be balanced and sinusoidal.

The following different operating conditions are compared in Table 3.4.

Case (i): $R_a = 0\ \Omega$, $R_b = 5\ \Omega$, $R_c = 25\ \Omega$, without neutral wire.
Case (ii): $R_a = 5\ \Omega$, $R_b = 5\ \Omega$, $R_c = 25\ \Omega$, without neutral wire.
Case (iii): $R_a = 5\ \Omega$, $R_b = 5\ \Omega$, $R_c = 25\ \Omega$, with neutral wire.
Case (iv): $R_a = 5\ \Omega$, $R_b = 5\ \Omega$, $R_c = 5\ \Omega$, without neutral wire.

As shown in Table 3.4, in case (i), phases a and c generate reactive power, whilst phase b absorbs reactive power, even though the overall reactive power requirement is zero. The effect is attenuated when a load is connected to phase a (case (ii)), but there is still reactive power in two phases. When the circuit topology is changed by

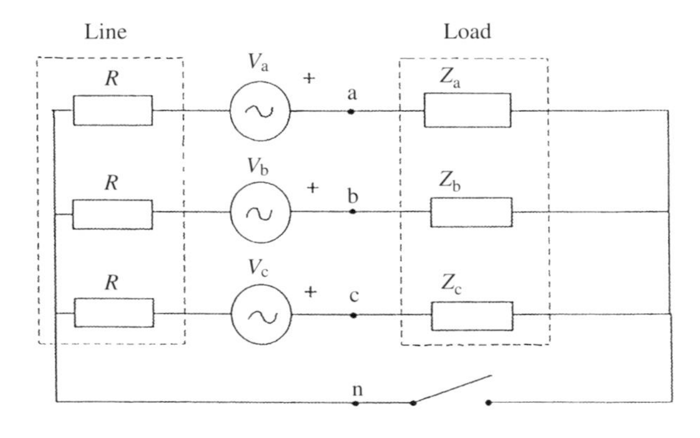

Figure 3.2 Three-phase test system

Table 3.4 Apparent powers and power factors for test cases (i) to (iv)

Case	SV_a	SV_b	SV_c	SV	$S_a = S_e$	S_s	PF_ν	$PF_a = PF$	PF_s
i	32.4 − j12.46	27 + j15.58	5.4 − j3.11	64.8	72.12	81.52	1	0.898	0.794
ii	14.7 − j5.7	14.7 + j5.7	4.9	34.36	36.47	39.58	1	0.942	0.868
iii	18	18	3.6	39.6	39.6	44.53	1	1	0.889
iv	18	18	18	54	54	54	1	1	1

Note: All the powers are expressed in kVA.

connecting a neutral wire (case (iii)) the generating source stops generating reactive power; the reactive power generation is also absent when phase c is made equal to those of phases a and b (case (iv)), i.e. when the network is perfectly balanced and this is the case of maximum efficiency.

Table 3.4 shows for each case the apparent powers resulting from the different definitions as well as the corresponding power factors. The table also shows the complex apparent powers generated by the sources V_a, V_b, V_c, respectively. It can be seen that the most pessimistic power factor is PF_s, which only gives the value of one when the resistive load is perfectly balanced; on the other hand PF_ν remains at unity in all cases since the total load demand for reactive power, being resistive, is zero.

Next, a set of test cases (v) to (x) involves the circuit of Figure 3.2 with a perfectly balanced source feeding either linear or non-linear loads. The calculations performed in each case involve the four power factors defined above, the resistive losses in the line and the line loss ratios of each of the cases to that of case v (the balanced case), which is used as a reference; the latter will show that only one of the definitions coincides with that ratio.

As the powers consumed by the load are different in all cases, for the purpose of comparison the power of case v is also used as a reference, i.e. 54 kW in the test system.

Case (v): V_{ns} (non-sinusoidal voltages), balanced load, 3 or 4 wires:

$$R_a = R_b = R_c = 5\ \Omega,\ R = 0.2\ \Omega/\text{phase},$$
$$\nu_a = 298.51\sqrt{2}\sin(\omega t) + 29.85\sqrt{2}\sin(5\omega t),\ V_a = 300\ \text{V},$$
$$i_a = 59.70\sqrt{2}\sin(\omega t) + 5.97\sqrt{2}\sin(5\omega t),\ I_a = 60\ \text{A},$$
$$P = 3 * 298.51 * 59.70 + 3 * 29.85 * 5.97 = 54000\ \text{W}.$$

The power loss in the line is: $P_\nu = 3 * (60)^2 * 0.2 = 2160$ W. It should be noted that this value is the same as case (iv) (the balanced sinusoidal circuit) for the same line resistance.

Case (vi)(a): V_{ns}, unbalanced load, 4 wires:

$$R_a = R_b = 5\ \Omega,\ R_c = 25\ \Omega,$$
$$\nu_a = 298.51\sqrt{2}\sin(\omega t) + 29.85\sqrt{2}\sin(5\omega t),$$
$$\text{i}_a = 59.70\sqrt{2}\sin(\omega t) + 5.97\sqrt{2}\sin(5\omega t),\quad I_a = 60\ \text{A},$$

$$i_b = 59.70\sqrt{2}\sin(\omega t - 120) + 5.97\sqrt{2}\sin(5\omega t + 120), \quad I_b = 60 \text{ A},$$
$$i_c = 11.94\sqrt{2}\sin(\omega t + 120) + 1.19\sqrt{2}\sin(5\omega t - 120), \quad I_c = 12 \text{ A}.$$

The topology of this case does permits neither distorted nor reactive power.

To calculate the line loss in relation to the power base, the calculated currents are multiplied by factor K, which is the ratio of the power in case (v) to that of (vi)(a); therefore

$$K = 54000/39600 = 1.36,$$
$$P_{vi} = [(60 * 1.3636)^2 + (60 * 1.3636)^2 + (12 * 1.3636)^2] * 0.2 = 2731.23 \text{ W}.$$

The ratio $P_v/P_{vi} = 0.79$ is the square of PF_s. It is observed that although the consumed power has no reactive or distorted component, due to load unbalance, line losses are larger and the power factors PF_v, PF_a and PF_e do not reflect that loss of network efficiency; on the other hand PF_s represents this increment faithfully. It should be noted that the value P_{iii} (of case (iii)) would be identical to P_{vi} if the line resistance (i.e. 0.2 Ω/phase) had been represented.

Case (vi)(b): V_{ns}, unbalanced load, 3 wires:

$$R_a = R_b = 5\ \Omega,\ R_c = 25\ \Omega,\ R = 0.2\ \Omega/\text{phase},$$
$$v_a = 298.51\sqrt{2}\sin(\omega t) + 29.85\sqrt{2}\sin(5\omega t),$$
$$i_a = 52.34\sqrt{2}\sin(\omega t + 21.05) + 5.234\sqrt{2}\sin(5\omega t - 21.05), \quad I_a = 52.60 \text{ A},$$
$$i_b = 52.34\sqrt{2}\sin(\omega t - 141.05) + 5.234\sqrt{2}\sin(5\omega t + 141.05), \quad I_b = 52.60 \text{ A},$$
$$i_c = 16.34\sqrt{2}\sin(\omega t + 120) + 1.628\sqrt{2}\sin(5\omega t - 120), \quad I_c = 16.36 \text{ A}.$$

The different behaviour in this case with respect to reactive and distorted power is due to the new topology, which excludes the neutral wire.

The line losses, normalised to 54 kW are $P_{vi}(b) = 2865.3$ and the ratio $P_v/P_{vi} = 0.75$ is again the square of PF_s.

Case (vii): V_s (sinusoidal), non-linear load, I_{ns} (non-sinusoidal) and balanced:

$$R = 0.2\ \Omega/\text{phase},$$
$$v_a = 300\sqrt{2}\sin(\omega t),$$
$$i_a = 60\sqrt{2}\sin(\omega t) + 6\sqrt{2}\sin(5\omega t), \quad I_a = I_b = I_c = 60.3 \text{ A}.$$

In this case, the load non-linearity produces a harmonic component not present in the voltage source. The power consumed is 54 kW and the values of the various power factors are identical and close to unity, in spite of the fifth harmonic current, because the fundamental component of the current is balanced and its power factor is unity.

Case (viii): V_s, non-linear load, I_{ns} unbalanced:

$$R = 0.2\ \Omega/\text{phase},$$
$$v_a = 300\sqrt{2}\sin(\omega t).$$

This case has the same currents as in vi(b), yielding lower power factors, due to the fact that the load active power decreases in the absence of fifth harmonic voltage.

Case (ix): V_{ns}, non-linear load, I_{ns} balanced:

$$R = 0.2\ \Omega/\text{phase},$$
$$v_{\mathrm{a}} = 298.51\sqrt{2}\sin(\omega t) + 29.85\sqrt{2}\sin(5\omega t),$$
$$i_{\mathrm{a}} = 59.70\sqrt{2}\sin(\omega t) + 5.97\sqrt{2}\sin(5\omega t) + 10\sqrt{2}\sin(7\omega t),\ I_{\mathrm{a}} = 60.83\ \text{A}.$$

In this case, the currents are as in case (v), but the load non-linearity injects the seventh harmonic which results in a reduction of power factors with respect to the base case.

Case (x): V_{ns}, non-linear load, I_{ns} unbalanced:

$$R = 0.2\ \Omega/\text{phase},$$
$$v_{\mathrm{a}} = 298.51\sqrt{2}\sin(\omega t) + 29.85\sqrt{2}\sin(5\omega t),$$
$$i_{\mathrm{a}} = 59.70\sqrt{2}\sin(\omega t),\ I_{\mathrm{a}} = 59.70\ \text{A},$$
$$i_{\mathrm{b}} = 59.70\sqrt{2}\sin(\omega t - 120) + 5.97\sqrt{2}\sin(5\omega t + 120),\ I_{\mathrm{b}} = 60\ \text{A},$$
$$i_{\mathrm{c}} = 59.70\sqrt{2}\sin(\omega t + 120) + 11.94\sqrt{2}\sin(5\omega t - 120),\ I_{\mathrm{b}} = 60.88\ \text{A}.$$

The fundamental component is as in case (v); the non-linear load contains the unbalanced fifth harmonic, chosen to ensure that the consumed active power is the same as that of the base circuit; this case also shows a decrease of the power factors with respect to the base case.

The main characteristics of the seven cases considered in this section are shown in Table 3.5. Table 3.6 illustrates, for each case, the magnitudes of the apparent powers; P_{j} is the power loss in the line, normalised to the base power of case (v), P_v/P_{j} is the ratio of the line power loss for the balanced line and that corresponding to each case. Finally, the table lists the magnitude of PF_{s} (the square of the system power factor), which coincides always with P_v/P_{j}.

Table 3.5 Circuit characteristics for test cases (v) to (x)

Case	Volt components	Impedance	Neutral	Current components
(v)	$\omega 1$, $\omega 5$	$R_{\mathrm{a}} = R_{\mathrm{b}} = R_{\mathrm{c}} = 5\ \Omega$	No	$\omega 1$, $\omega 5$, balanced
(vi)(a)	$\omega 1$, $\omega 5$	$R_{\mathrm{a}} = R_{\mathrm{b}} = 5$; $R_{\mathrm{c}} = 25\ \Omega$	Yes	$\omega 1$, $\omega 5$, unbalanced
(vi)(b)	$\omega 1$, $\omega 5$	$R_{\mathrm{a}} = R_{\mathrm{b}} = 5$; $R_{\mathrm{c}} = 25\ \Omega$	No	$\omega 1$, $\omega 5$, unbalanced
(vii)	$\omega 1$	non-linear	Yes	$\omega 1$, $\omega 5$, balanced
(viii)	$\omega 1$	non-linear	Yes	$\omega 1$, $\omega 5$, unbalanced
(ix)	$\omega 1$, $\omega 5$	non-linear	Yes	$\omega 1$, $\omega 5$, balanced
(x)	$\omega 1$, $\omega 5$	non-linear	Yes	$\omega 1$, balanced; $\omega 5$, unbalanced

Table 3.6 Apparent powers and power factors of test cases (v) to (x)

Case	Sv	$S_a = S_e$	S_s	PF_v	$PF_a = PF_e$	PF_s	P_1	P_v/P_1	PF_s^2
(v)	54 000	54 000	54 000	1	1	1	2160	1	1
(vi)(a)	39 600	39 600	44 529.5	1	1	0.889	2731.2	0.790	0.790
(vi)(b)	34 436.8	36 470.1	39 578.3	0.997	0.942	0.868	2865.3	0.753	0.753
(vii)	54 269.3	54 269.3	54 269.3	0.995	0.995	0.995	2181.6	0.990	0.990
(viii)	34 385.1	36 470.1	39 578.3	0.994	0.937	0.863	2893.9	0.746	0.746
(ix)	54 744.8	54 744.8	54 744.8	0.986	0.986	0.986	2220	0.972	0.972
(x)	54 117.5	54 176.0	54 177.9	0.997	0.996	0.996	2174.2	0.993	0.993

3.4 Summary

During this century, many alternative definitions have been proposed for reactive power in the presence of waveform distortion. Among them, Budeanu's proposal is still the most commonly accepted and quoted in international circles; it is, however, questioned by many specialists. The main effort in recent times has gone into reducing the difference between active and apparent power; the emphasis is changing from a direct optimisation of the power factor at the point of common coupling, to a global minimisation of the power loss throughout the system.

Four different definitions have been considered for apparent power in three-phase networks, the varied criteria affecting the measuring technology (power, energy and the power factor) and therefore the philosophy of tariffs.

The system equivalent apparent power S_s, being the largest of the four, results in the lowest power factor PF_s. The square of this factor is found to be equal to the ratio of the balanced line power loss to that of any other unbalanced operating condition; therefore, the larger the current unbalance is, either the fundamental or harmonic components, the smaller will be the energy efficiency, which can be represented by PF_s. Therefore, the latter index includes the three factors that affect line losses, i.e. unbalanced operation, non-linear load harmonic generation and reactive power demand.

It is possible for some of the phases of a perfectly sinusoidal and balanced power source to generate reactive power even when the load is purely resistive. In such cases, there is an interchange of reactive energy between the different phases, even though the load has no energy-storing elements.

Finally, it should be emphasised that, regardless of the definition used for reactive power, the important thing is to minimise the difference ($S^2 - P^2$) because the effect of distortion on losses cannot be neglected; this is why Fryze's theory, adapted by Emanuel, appears to be gaining acceptance.

3.5 References

1. Antoniu, S, (1984). Le régime énergétique déformant. Une question de priorite, *RGE*, **6/84**, pp. 357–362.
2. Fryze, S, (1932). Wirk-, Blind- und Scheinleistung in Elektrischen Stromkreisen mit nichtsinusförmigen Verlauf von Strom und Spannung, *Elektrotecnische Zeitschrift*, pp. 596–599.
3. IEEE, (1993) *The New IEEE Standard Dictionary of Electrical and Electronics Terms*, 5th edition, IEEE.

4. Shepherd, W and Zakikhani, P, (1972). Suggested definition of reactive power for nonsinusoidal systems, *Proc. IEE*, **119**, pp. 1361–1362.
5. Sharon, D, (1973). Reactive power definition and power factor improvement in nonlinear systems, *Proc. IEE*, **120**, pp. 704–706.
6. Emanuel, A E, (1977). Energetical factors in power systems with nonlinear loads, *Archiv für Elektrotechnik*, **59**, pp. 183–189.
7. Kusters, N L and Moore, W J M, (1980). On definition of reactive power under nonsinusoidal conditions, *IEEE Transactions on Power Apparatus and Systems*, **PAS-99**, September 1980, pp. 1845–1850.
8. Emanuel, A E, (1990). 'Power in nonsinusoidal situations. A review of definitions and physical meaning', *IEEE Transactions on Power Delivery*, **PWRD-5**, July 1990, pp. 1377–1383.
9. Czarnecki, L S, (1987). What is wrong with the Budeanu concept of reactive and distortion power and why it should be abandoned, *IEEE Transactions on Instrumentation and Measurement*, **IM-36**(3), pp. 834–837.
10. Slonin, M A and Van Wyk, J D, (1988). Power components in a system with sinusoidal and nonsinusoidal voltages and/or currents, *IEE Proceedings of the* **135**, pp. 76–84.
11. Emanuel, A E, (1998). Apparent power: Components and physical interpretation, *International Conference on Harmonics and Quality of Power (ICHQP'98)*, Athens, pp. 1–13.
12. Eguiluz, L I and Arrillaga, J, (1995). Comparison of power definitions in the presence of waveform distortion, *International Journal of Electrical Engineering Education*, **32**, pp. 141–153.
13. Filipski, P S, (1991). Polyphase apparent power and power factor under distorted waveform conditions, *IEEE Transactions on Power Delivery*, **6**(3).
14. Eguiluz, L I, Benito, P and Arrillaga, J, (1996). Power factor and efficiency in three-phase distorted circuits, *IPENZ Proceedings of the* Dunedin, New Zealand, pp. 131–140.
15. Eguiluz, L I, Mañana, M, Benito P and Lavandero, J C, (1995). El *PFs* un factor de potencia que relaciona las perdidas en la linea en circuitos trifasicos distorsionados, *4as J Luso-Espanholas de Engenharia Electrotecnica*, **3**, pp. 1212–1220.

4

WAVEFORM PROCESSING TECHNIQUES

4.1 Introduction

The voltage and current waveforms at points of connection of non-linear devices can either be obtained from appropriate transducers or calculated for a given operating condition from knowledge of the network impedances and of the device non-linear characteristics.

It is then necessary to extract relevant information from these waveforms to calculate the power quality indices described in Chapter 2. To derive such information, the waveforms are subjected to different levels of processing as follows.

First is the detection of non-repetitive subcycle transients, a straightforward operation not requiring spectral processing.

Next is the extraction of the fundamental frequency voltage and current phasors, i.e. their magnitude and phase. In this respect, techniques are needed to extract efficiently and accurately those components in the presence of transients and waveform distortion.

Finally, a higher level of processing complexity is required to obtain the complete frequency spectrum of periodic and non-periodic voltage and current waveforms. Traditionally, this has been achieved exclusively by Fourier techniques but more recently alternatives such as wavelets, neural networks and fuzzy logic are also being considered.

4.2 Fundamental Frequency Characterisation

Fundamental voltage magnitude and phase estimates are required in a variety of power systems applications.

Among them are the need to derive a reference signal for active filtering, the detection of voltage peaks for VAR compensation and the need to extract fundamental frequency power for transient stability assessment. Fast Fourier Transform (FFT)-based techniques are in general use to obtain spectra from discrete time samples. However, aliasing, spectral leakage and picket-fence effects may lead to inaccurate estimates if the FFT is misapplied. Special care is thus required when the FFT is applied to transient waveforms.

An alternative approach is the use of an estimator that minimises the least square error. This technique, described in Appendix III, is a static solution that performs post-processing of the given data. Although normally listed separately as a different method, Kalman filtering is a type of least error square estimator of the dynamic system parameters. The Kalman approach, described in Appendix IV, is recursive and thus allows each new sample to be efficiently incorporated into the estimation.

The traditional least square error technique penalises errors heavily and, therefore, bad data (errors) have a strong effect on the solution. To reduce this problem, a method is described next that minimises the non-sinusoidal root mean square content of the waveform.

4.2.1 Curve-Fitting Algorithm (CFA)

The curve-fitting approach detailed here uses least square error estimation to find the magnitude and phase of the fundamental component; however, the method is generally applicable to any other frequency [1–3].

Curve fitting selects the best fit of a curve to a waveform and measures the discrete residual values between the waveform and the fitted curve. In the least squares method, the size of these residuals is measured by the sum of their squared values. This is then minimised to obtain the least squared error, and the amplitude and phase of the best fitted curve calculated. Least squares curve fitting has both computational and theoretical advantages over Fourier processing. A least squares curve-fitted approach is most useful when periodicity clearly exists in the data. This is normally the case in the power system field since the fundamental component is most commonly predominant. Also of particular advantage is that in a curve-fitted approach, it is not necessary to truncate data exactly every period as with the Fourier Transform.

The formulation of the curve-fitting algorithm is as follows.

Assume a sinewave signal with a frequency of ω rad/s and a phase shift of ψ relative to some arbitrary time T_0:

$$y(t) = A\sin(\omega t - \psi), \tag{4.1}$$

where $\psi = \omega T_0$.

This can be written

$$y(t) = A\sin(\omega t)\cos(\omega T_0) - A\cos(\omega t)\sin(\omega T_0). \tag{4.2}$$

Letting $C_1 = A\cos(\omega T_0)$ and $C_2 = A\sin(\omega T_0)$ and if $\sin(\omega t)$ and $\cos(\omega t)$ are represented by functions $F_1(t)$ and $F_2(t)$, respectively, then

$$y(t) = C_1F_1(t) + C_2F_2(t). \tag{4.3}$$

$F_1(t)$ and $F_2(t)$ are known if the fundamental frequency ω is known. However, the amplitude and phase of this frequency generally need to be found, so the equation has to be solved for C_1 and C_2. If the signal $y(t)$ is distorted, then its deviation from a sinusoid can be described by an error function E:

$$x(t) = y(t) + E. \tag{4.4}$$

For a least squares method of curve fitting, the size of the error function is measured by the sum of the individual residual squared values such that

$$E = \sum_{i=1}^{n} \{x_i - y_i\}^2, \tag{4.5}$$

where $x_i = x(t_0 + i\Delta t)$ and $y_i = y(t_0 + i\Delta t)$.

From Equation (4.3)

$$E = \sum_{i=1}^{n} \{x_i - C_1 F_1(t_i) - C_2 F_2(t_i)\}^2, \tag{4.6}$$

where the residual value r at each discrete step is defined as

$$r_i = x_i - C_1 F_1(t_i) - C_2 F_2(t_i). \tag{4.7}$$

In matrix form

$$\begin{bmatrix} r_1 \\ r_2 \\ \vdots \\ r_n \end{bmatrix} = \begin{bmatrix} x_1 \\ x_2 \\ \vdots \\ x_n \end{bmatrix} - \begin{bmatrix} F_1(t_1) & F_2(t_1) \\ F_1(t_2) & F_2(t_2) \\ \vdots & \vdots \\ F_1(t_n) & F_2(t_n) \end{bmatrix} \begin{bmatrix} C_1 \\ C_2 \end{bmatrix}, \tag{4.8}$$

$$[r] = [X] - [F][C]. \tag{4.9}$$

The error component can be described in terms of the residual matrix as follows

$$\begin{aligned} E &= [r]^{\mathrm{T}}[r] \\ &= [r_1 \quad r_2 \ldots r_n] \begin{bmatrix} r_1 \\ r_2 \\ \vdots \\ r_n \end{bmatrix} \\ &= r_1^2 + r_2^2 + \cdots + r_n^2 \\ &= [[X] - [F][C]]^{\mathrm{T}}[[X] - [F][C]] \\ &= [X]^T[X] - [C]^{\mathrm{T}}[F]^{\mathrm{T}}[X] - [X]^{\mathrm{T}}[F][C] + [C]^{\mathrm{T}}[F]^{\mathrm{T}}[F][C]. \end{aligned} \tag{4.10}$$

This error can be minimised as shown in Appendix III, i.e.

$$\begin{aligned} \frac{\partial E}{\partial C} &= -2[F]^{\mathrm{T}}[X] + 2[F]^{\mathrm{T}}[F][C] = 0, \\ [F]^{\mathrm{T}}[F][C] &= [F]^{\mathrm{T}}[X], \\ [C] &= [[F]^{\mathrm{T}}[F]]^{-1}[F]^{\mathrm{T}}[X]. \end{aligned} \tag{4.11}$$

If $[A] = [F]^{\mathrm{T}}[F]$ and $[B] = [F]^{\mathrm{T}}[X]$, then

$$[C] = [A]^{-1}[B] \tag{4.12}$$

and the following recursive solution is possible:

$$
\begin{aligned}
[A] &= \begin{bmatrix} F_1 \\ F_2 \end{bmatrix} [\, F_1 \quad F_2 \,] \\
&= \begin{bmatrix} F_1 F_1(t_i) & F_1 F_2(t_i) \\ F_2 F_1(t_i) & F_2 F_2(t_i) \end{bmatrix} \\
&= \begin{bmatrix} a_{11} & a_{12} \\ a_{21} & a_{22} \end{bmatrix}.
\end{aligned}
$$

Elements of matrix $[A]$ can then be derived as shown:

$$
\begin{aligned}
a_{11n} &= \begin{bmatrix} F_1(t_1) \\ \vdots \\ F_1(t_n) \end{bmatrix}^{\mathrm{T}} \begin{bmatrix} F_1(t_1) \\ \vdots \\ F_1(t_n) \end{bmatrix} \\
&= \sum_{i=1}^{n-1} F_1^2(t_i) + F_1^2(t_n) \\
&= a_{11n-1} + F_1^2(t_n), \quad \text{etc.}
\end{aligned}
\tag{4.13}
$$

Since a recursive technique is used, there is no need to store all the sample points when increasing the window size.

Similarly,

$$
\begin{aligned}
[B] &= \begin{bmatrix} F_1(t_i)x(t_i) \\ F_2(t_i)x(t_i) \end{bmatrix} \\
&= \begin{bmatrix} b_1 \\ b_2 \end{bmatrix}
\end{aligned}
$$

and

$$
b_{1n} = b_{1n-1} + F_1(t_n)x(t_n), \tag{4.14}
$$

$$
b_{2n} = b_{2n-1} + F_2(t_n)x(t_n). \tag{4.15}
$$

From these matrix element equations, C_1 and C_2 can be calculated recursively using sequential data.

To give more weight to the recent samples, a *forgetting factor* (less than 1.0) can be used to multiply the a_{xxn-1} terms in Equation (4.13) and the b_{xn-1} terms in Equations (4.14) and (4.15).

4.2.2 Curve-Fitting Implementation

Four different test waveforms are shown in Figure 4.1, representing cases of (i) strong waveform amplitude modulation, (ii) heavy harmonic distortion, (iii) high frequency ringing and (iv) d.c. offset.

These four test cases were subjected to CFA and FFT processing. The CFA was applied using two different unit magnitude window sizes corresponding to one fundamental period and half of this period. Also, a staggered approach was used with the full period window in an overlapping manner to produce results twice every period.

The curve-fitting algorithm compares extremely well to Fourier analysis when a fundamental period rectangular window is used. Responses to amplitude variations, harmonic interference, and the addition of a d.c. component are then practically identical. Given the computational advantage of the CFA over the FFT, it is considered to be a better choice when only the fundamental frequency is required.

Although window sizes of less than a fundamental period respond quicker to amplitude variation, they suffer from interference in the presence of d.c. For this reason, it is not an advisable method for extracting accurate fundamental frequency data.

Staggering a full fundamental period window so that results are obtainable twice every cycle give very good results. The generation of results every half cycle gives a better general tracking response than using a non-staggered full period approach.

When only periodic information is required, CFA should avoid regions with aperiodic subcycle (fast transient) events and discontinuities, because these affect the response across the boundaries of such disturbances.

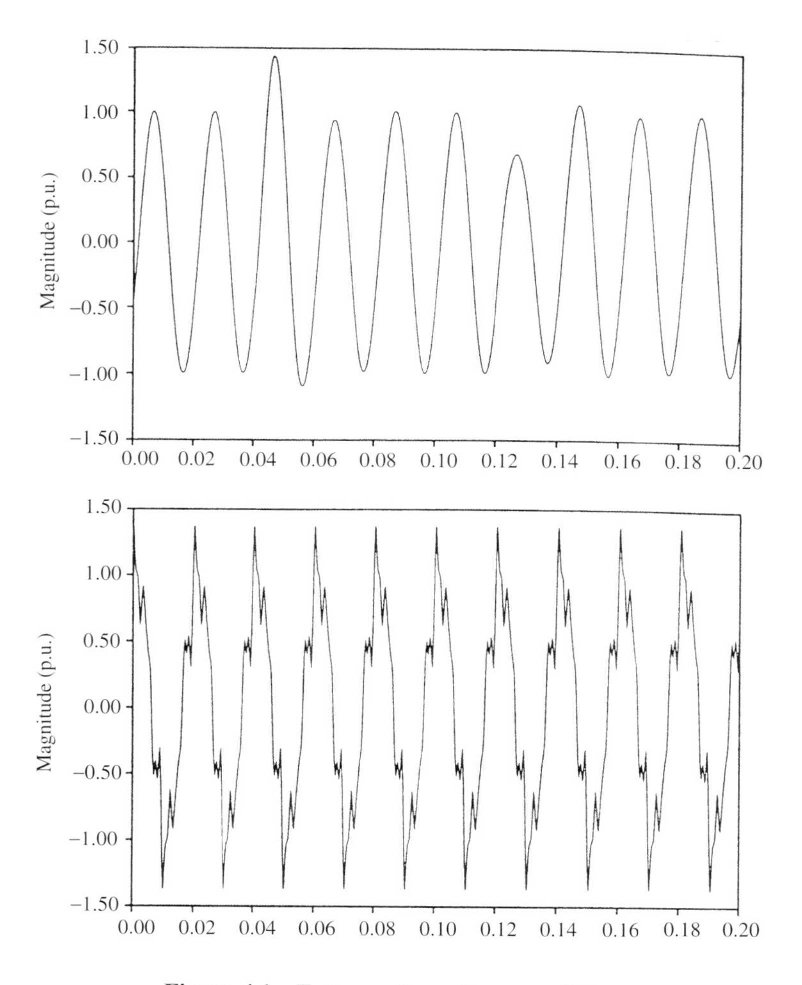

Figure 4.1 Test waveforms for curve fitting

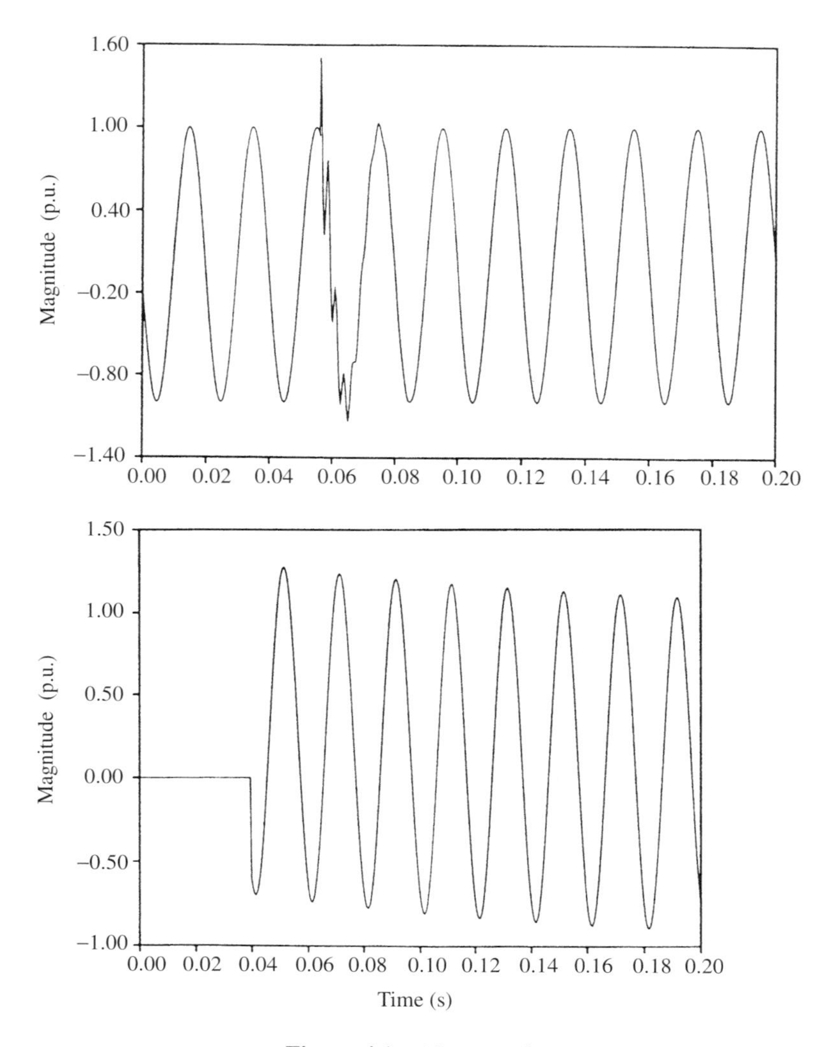

Figure 4.1 *(Continued)*

Electromagnetic transient information is obtained in steps of the order of 50 μs to 100 μs. Using these time steps, and with an analysis window of a period length, there is no risk of violating the Nyquist criteria for the fundamental frequency.

If fundamental frequency data is required at discrete intervals of a fundamental cycle, then a CFA analysis of a non-staggered fundamental period length is considered the best choice. When data is required at discrete intervals of less than a cycle, a CFA analysis with a similar length of fundamental period used in a staggered manner will, instead, be the optimum choice. The CFA technique gives very good results for all disturbing effects and provides computational efficiency with its curve-fitted approach. A further advantage is that the window analysis length is not restricted to precisely one period of the fundamental frequency but can be extended beyond this if necessary.

Finally, in the presence of a d.c. component the window size should not be reduced below the width of a fundamental period.

4.2.3 Frequency Estimates

Estimation of the main power frequency is an important parameter for the control and protection of power systems. As frequency estimates are normally required on-line, the processing time is critical and hence computationally expensive algorithms and time delays due to pre- and post-filtering, to remove distortion, are undesirable. The zero-crossing methods and Discrete Fourier Transform (DFT) techniques often used assume sinusoidal waveforms of one frequency and are unreliable in the presence of distortion. Instead, techniques utilising least square estimation [4] and, more recently, Newton techniques [5] are recommended.

4.2.4 R.M.S. Error Assessment

The difference between the actual waveform and an ideal sine wave is best expressed by the r.m.s. error, which contains all the *unwanted* non-sinusoidal content of the waveform.

The curve-fitting techniques described in the previous section can be used to derive the reference ideal fundamental component. A step-by-step comparison is then made between the reference and actual waveform and their difference can be compounded over any required time period.

As an illustration, a typical simulated voltage waveform following the occurrence of a three-cycle line short circuit is shown in Figure 4.2, together with its corresponding reference voltage derived from the curve-fitting technique.

The r.m.s.error is derived from the expression

$$\text{r.m.s.(error)} = \sqrt{\frac{\sum(\Delta e)^2}{N}}, \tag{4.16}$$

where Δe are the differences between the expected (reference) and actual voltage samples, and N is the number of sample points.

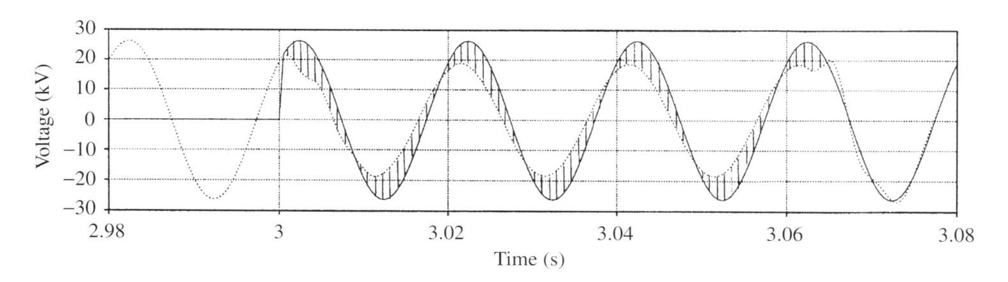

Figure 4.2 Waveforms used for the r.m.s. error calculation: ______Vreference . . . . volt Invercargill 033

4.3 Fourier Analysis

Fourier analysis is used to convert time domain waveforms into their frequency components and vice-versa [6]. When the waveform is periodical, the Fourier series can be used to calculate the magnitudes and phases of the fundamental and its harmonic components.

More generally, the Fourier Transform and its inverse are used to map any function in the interval $-\infty$ to ∞ in either the time or frequency domain into a continuous function in the inverse domain. The Fourier series, therefore, represents the special case of the Fourier Transform applied to a periodic signal.

In practice, data are often available in the form of a sampled time function, represented by a time series of amplitudes, separated by fixed time intervals of limited duration. When dealing with such data, a modification of the Fourier Transform, the DFT, is used. The implementation of the DFT, by means of the FFT algorithm, forms the basis of most modern spectral and harmonic analysis systems. The development of the Fourier Transform and DFTs is also examined in this section along with the implementation of the FFT.

4.3.1 Fourier Series and Coefficients [7,8]

The Fourier series of a periodic function $x(t)$ has the expression

$$x(t) = a_0 + \sum_{n=1}^{\infty} \left(a_n \cos\left(\frac{2\pi nt}{T}\right) + b_n \sin\left(\frac{2\pi nt}{T}\right)\right). \tag{4.17}$$

This constitutes a frequency domain representation of the periodic function.

In this expression, a_0 is the average value of the function $x(t)$, whilst a_n and b_n, the coefficients of the series, are the rectangular components of the nth harmonic. The corresponding nth harmonic vector is

$$A_n/\phi_n = a_n + jb_n \tag{4.18}$$

with a magnitude

$$A_n = \sqrt{a_n^2 + b_n^2}$$

and a phase angle

$$\phi_n = \tan^{-1}\left(\frac{b_n}{a_n}\right).$$

For a given function $x(t)$, the constant coefficient, a_0, can be derived by integrating both sides of Equation (4.17) from $-T/2$ to $T/2$ (over a period T), i.e.

$$\int_{-T/2}^{T/2} x(t)\,\mathrm{d}t = \int_{-T/2}^{T/2} \left[a_0 + \sum_{n=1}^{\infty} \left[a_0 \cos\left(a_n \cos\left(\frac{2\pi nt}{T}\right) + b_n \sin\left(\frac{2\pi nt}{T}\right)\right)\right]\right] \mathrm{d}t. \tag{4.19}$$

The Fourier series of the right-hand side can be integrated term by term, giving

$$\int_{-T/2}^{T/2} x(t)\,\mathrm{d}t = a_0 \int_{-T/2}^{T/2} \mathrm{d}t + \sum_{n=1}^{\infty} \left[a_n \int_{-T/2}^{T/2} \cos\left(\frac{2\pi n t}{T}\right) \mathrm{d}t + b_n \int_{-T/2}^{T/2} \sin\left(\frac{2\pi n t}{T}\right) \mathrm{d}t \right]. \tag{4.20}$$

The first term on the right-hand side equals Ta_0, while the other integrals are zero. Hence, the constant coefficient of the Fourier series is given by

$$a_0 = 1/T \int_{-T/2}^{T/2} x(t)\,\mathrm{d}t, \tag{4.21}$$

which is the area under the curve of $x(t)$ from $-T/2$ to $T/2$, divided by the period of the waveform, T.

The a_n coefficients can be determined by multiplying Equation (4.17) by $\cos(2\pi mt/T)$, where m is any fixed positive integer, and integrating between $-T/2$ and $T/2$, as previously, i.e

$$\begin{aligned}
\int_{-T/2}^{T/2} x(t) \cos\left(\frac{2\pi m t}{T}\right) \mathrm{d}t
&= \int_{-T/2}^{T/2} \left[a_0 + \sum_{n=1}^{\infty} \left[a_n \cos\left(\frac{2\pi n t}{T}\right) + b_n \sin\left(\frac{2\pi n t}{T}\right) \right] \right] \cos\left(\frac{2\pi m t}{T}\right) \mathrm{d}t \\
&= a_o \int_{-T/2}^{T/2} \cos\left(\frac{2\pi m t}{T}\right) \mathrm{d}t + \sum_{n=1}^{\infty} \left[a_n \int_{-T/2}^{T/2} \cos\left(\frac{2\pi n t}{T}\right) \right. \\
&\quad \left. \times \cos\left(\frac{2\pi m t}{T}\right) \mathrm{d}t + b_n \int_{-T/2}^{T/2} \sin\left(\frac{2\pi n t}{T}\right) \cos\left(\frac{2\pi m t}{T}\right) \mathrm{d}t \right].
\end{aligned} \tag{4.22}$$

The first term on the right-hand side is zero, as are all the terms in b_n, since $\sin(2\pi nt/T)$ and $\cos(2\pi mt/T)$ are orthogonal functions for all n and m.

Similarly, the terms in a_n are zero, being orthogonal, unless $m = n$. In this case, Equation (4.22) becomes

$$\begin{aligned}
\int_{-T/2}^{T/2} x(t) \cos\left(\frac{2\pi m t}{T}\right) \mathrm{d}t &= a_n \int_{-T/2}^{T/2} \cos\left(\frac{2\pi n t}{T}\right) \mathrm{d}t \\
&= \frac{a_n}{2} \int_{-T/2}^{T/2} \cos\left(\frac{4\pi n t}{T}\right) \mathrm{d}t + \frac{a_n}{2} \int_{-T/2}^{T/2} \mathrm{d}t.
\end{aligned} \tag{4.23}$$

The first term on the right-hand side is zero, while the second term equals $a_n T/2$. Hence, the coefficients a_n can be obtained from

$$a_n = \frac{2}{T} \int_{-T/2}^{T/2} x(t) \cos\left(\frac{2\pi n t}{T}\right) \mathrm{d}t, \quad \text{for } n = 1 \to \infty. \tag{4.24}$$

To determine the coefficients b_n, Equation (4.17) is multiplied by $\sin(2\pi mt/T)$ and, by a similar argument to the above,

$$b_n = \frac{2}{T} \int_{-T/2}^{T/2} x(t) \sin\left(\frac{2\pi n t}{T}\right) \mathrm{d}t, \quad \text{for } n = 1 \to \infty. \tag{4.25}$$

It should be noted that because of the periodicity of the integrands in Equations (4.21), (4.24) and (4.25), the interval of integration can be taken more generally as t and $t + T$.

If the function $x(t)$ is piecewise continuous (i.e. has a finite number of vertical jumps) in the interval of integration, the integrals exist and Fourier coefficients can be calculated for this function. Equations (4.20), (4.24) and (4.25) are often expressed in terms of the angular frequency as follows

$$a_0 = \frac{1}{2\pi}\int_{-\pi}^{\pi} x(\omega t)\,\mathrm{d}(\omega t), \tag{4.26}$$

$$a_n = \frac{1}{\pi}\int_{-\pi}^{\pi} x(\omega t)\cos(n\omega t)\,\mathrm{d}(\omega t), \tag{4.27}$$

$$b_n = \frac{1}{\pi}\int_{-\pi}^{\pi} x(\omega t)\sin(n\omega t)\mathrm{d}(\omega \mathrm{t}), \tag{4.28}$$

so that

$$x(t) = a_o + \sum_{n=1}^{\infty}[a_n \cos(n\omega t) + b_n \sin(n\omega t)]. \tag{4.29}$$

4.3.2 Simplifications Resulting from Waveform Symmetry [7,8]

Equations (4.21), (4.24) and (4.25), the general formulas for the Fourier coefficients, can be represented as the sum of two separate integrals, i.e.

$$a_n = \frac{2}{T}\int_{0}^{T/2} x(t)\cos\left(\frac{2\pi nt}{T}\right)\,\mathrm{d}t + \frac{2}{T}\int_{-T/2}^{0} x(t)\cos\left(\frac{2\pi nt}{T}\right)\,\mathrm{d}t, \tag{4.30}$$

$$b_n = \frac{2}{T}\int_{0}^{T/2} x(t)\sin\left(\frac{2\pi nt}{T}\right)\,\mathrm{d}t + \frac{2}{T}\int_{-T/2}^{0} x(t)\sin\left(\frac{2\pi nt}{T}\right)\,\mathrm{d}t. \tag{4.31}$$

Replacing t by $-t$ in the second integral of Equation (4.30), with limits $(-T/2, 0)$:

$$\begin{aligned} a_n &= \frac{2}{T}\int_{0}^{T/2} x(t)\cos\left(\frac{2\pi nt}{T}\right)\,\mathrm{d}t + \frac{2}{T}\int_{+T/2}^{0} x(-t)\cos\left(\frac{-2\pi nt}{T}\right)\,\mathrm{d}(-t) \\ &= \frac{2}{T}\int_{0}^{T/2} [x(t) + x(-t)]\cos\left(\frac{2\pi nt}{T}\right)\,\mathrm{d}t. \end{aligned} \tag{4.32}$$

Similarly,

$$b_n = \frac{2}{T}\int_{0}^{T/2} [x(t) - x(-t)]\sin\left(\frac{2\pi nt}{T}\right)\,\mathrm{d}t. \tag{4.34}$$

Odd Symmetry The waveform has odd symmetry if

$$x(t) = -x(-t).$$

Then the a_n terms become zero for all n, while

$$b_n = \frac{4}{T}\int_0^{T/2} x(t)\sin\left(\frac{2\pi nt}{T}\right)\,\mathrm{d}t. \tag{4.35}$$

The Fourier series for an odd function will, therefore, contain only sine terms.

Even Symmetry The waveform has even symmetry if

$$x(t) = x(-t).$$

In this case

$$b_n = 0, \quad \text{for all } n,$$

and

$$a_n = \frac{4}{T}\int_0^{T/2} x(t)\cos\left(\frac{2\pi nt}{T}\right)\,\mathrm{d}t. \tag{4.36}$$

The Fourier series for an even function will, therefore, contain only cosine terms.

Certain waveforms may be odd or even depending on the time reference position selected. For instance, the square wave of Figure 4.3, drawn as an odd function, can be transformed into an even function simply by shifting the origin (vertical axis) by $T/2$.

Halfwave Symmetry A function $x(t)$ has halfwave symmetry if

$$x(t) = -x(t + T/2), \tag{4.37}$$

i.e. the shape of the waveform over a period $t + T/2$ to $t + T$ is the negative of the shape of the waveform over the period t to $t + T/2$. Consequently, the square wave function of Figure 4.3 has halfwave symmetry.

Using Equation (4.24) and replacing (t) by $(t + T/2)$ in the interval $(-T/2, 0)$

$$a_n = \frac{2}{T}\int_0^{T/2} x(t)\cos\left(\frac{2\pi nt}{T}\right)\mathrm{d}t + \frac{2}{T}\int_{-T/2+T/2}^{0+T/2} x(t+T/2)\cos\left(\frac{2\pi n(t+T/2)}{T}\right)\mathrm{d}t$$

$$= \frac{2}{T}\int_0^{T/2} x(t)\left[\cos\left(\frac{2\pi nt}{T}\right) - \cos\left(\frac{2\pi nt}{T} + n\pi\right)\right]\mathrm{d}t \tag{4.38}$$

since by definition $x(t) = -x(t + T/2)$.

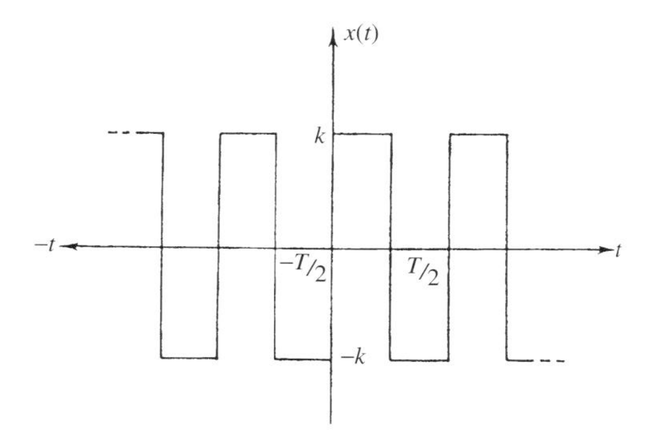

Figure 4.3 Square wave function

If n is an odd integer, then

$$\cos\left(\frac{2\pi nt}{T} + n\pi\right) = -\cos\left(\frac{2\pi nt}{T}\right)$$

and

$$a_n = \frac{4}{T}\int_0^{T/2} x(t)\cos\left(\frac{2\pi nt}{T}\right)\,\mathrm{d}t. \tag{4.39}$$

However, if n is an even integer, then

$$\cos\left(\frac{2\pi nt}{T} + n\pi\right) = \cos\left(\frac{2\pi nt}{T}\right)$$

and

$$a_n = 0.$$

Similarly,

$$\begin{aligned} b_n &= \frac{4}{T}\int_0^{T/2} x(t)\sin\left(\frac{2\pi nt}{T}\right)\,\mathrm{d}t, && \text{for } n \text{ odd,} \\ &= 0, && \text{for } n \text{ even.} \end{aligned} \tag{4.40}$$

Thus, waveforms, which have halfwave symmetry, contain only odd order harmonics.

The square wave of Figure 4.3 is an odd function with halfwave symmetry. Consequently, only the b_n coefficients and odd harmonics will exist. The expression for the coefficients taking into account these conditions is

$$b_n = \frac{8}{T}\int_0^{T/4} x(t)\sin\left(\frac{2\pi nt}{T}\right)\,\mathrm{d}t, \tag{4.41}$$

which can be represented by a line spectrum of amplitudes inversely proportional to the harmonic order, as shown in Figure 4.4.

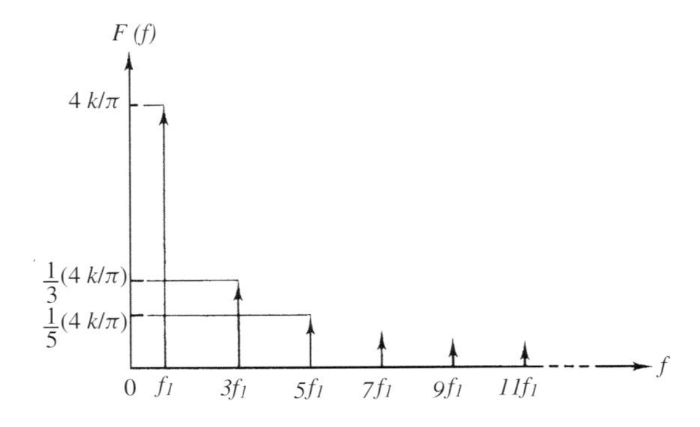

Figure 4.4 Line spectrum representation of a square wave

4.3.3 Complex Form of the Fourier Series

The representation of the frequency components as rotating vectors in the complex plane gives a geometrical interpretation of the relationship between waveforms in the time and frequency domains.

A uniformly rotating vector $A/2e^{j\phi}(X(f_n))$ has a constant magnitude $A/2$, and a phase angle ϕ, which is time varying according to

$$\phi = 2\pi ft + \theta, \tag{4.42}$$

where θ is the initial phase angle when $t = 0$.

A second vector $A/2e^{-j\phi}(X(-f_n))$ with magnitude $A/2$ and phase angle $-\phi$, will rotate in the opposite direction to $A/2e^{+j\phi}(X(f_n))$. This negative rate of change of phase angle can be considered as a negative frequency.

The sum of the two vectors will always lie along the real axis, the magnitude oscillating between A and $-A$ according to

$$\frac{A}{2}e^{j\phi} + \frac{A}{2}e^{-j\phi} = A\cos\phi. \tag{4.43}$$

Thus, each harmonic component of a real-valued signal can be represented by two half amplitude contra-rotating vectors as shown in Figure 4.5, such that

$$X(f_n) = X^*(-f_n), \tag{4.44}$$

where $X^*(-f_n)$ is the complex conjugate of $X(-f_n)$.

The sine and cosine terms of Equations (4.27) and (4.28) may, therefore, be solved into positive and negative frequency terms using the trigonometric identities

$$\cos(n\omega t) = \frac{e^{jn\omega t} + e^{-jn\omega t}}{2}, \tag{4.45}$$

$$\sin(n\omega t) = \frac{e^{jn\omega t} - e^{-jn\omega t}}{2j}. \tag{4.46}$$

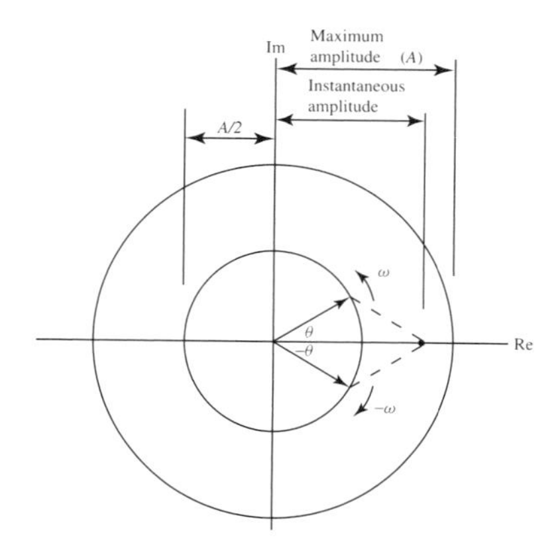

Figure 4.5 Contra-rotating vector pair producing a varying amplitude (pulsating) vector

Substituting into Equation (4.29) and simplifying yields

$$x(t) = \sum c_n \mathrm{e}^{jn\omega t}, \tag{4.47}$$

where

$$\begin{aligned} c_n &= \tfrac{1}{2}(a_n - jb_n), \quad n > 0, \\ c_{-n} &= c_n, \\ c_0 &= a_0. \end{aligned}$$

The c_n terms can also be obtained by complex integration:

$$c_n = \frac{1}{\pi}\int_{-\pi}^{\pi} x(\omega t)\mathrm{e}^{-jn\omega t}\,\mathrm{d}(\omega t), \tag{4.48}$$

$$c_0 = \frac{1}{2\pi}\int_{-\pi}^{\pi} x(\omega t)\,\mathrm{d}(\omega t). \tag{4.49}$$

If the time-domain signal $x(t)$ contains a component rotating at a single frequency nf, then multiplication by the unit vector $\mathrm{e}^{-j2\pi nft}$, which rotates at a frequency $-nf$, annuls the rotation of the component, such that the integration over a complete period has a finite value. All components at other frequencies will continue to rotate after multiplication by $\mathrm{e}^{-j2\pi nft}$, and will thus integrate to zero.

The Fourier series is most generally used to approximate a periodic function by truncation of the series. In this case, the truncated Fourier series is the best trigonometric series expression of the function, in the sense that it minimises the square error between the function and the truncated series. The number of terms required depends upon the magnitude of repeated derivatives of the function to be approximated. Repeatedly differentiating Equation (4.48) by parts, it can readily be shown that

$$c_n = \frac{1}{2\pi}\frac{1}{n^{m+1}}\int_{-\pi}^{\pi} f^{(m+1)}(\omega t)\,\mathrm{d}(\omega t). \tag{4.50}$$

Consequently, the Fourier series for repeatedly differentiated functions will converge faster than that for functions with low-order discontinuous derivatives.

The complex Fourier series expansion is compatible with the FFT, the method of choice for converting time-domain data samples into a Nyquist rate limited frequency spectrum. The trigonometric Fourier expression can also be written as a series of phase shifted sine terms by substituting

$$a_n \cos n\omega t + b_n \sin n\omega t = d_n \sin(n\omega t + \psi_n) \tag{4.51}$$

into Equation (4.29), where

$$\begin{aligned} d_n &= \sqrt{a_n^2 + b_n^2}, \\ \psi_n &= \tan^{-1}\frac{b_n}{a_n}. \end{aligned} \tag{4.52}$$

Finally, the phase-shifted sine terms can be represented as peak value phasors by setting

$$\Psi_n = d_n \mathrm{e}^{j\psi_n}, \tag{4.53}$$

so that

$$d_n \sin(n\omega t + \psi_n) = I\{\Psi_n e^{jn\omega t}\}$$
$$= |\Psi_n| \sin(n\omega t + \angle\Psi_n). \quad (4.54)$$

The harmonic phasor Fourier series is, therefore,

$$f(t) = \sum_{n=0}^{\infty} I\{\Psi_n e^{jn\omega t}\}, \quad (4.55)$$

which does not contain negative frequency components. Note that the d.c. term becomes

$$\Psi_0 = \frac{a_0}{2} e^{j\pi/2}$$
$$= j\frac{a_0}{2}. \quad (4.56)$$

In practice, the upper limit of the summation is set to n_h, the highest harmonic order of interest.

4.3.4 Convolution of Harmonic Phasors

Viewing a continuous function through a sampling period of interval T s is equivalent to multiplying the signal in the time domain by a rectangular pulse of length T (Figure 4.6). This corresponds to the convolution in the frequency domain of their respective frequency spectra.

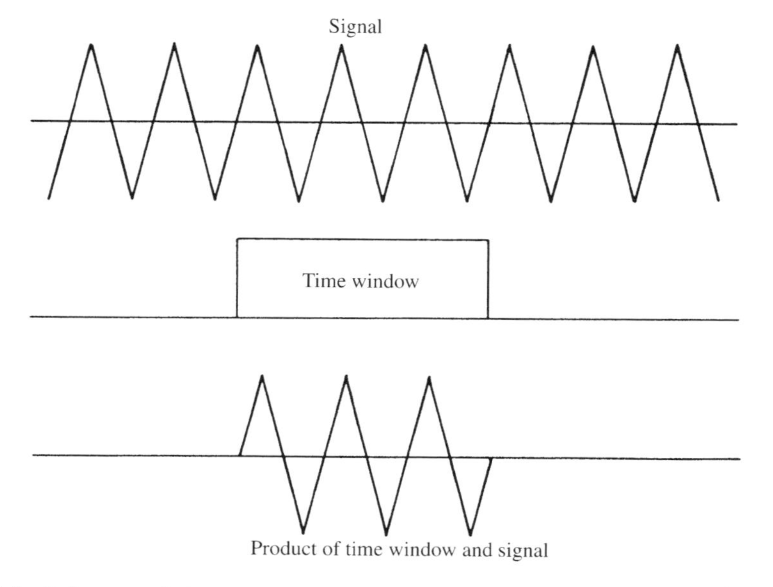

Figure 4.6 Influence of viewing a continuous function through a rectangular time window

In general, the point-by-point multiplication of two time-domain waveforms is expressed in the harmonic domain by a discrete convolution of their Fourier series. When two harmonic phasors of different frequencies are convolved, the results are harmonic phasors at sum and difference harmonics. This is best explained by multiplying the corresponding sinusoids using the trigonometric identity for the product of sine waves, and then converting back to phasor form. Given two phasors, A_k and B_m, of harmonic orders k and m, the trigonometric identity for their time-domain multiplication is:

$$\begin{aligned}|A_k| \sin(k\omega t + \angle A_k)|B_m| \sin(m\omega t + \angle B_m) \\ = \frac{1}{2}|A_k||B_m| \left[\sin\left((k-m)\omega t + \angle A_k - \angle B_m + \frac{\pi}{2}\right)\right. \\ \left. - \sin\left((k+m)\omega t + \angle A_k + \angle B_m + \frac{\pi}{2}\right)\right]. \end{aligned} \tag{4.57}$$

Converting to phasor form

$$\begin{aligned} A_k \otimes B_m &= \tfrac{1}{2}|A_k||B_m| \left[e^{j(\angle A_k - \angle B_m + \pi/2)}|_{k-m} - e^{j(\angle A_k - \angle B_m + \pi/2)}|_{k+m} \right] \\ &= \tfrac{1}{2}\left[(|A_k|e^{j\angle A_k}|B_m|e^{-j\angle B_m}e^{j\pi/2})|_{k-m} - (|A_k|e^{j\angle A_k}|B_m|e^{j\angle B_m}e^{j+\pi/2})_{k+m} \right] \\ &= \tfrac{1}{2}j\left[(A_k B_m^*)_{k-m} - (A_k B_m)_{k+m} \right]. \end{aligned} \tag{4.58}$$

If k is less than m, then a negative harmonic can be avoided by conjugating the difference term. This leads to the overall equation

$$A_k \otimes B_m = \begin{cases} \frac{1}{2}j(A_k B_m^*)_{k-m} - \frac{1}{2}j(A_k B_m)_{k+m}, & \text{if } k \geq m, \\ \frac{1}{2}j(A_k B_m^*)_{m-k}^* - \frac{1}{2}j(A_k B_m)_{k+m}, & \text{otherwise.} \end{cases} \tag{4.59}$$

The multiplication of two non-sinusoidal periodic waveforms leads to a discrete convolution of their harmonic phasor Fourier series

$$\begin{aligned} f_a(t) f_b(t) &= \sum_{k=0}^{n_h} |A_k| \sin(k\omega t + \angle A_k) \sum_{m=0}^{n_h} |B_k| \sin(m\omega t + \angle B_m) \\ &= \sum_{k=0}^{nh}\sum_{m=0}^{nh} |A_k| \sin(k\omega t + \angle A_k)|B_k| \sin(m\omega t + \angle B_m). \end{aligned} \tag{4.60}$$

Rewriting this in terms of phasors yields

$$\mathbf{F}_A \otimes \mathbf{F}_B = \sum_{k=0}^{nh}\sum_{m=0}^{nh} A_k \otimes B_m. \tag{4.61}$$

Equation (4.61) generates harmonic phasors of order up to $2n_h$, due to the sum terms. Substituting the equation for the convolution of two phasors, Equation (4.59), into (4.61) and solving for the lth order component yields

$$(A \otimes B)_l = \frac{1}{2}j\left[\sum_{k=l}^{n_h} A_k B_{k-l}^* + \sum_{k=1}^{n_h} (A_k B_{k+l}^*)^* - \sum_{k=0}^{n_h} A_k B_{l-k}^*\right], \quad l > 0, \tag{4.62}$$

$$(A \otimes B)_l = \frac{1}{2} j \left[-A_0 B_0 + \sum_{k=0}^{n_h} A_k B_k^* \right], \quad l = 0. \tag{4.63}$$

The convolution equations are non-analytic in the complex plane but are differentiable by decomposing into two real-valued components (typically rectangular).

If negative frequencies are retained, the convolution is just the multiplication of two series

$$\begin{aligned} f_{a(t)} f_{b(t)} &= \sum_{n=-n_h}^{n_h} c_{an} e^{jn\omega t} \sum_{l=-n_h}^{n_h} c_{bl} e^{jl\omega t} \\ &= \sum_{n=-n_h}^{n_h} \sum_{l=-n_h}^{n_h} c_{an} c_{bl} e^{j(l+n)\omega t}. \end{aligned} \tag{4.64}$$

In practice, the discrete convolution can be evaluated faster using FFT methods.

4.3.5 The Fourier Transform [8,9]

Fourier analysis, when applied to a continuous, periodic signal in the time domain, yields a series of discrete frequency components in the frequency domain.

By allowing the integration period to extend to infinity, the spacing between the harmonic frequencies, ω, tends to zero and the Fourier coefficients, c_n, of Equation (4.48) become a continuous function, such that

$$X(f) = \int_{-\infty}^{\infty} x(t) e^{-j2\pi f t} \, dt. \tag{4.65}$$

The expression for the time-domain function $x(t)$, which is also continuous and of infinite duration, in terms of $X(f)$ is then

$$x(t) = \int_{-\infty}^{\infty} X(f) e^{j2\pi f t} \, df. \tag{4.66}$$

$X(f)$ is known as the spectral density function of $x(t)$.

Equations (4.65) and (4.66) form the Fourier Transform Pair. Equation (4.65) is referred to as the 'Forward Transform' and Equation (4.66) as the 'Reverse' or 'Inverse Transform'.

In general, $X(f)$ is complex and can be written as

$$X(f) = \operatorname{Re} X(f) + j \operatorname{Im} X(f). \tag{4.67}$$

The real part of $X(f)$ is obtained from

$$\begin{aligned} \operatorname{Re} X(f) &= \tfrac{1}{2}[X(f) + X(-f)] \\ &= \int_{-\infty}^{\infty} x(t) \cos 2\pi f t \, dt. \end{aligned} \tag{4.68}$$

Similarly, for the imaginary part of $X(f)$

$$\begin{aligned}\operatorname{Im} X(f) &= \tfrac{1}{2} j[X(f) - X(-f)] \\ &= -\int_{-\infty}^{\infty} x(t) \sin 2\pi f t \, \mathrm{d}t. \qquad (4.69)\end{aligned}$$

The amplitude spectrum of the frequency signal is obtained from

$$|X(f)| = [(\operatorname{Re} X(f))^2 + (\operatorname{Im} X(f))^2]^{1/2}. \qquad (4.70)$$

The phase spectrum is

$$\phi(f) = \tan^{-1}\left[\frac{\operatorname{Im} X(f)}{\operatorname{Re} X(f)}\right]. \qquad (4.71)$$

Using Equations (4.67)–(4.71), the Inverse Fourier Transform can be expressed in terms of the magnitude and phase spectra components.

$$x(t) = \int_{-\infty}^{\infty} |X(f)| \cos[2\pi f t - \phi(f)] \, \mathrm{d}f. \qquad (4.72)$$

As an example, let us consider a rectangular function such as Figure 4.7 defined by

$$\begin{aligned} x(t) &= K, \quad \text{for } |t| \le T/2, \\ &= 0, \quad \text{for } |t| > T/2, \end{aligned}$$

i.e. the function is continuous over all t but is zero outside the limits $(-T/2, T/2)$.

Its Fourier Transform is

$$\begin{aligned} X(f) &= \int_{-\infty}^{\infty} x(t) \mathrm{e}^{-j2\pi f t} \, \mathrm{d}t \\ &= \int_{-T/2}^{T/2} K e^{-j2\pi f t} \, \mathrm{d}t \\ &= \frac{-K}{\pi f} \cdot \frac{1}{2j} \left[\mathrm{e}^{-j\pi f T} - \mathrm{e}^{j\pi f T}\right] \qquad (4.73) \end{aligned}$$

and using the identity

$$\sin \phi = \frac{1}{2j}(\mathrm{e}^{j\phi} - \mathrm{e}^{-j\phi})$$

yields the following expression for the Fourier Transform:

$$X(f) = \frac{K}{\pi f} \sin(\pi f T) = KT\left[\frac{\sin(\pi f T)}{\pi f T}\right]. \qquad (4.74)$$

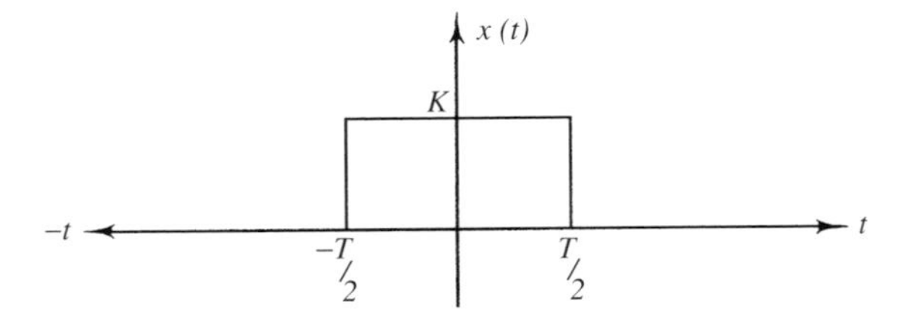

Figure 4.7 Rectangular function

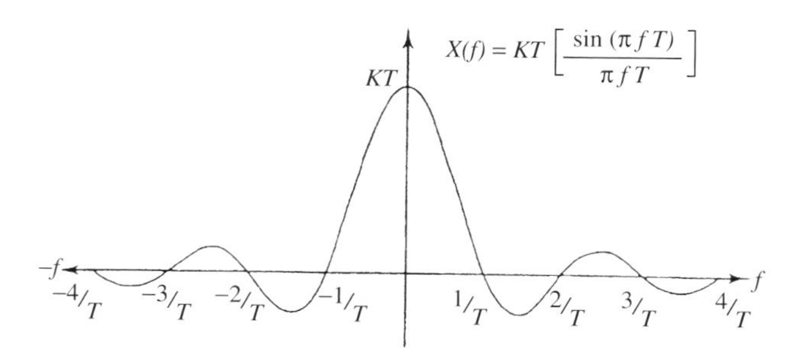

Figure 4.8 The sinc function, $\sin(\pi f T)/(\pi f T)$

The term in brackets, known as the sinc function, is shown in Figure 4.8.

While the function is continuous, it has zero value at the points $f = n/T$ for $n = \pm 1, \pm 2, \ldots$ and the side lobes decrease in magnitude as $1/T$. This should be compared to the Fourier series of a periodic square wave which has discrete frequencies at odd harmonics. The interval $1/T$ is the effective bandwidth of the signal.

4.3.6 Sampled Time Functions [9,10]

With an increase in the digital processing of data, functions are often recorded by samples in the time domain. Thus, the signal can be represented as in Figure 4.9, where $f_s = 1/t_1$ is the frequency of the sampling. In this case, the Fourier Transform of the signal is expressed as the summation of the discrete signal where each sample is multiplied by $e^{-j2\pi f n t_1}$; i.e.

$$X(f) = \sum_{n=-\infty}^{\infty} x(nt_1)e^{-j2\pi f n t_1}. \tag{4.75}$$

The frequency-domain spectrum, shown in Figure 4.10, is periodic and continuous.

The Inverse Fourier Transform is thus

$$x(t) = 1/f_s \int_{-f_s/2}^{f_s/2} X(f)e^{j2\pi f n t_1}\,df. \tag{4.76}$$

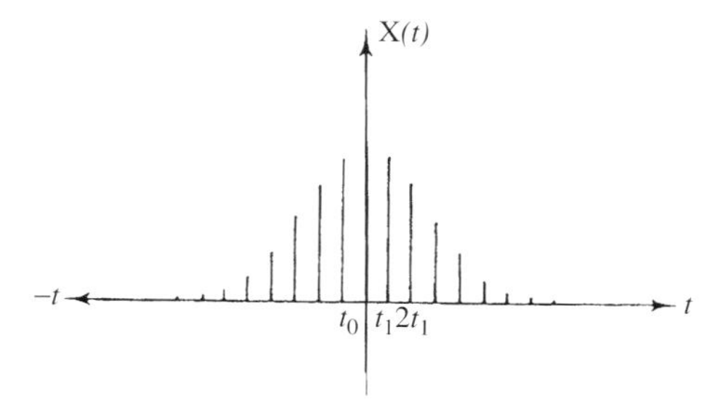

Figure 4.9 Sampled time-domain function

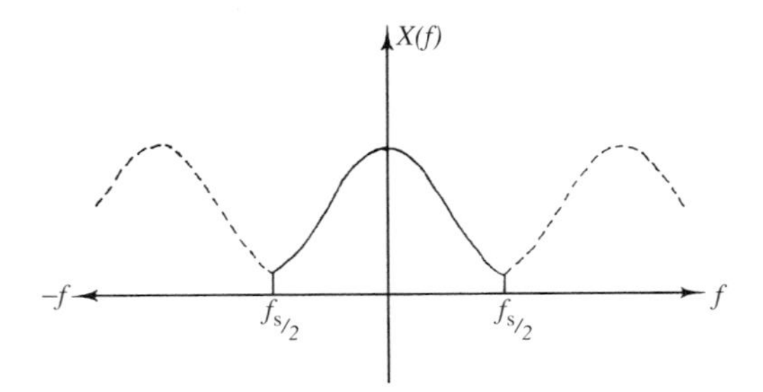

Figure 4.10 Frequency spectrum for discrete time-domain function

4.3.7 Discrete Fourier Transform [9,10]

In the case where the frequency-domain spectrum is a sampled function, as well as the time-domain function, we obtain a Fourier Transform Pair made up of discrete components:

$$X(f_k) = 1/N \sum_{n=0}^{N-1} x(t_n) e^{-j2\pi kn/N} \tag{4.77}$$

and

$$x(t_n) = \sum_{k=0}^{N-1} X(f_k) e^{j2\pi kn/N}. \tag{4.78}$$

Both the time-domain function and the frequency-domain spectrum are assumed periodic as in Figure 4.11, with a total of N samples per period. It is in this discrete form that the Fourier Transform is most suited to numerical evaluation by digital computation.

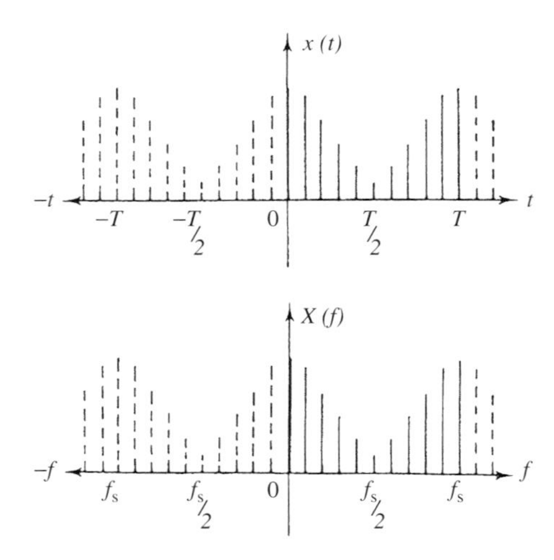

Figure 4.11 Discrete time- and frequency-domain function

Consider Equation (4.77) rewritten as

$$X(f_k) = \frac{1}{N}\sum_{n=0}^{N-1} x(t_n)W^{\mathrm{kn}}, \tag{4.79}$$

where $W = \mathrm{e}^{-j2\pi/N}$.

Over all the frequency components, Equation (4.79) becomes a matrix equation.

$$\begin{bmatrix} X(f_0) \\ X(f_1) \\ \cdot \\ \cdot \\ \cdot \\ X(f_k) \\ \cdot \\ \cdot \\ X(f_{N-1}) \end{bmatrix} = 1/N \begin{bmatrix} 1 & 1 & \cdot\;\cdot\;\cdot & 1 & \cdot\;\cdot\;\cdot & 1 \\ 1 & W & \cdot\;\cdot\;\cdot & W^k & \cdot\;\cdot\;\cdot & W^{N-1} \\ \cdot & \cdot & \cdot & & & \cdot \\ \cdot & \cdot & & \cdot & & \cdot \\ \cdot & \cdot & & & \cdot & \cdot \\ 1 & W^k & \cdot\;\cdot\;\cdot & W^{k^2} & \cdot\;\cdot\;\cdot & W^{k(N-1)} \\ \cdot & \cdot & & \cdot & \cdot & \cdot \\ \cdot & \cdot & & \cdot & \cdot & \cdot \\ 1 & W^{N-1} & \cdot\;\cdot\;\cdot & W^{(N-1)k} & \cdot\;\cdot\;\cdot & W^{(N-1)^2} \end{bmatrix} \cdot \begin{bmatrix} x(t_0) \\ x(t_1) \\ \cdot \\ \cdot \\ \cdot \\ x(t_k) \\ \cdot \\ \cdot \\ x(t_{N-1}) \end{bmatrix} \tag{4.80}$$

or in a condensed form

$$[X(f_k)] = \frac{1}{N}[W^{kn}][x(t_n)]. \tag{4.81}$$

In these equations, $[X(f_k)]$ is a vector representing the N components of the function in the frequency domain, while $[x(t_n)]$ is a vector representing the N samples of the function in the time domain.

Calculation of the N frequency components from the N time samples, therefore, requires a total of N^2 complex multiplications to implement in the above form.

Each element in the matrix $[W^{kn}]$ represents a unit vector with a clockwise rotation of $2n\pi/N(n = 0, 1, 2, \ldots, (N-1))$ introduced between successive components. Depending on the value of N, a number of these elements are the same.

For example, if $N = 8$, then

$$\begin{aligned} W &= \mathrm{e}^{-j2\pi/8} \\ &= \cos\frac{\pi}{4} - j\sin\frac{\pi}{4}. \end{aligned}$$

As a consequence,

$$\begin{aligned} W^0 &= -W^4 = 1, \\ W^1 &= -W^5 = \left(\frac{1}{\sqrt{2}} - j\frac{1}{\sqrt{2}}\right), \\ W^2 &= -W^6 = -j, \\ W^3 &= -W^7 = -\left(\frac{1}{\sqrt{2}} + j\frac{1}{\sqrt{2}}\right). \end{aligned}$$

These can also be thought of as unit vectors rotated through $\pm 0^\circ$, $\pm 45^\circ$, $\pm 90^\circ$ and $\pm 135^\circ$, respectively.

Further more, W^8 is a complete rotation and hence equal to 1. The value of the elements of W^{kn} for $kn > 8$ can thus be obtained by subtracting full rotations, to leave only a fraction of a rotation, the values for which are shown above. For example, if $k = 5$ and $n = 6$, then $kn = 30$ and $W^{30} = W^{3\times 8+6} = W^6 = j$.

Thus, there are only four unique absolute values of W^{kn} and the matrix $[W^{kn}]$, for the case $N = 8$, becomes

$$\begin{bmatrix} 1 & 1 & 1 & 1 & 1 & 1 & 1 & 1 \\ 1 & W & -j & W^3 & -1 & -W & j & -W^3 \\ 1 & -j & -1 & j & 1 & -j & -1 & j \\ 1 & W^3 & j & W & -1 & -W^3 & -j & -W \\ 1 & -1 & 1 & -1 & 1 & -1 & 1 & -1 \\ 1 & -W & -j & -W^3 & -1 & W & j & W^3 \\ 1 & j & -1 & -j & 1 & j & -1 & -j \\ 1 & -W^3 & j & -W & -1 & W^3 & -j & W \end{bmatrix}.$$

It can be observed that the d.c. component of the frequency spectrum, $X(f_0)$, obtained by the algebraic addition of all the time-domain samples, divided by the number of samples, is the average value of all the samples.

Subsequent rows show that each time sample is weighted by a rotation dependent on the row number. Thus, for $X(f_1)$ each successive time sample is rotated by $1/N$ of a revolution; for $X(f_2)$ each sample is rotated by $2/N$ revolutions, and so on.

4.3.8 The Nyquist Frequency and Aliasing [9]

With regard to Equation (4.80) for the DFT and the matrix $[W^{kn}]$ it can be observed that for the rows $N/2$ to N, the rotations applied to each time sample are the negative of those in rows $N/2$ to 1. Frequency components above $k = N/2$ can be considered as negative frequencies, since the unit vector is being rotated through increments greater than π between successive components. In the example of $N = 8$, the elements of row 3 are successively rotated through $-\pi/2$. The elements of row 7 are similarly rotated through $-3\pi/2$; or in negative frequency form through $\pi/2$. More generally, a rotation through

$$2\pi(N/2 + p)/N \text{ rad}, \quad \text{for } p = 1, 2, 3 \ldots, (N/2 - 1) \quad [\text{with } N \text{ even}],$$

corresponds to a negative rotation of

$$-2\pi(N/2 - p)/N \text{ rad}.$$

Hence, $-X(k)$ corresponds to $X(N - k)$ for $k = 1$ to $N/2$ as shown by Figure 4.12.

This is an interpretation of the sampling theorem which states that the sampling frequency must be at least twice the highest frequency contained in the original signal for a correct transfer of information to the sampled system.

The frequency component at half the sampling frequency is referred to as the Nyquist frequency.

The representation of frequencies above the Nyquist frequency as negative frequencies means that should the sampling rate be less than twice the highest frequency present in the sampled waveform, then these higher frequency components can mimic components below the Nyquist frequency, introducing error into the analysis.

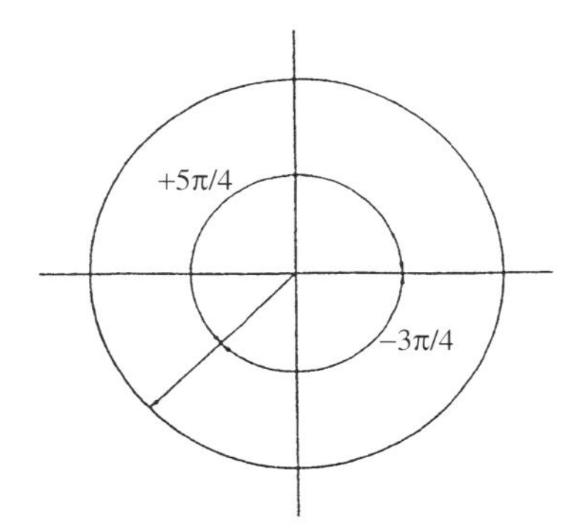

Figure 4.12 Correspondence of positive and negative angles

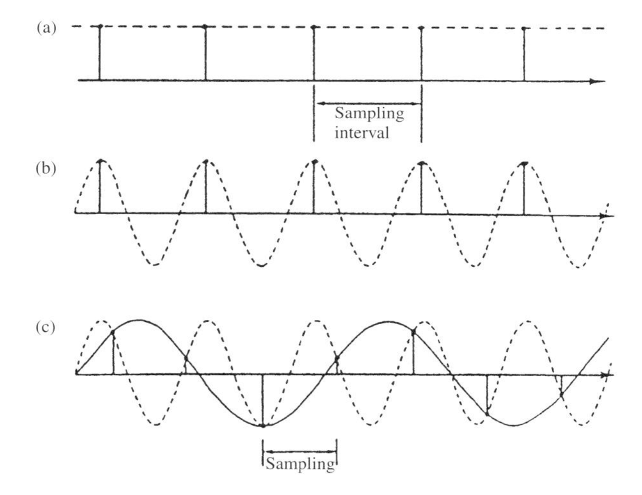

Figure 4.13 The effect of aliasing: (a) $x(t) = k$; (b) $x(t) = k \cos 2\pi n f t$. For (a) and (b) both signals are interpreted as being d.c. In (c) the sampling can represent two different signals with frequencies above and below the Nyquist or sampling rate

It is possible for high-frequency components to complete many revolutions between samplings; however, since they are only sampled at discrete points in time, this information is lost.

This misinterpretation of frequencies above the Nyquist frequency as being lower frequencies, is called *aliasing* and is illustrated in Figure 4.13.

To prevent aliasing it is necessary to pass the time-domain signal through a band-limited low-pass filter, the ideal characteristic of which is shown in Figure 4.14, with a cut-off frequency, f_c, equal to the Nyquist frequency.

Thus, if sampling is undertaken on the filtered signal and the DFT applied, the frequency spectrum has no aliasing effect and is an accurate representation of the frequencies in the original signal that are below the Nyquist frequency. However, information on those frequencies above the Nyquist frequency is lost due to the filtering process.

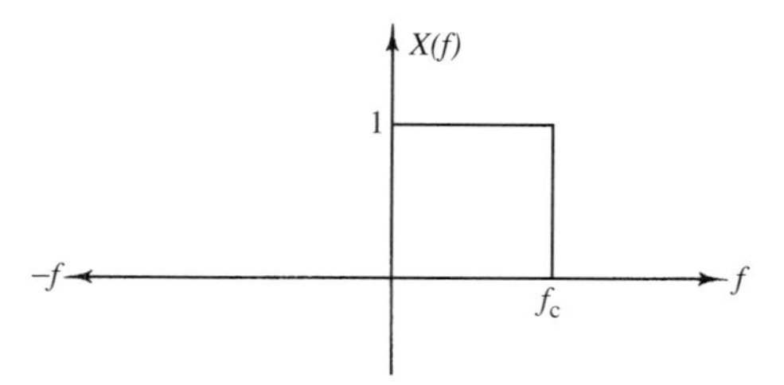

Figure 4.14 Frequency domain characteristics of an ideal low-pass filter with cut-off frequency f_c

4.3.9 Fast Fourier Transform [9,10,11,12]

For large values of N, the computational time and cost of executing the N^2 complex multiplications of the DFT can become prohibitive.

Instead, a calculation procedure known as the FFT, which takes advantage of the similarity of many of the elements in the matrix $[W^{kn}]$, produces the same frequency components using only $N/2 \log_2 N$ multiplications to execute the solution of Equation (4.81). Thus, for the case $N = 1024 = 2^{10}$, there is a saving in computation time by a factor of over 200. This is achieved by factorising the $[W^{kn}]$ matrix of equation (4.81) into $\log_2 N$ individual or factor matrices such that there are only two non-zero elements in each row of these matrices, one of which is always unity. Thus, when multiplying by any factor matrix, only N operations are required.

The reduction in the number of multiplications required, to $(N/2)\log_2 N$, is obtained by recognising that

$$W^{N/2} = -W^0,$$
$$W^{(N+2)/2} = -W^1, \text{ etc.}$$

To obtain the factor matrices, it is first necessary to re-order the rows of the full matrix. If rows are denoted by a binary representation, then the re-ordering is by bit reversal.

For the example where $N = 8$; row 5, represented as 100 in binary (row 1 is 000), now becomes row 2, or 001 in binary. Thus, rows 2 and 5 are interchanged. Similarly, rows 4 and 7, represented as 011 and 110, respectively, are also interchanged. Rows 1, 3, 6 and 8 have binary representations which are symmetrical with respect to bit reversal and hence remain unchanged.

The corresponding matrix is now

$$\begin{bmatrix} 1 & 1 & 1 & 1 & 1 & 1 & 1 & 1 \\ 1 & -1 & 1 & -1 & 1 & -1 & 1 & -1 \\ 1 & -j & -1 & j & 1 & -j & -1 & j \\ 1 & j & -1 & -j & 1 & j & -1 & -j \\ 1 & W & -j & W^3 & -1 & -W & j & -W^3 \\ 1 & -W & -j & -W^3 & -1 & W & j & W^3 \\ 1 & W^3 & j & W & -1 & -W^3 & -j & -W \\ 1 & -W^3 & j & -W & -1 & W^3 & -j & W \end{bmatrix}.$$

This new matrix can be separated into $\log_2 8 (= 3)$ factor matrices.

$$
\begin{bmatrix}
1 & 1 & & & & & & \\
1 & -1 & & & & & & \\
& & 1 & -j & & & & \\
& & 1 & j & & & & \\
& & & & 1 & W & & \\
& & & & 1 & -W & & \\
& & & & & & 1 & W^3 \\
& & & & & & 1 & -W^3
\end{bmatrix}
\begin{bmatrix}
1 & & 1 & & & & & \\
& 1 & & 1 & & & & \\
1 & & -1 & & & & & \\
& 1 & & -1 & & & & \\
& & & & 1 & & -j & \\
& & & & & 1 & & -j \\
& & & & 1 & & j & \\
& & & & & 1 & & j
\end{bmatrix}
$$

$$
\times
\begin{bmatrix}
1 & & & & 1 & & & \\
& 1 & & & & 1 & & \\
& & 1 & & & & 1 & \\
& & & 1 & & & & 1 \\
1 & & & & -1 & & & \\
& 1 & & & & -1 & & \\
& & 1 & & & & -1 & \\
& & & 1 & & & & -1
\end{bmatrix}
$$

As previously stated, each factor matrix has only two non-zero elements per row, the first of which is unity.

The re-ordering of the $[W^{kn}]$ matrix results in a frequency spectrum which is also re-ordered. To obtain the natural order of frequencies, it is necessary to reverse the previous bit-reversal.

In practice, a mathematical algorithm implicitly giving factor matrix operations is used for the solution of an FFT [13].

Using $N = 2^m$, it is possible to represent n and k by m bit binary numbers such that

$$n = n_{\mathrm{m}-1}2^{\mathrm{m}-1} + n_{\mathrm{m}-2}2^{\mathrm{m}-2} + \cdots + 4n_2 + 2n_1 + n_0, \tag{4.82}$$

$$k = k_{\mathrm{m}-1}2^{\mathrm{m}-1} + k_{\mathrm{m}-2}2^{\mathrm{m}-2} + \cdots + 4k_2 + 2k_1 + k_0, \tag{4.83}$$

where

$$n_{\mathrm{i}} = 0, 1 \quad \text{and} \quad k_{\mathrm{i}} = 0, 1.$$

For $N = 8$:

$$n = 4n_2 + 2n_1 + n_0$$

and

$$k = 4k_2 + 2k_1 + k_0,$$

where n_2, n_1, n_0 and k_2, k_1, k_0 are binary bits (n_2, k_2 most significant and n_0, k_0 least significant).

Equation (4.79) can now be re-written as

$$X(k_2, k_1, k_0) = \sum_{n_2=0}^{1} \sum_{n_1=0}^{1} \sum_{n_0=0}^{1} \frac{1}{N} x(n_2, n_1, n_0) W. \tag{4.84}$$

Defining n and k in this way enables the computation of Equation (4.79) to be performed in three independent stages computing in turn:

$$A_1(k_0, n_1, n_0) = \sum_{n_2=0}^{1} \frac{1}{N} x(n_2, n_1, n_0) W^{4k_0 n_2}, \tag{4.85}$$

$$A_2(k_0, k_1, n_0) = \sum_{n_1=0}^{1} A_1(k_0, n_1, n_0) W^{2(k_0+2k_1)n_1}, \tag{4.86}$$

$$A_3(k_0, k_1, k_2) = \sum_{n_0=0}^{1} A_2(k_0, k_1, n_0) W^{(k_0+2k_1+4k_2)n_0}. \tag{4.87}$$

From Equation (4.87) it is seen that the A_3 coefficients contain the required $X(k)$ coefficients but in reverse binary order:

$$\begin{aligned} &\text{Order of } A_3 \text{ in binary form is } k_0 k_1 k_2, \\ &\text{Order of } X(k) \text{ in binary form is } k_2 k_1 k_0. \end{aligned}$$

Hence

$$\begin{array}{ccccccc} & & \text{Binary} & & \text{Reversed} & & \\ A_3(3) & = & A_3(011) & = & X(110) & = & X(6) \\ A_3(4) & = & A_3(100) & = & X(001) & = & X(1) \\ A_3(5) & = & A_3(101) & = & X(101) & = & X(5). \end{array}$$

4.4 Window Functions [14]

In any practical measurement of a time-domain signal, it is normal to limit the time duration over which the signal is observed. This process is known as *windowing* and is particularly useful for the measurement of non-stationary signals which may be divided into short segments of a quasi-stationary nature with an implied infinite periodicity. Furthermore, in the digital analysis of waveforms, only a finite number of samples of the signal are recorded on which a spectral analysis is made. Thus, even stationary signals are viewed from limited time data and this can introduce errors in the frequency spectrum of the signal.

The effect of windowing can best be seen by defining a time-domain function which lies within finite time limits. Outside of these, the function is zero. The simplest window function is the rectangular window of Figure 4.15. The frequency spectrum of this function, obtained in Section 4.3.5, is also included.

The application of a window function has the effect of multiplying each point of a time-domain signal by the corresponding time point of the window function. Thus, within a rectangular window, the signal is just itself, but outside of this the signal is completely attenuated, although a periodicity of the signal within the window is implied outside the defined window. This time-domain multiplication has its equivalent in the frequency domain as the convolution of the spectra of the window function and the signal. This is illustrated for an infinite periodic function and a rectangular window

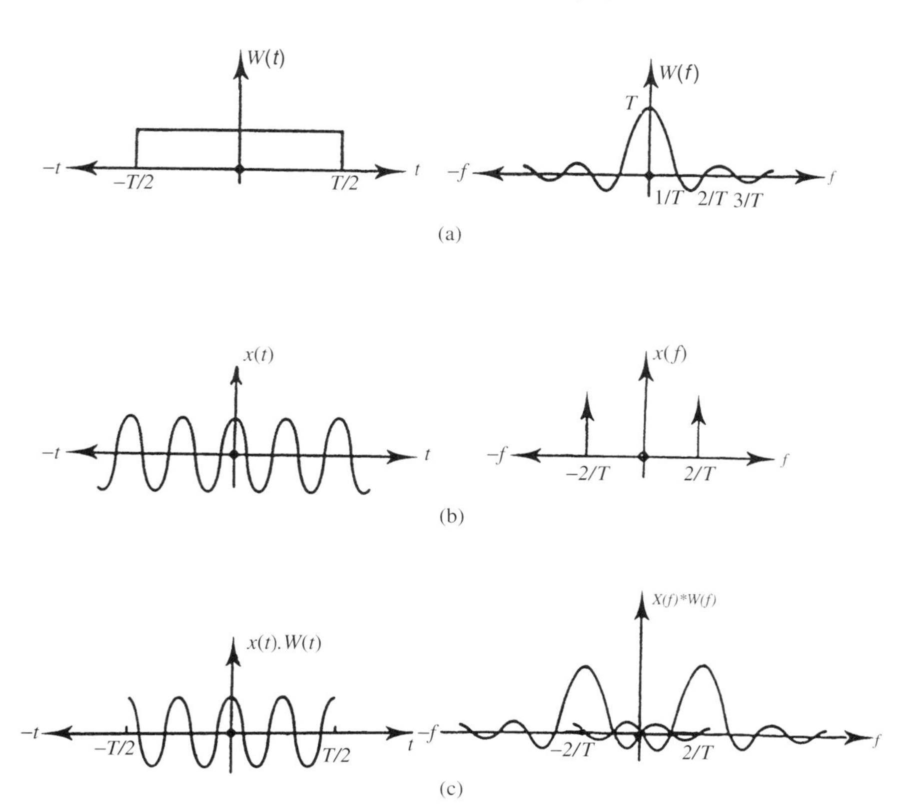

Figure 4.15 Infinite periodic function processed with a rectangular window function: (a) rectangular window function and frequency spectrum; (b) periodic function $x(t) = A\cos(4\pi t/T)$ and frequency spectrum; (c) infinite periodic function viewed through a rectangular time window

function in Figure 4.15(b) and (c). It can be observed that there is significant power in the frequencies of the sidelobes about the fundamental frequency, which is not present in the infinite fundamental frequency waveform. In this simple case, where the signal $x(t)$ is of fundamental frequency the only spectral component which contributes is that being evaluated, the component at f_1 in Figure 4.15(c).

However, it is highly likely that waveforms will be made up of many frequency components, not necessarily integer multiples of the fundamental window frequency f_1. Consequently, discontinuities will exist between the function at the start and finish of the window which will introduce uncertainty in the identification of the periodic components present, since Fourier analysis assumes periodicity of functions and continuity at the boundaries. The resulting error is known as spectral leakage and is the non-periodic noise contributing to each of the periodic spectral components present.

As an illustration of spectral leakage occurring when the duration of the rectangular time window is different from the fundamental of the actual waveform, consider a single-frequency periodic waveform where the worst-case sample gives the phase discontinuity as shown in Figure 4.16(b). Here, the spectrum shows the main frequency component, the existence of high side lobes and a d.c. component.

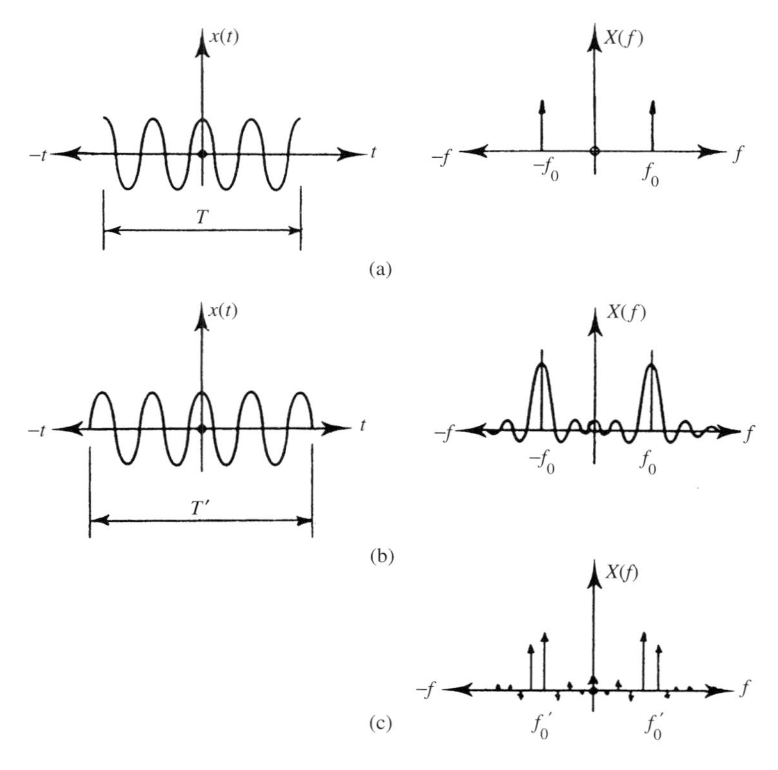

Figure 4.16 Infinite periodic signal viewed during different duration time windows with periodicity implied: (a) the time window is an exact multiple of the period of the waveform; (b) the time window is a $(2n + 1)/2$ multiple of the period of the waveform; (c) case (b) viewed as a discrete frequency spectrum

4.4.1 The Picket Fence

The combination of the DFT and window function produces a response equivalent to filtering the time-domain signal through a series of filters with centre frequencies at integer multiples of $1/T$, where T is the sampling period. The filter characteristic and the associated leakage is determined by the particular window function chosen. The resulting spectrum can therefore be considered as the true spectrum viewed through a picket fence with only frequencies at points corresponding to the gaps in the fence being visible.

When the signal being analysed is not one of these discrete, orthogonal frequencies, then, because of the non-ideal nature of the DFT filter, it will be seen by more than one such filter, but at a reduced level in each. The effect can be reduced by adding a number of zeros, usually equivalent to the original record length, to the data to be analysed. This is called zero padding. This effective increase in the sampling period T introduces extra DFT filters at points between the original filters. The bandwidth of the individual filters still depends upon the original sample period and is therefore unchanged.

4.4.2 Spectral Leakage Reduction

The effect of spectral leakage can be reduced by changing the form of the window function. In particular, if the magnitude of the window function is reduced towards zero at the boundaries, any discontinuity in the original waveform is weighted to a very small value and thus the signal is effectively continuous at the boundaries. This implies a more periodic waveform which has a more discrete frequency spectrum.

A number of window functions are shown in Figure 4.17, along with their Fourier Transforms, which give a measure of the attenuation of the side lobes that give rise to spectral leakage.

4.4.3 Choice of Window Function

The objective in choosing a window to minimise spectral leakage is to obtain a mainlobe width which is as narrow as possible, so that it only includes the spectral component of interest, with minimal sidelobe levels to reduce the contribution from interfering spectral components. These two specifications are inter-related for realisable windows and a compromise is made between mainlobe width compression and sidelobe level reduction.

The rectangular window function, defined by

$$W(t) = \begin{cases} 1, & \text{for } \dfrac{-T}{2} < t < \dfrac{T}{2}, \\ 0, & \text{otherwise}, \end{cases} \tag{4.88}$$

has a noise or effective bandwidth of $1/T$, where T is the window length; the sidelobe levels are large (-13 dB from the main lobe for the first sidelobe), and their rate of decay with frequency is slow (being 20 dB per decade). This means that when evaluating the fundamental component of a signal, interfering spectral components near to it will be weighted heavily, contributing greater interference to the fundamental than for the other windows illustrated in Figure 4.17.

However, as mentioned previously, there is one situation where the rectangular window ideally results in zero spectral leakage and high spectral resolution. This situation occurs when the duration of the rectangular window is equal to an integer multiple of the period of a periodic signal. When the rectangular window spans exactly one period, the zeros in the spectrum of the window coincide with all the harmonics except one. This results in no spectral leakage under ideal conditions. Consequently, spectrum analysers often incorporate the rectangular window function facility for the analysis of periodic waveforms, to which the duration of the window can be matched. This is achieved through the use of a phase-locked loop. This frequency matching gives the greatest resolution of the periodic frequency.

The triangular window, defined by

$$W(t) = \begin{cases} 1 + \dfrac{2t}{T}, & \text{for } \dfrac{-T}{2} < t < 0, \\ 1 - \dfrac{2t}{T}, & \text{for } 0 < t < \dfrac{T}{2}, \\ 0, & \text{otherwise}, \end{cases} \tag{4.89}$$

is a simple modification of the rectangular window, where the amplitude of the multiplying window is reduced linearly to zero from the window centre. The reduction

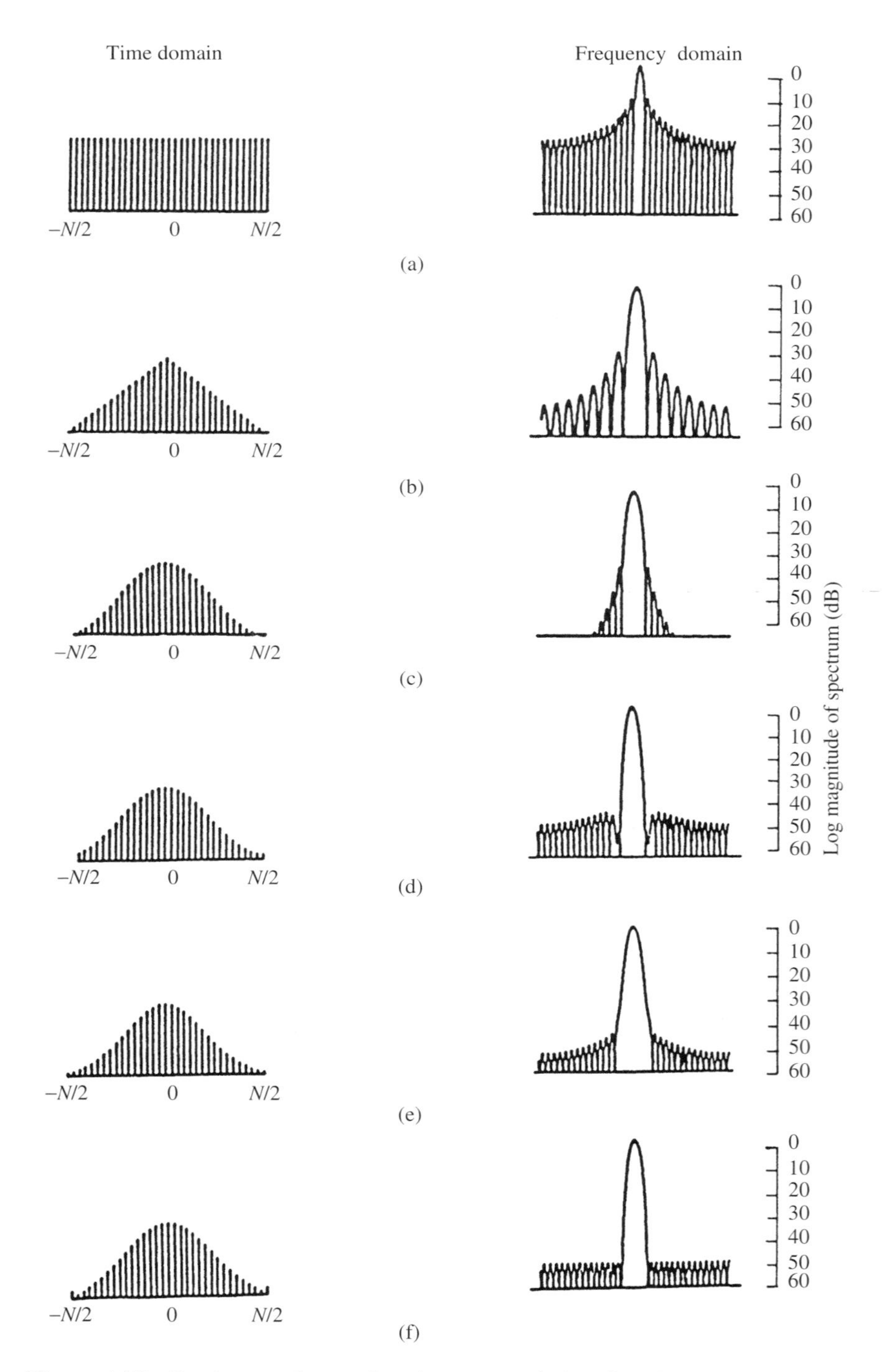

Figure 4.17 Fourier transform pairs of common window functions: (a) rectangular; (b) triangular; (c) cosine squared (hanning); (d) hamming; (e) gaussian; (f) dolph–chebyshev

of sidelobe is readily seen in Figure 4.17(b), but this is at the expense of mainlobe width and a consequent reduction in frequency resolution.

An international standard window function often incorporated into spectrum analysers is the cosine-squared or Hanning window, defined by

$$W(t) = \frac{1}{2}\left(1 - \cos\frac{2\pi t}{T}\right), \quad \text{for } -\frac{T}{2} < t < \frac{T}{2}, \tag{4.90}$$

and in which it is the power term that is cosine squared. This function is easily generated from sinusoidal signals and in FFT analysers a table of cosine values can be utilised for generating the window. The mainlobe noise bandwidth is greater than that for the rectangular window, being $1.5T$; however, the highest sidelobe is at -32 dB and the sidelobe fall-off rate is 60 dB per decade, thus reducing the effect of spectral leakage. This is illustrated in Figure 4.18 where the Hanning window is compared with the rectangular window for sidelobe-level reduction.

By mounting the Hanning window on a small rectangular pedestal (but limiting the maximum of the function to unity) the Hamming window is obtained. This is described in its amplitude form as

$$W(t) = 0.54 - 0.46\cos\frac{2\pi t}{T}, \quad \text{for } -\frac{T}{2} < t < \frac{T}{2}. \tag{4.91}$$

The second sidelobe of the rectangular function coincides with the first sidelobe of the Hanning function and since these are in opposite phase, they can be scaled to cancel each other. As a consequence, the highest sidelobe level is -42 dB (Figure 4.18). The

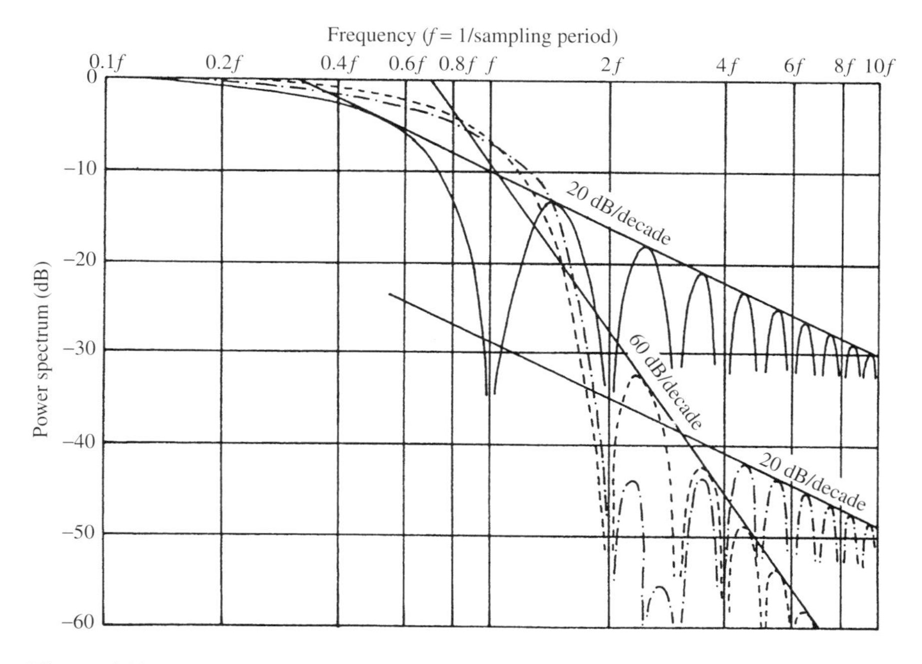

Figure 4.18 Power spectrum *versus* log(frequency) for selected window functions. ———, rectangular window; - - - , hanning window; . . . . , hamming window

remaining sidelobes are dominated by the rectangular function and have a fall-off rate of 20 dB per decade. A slight improvement in mainlobe noise bandwidth (to $1.4/T$) is observed also.

The ideal window which has a single mainlobe and no sidelobes is in the form of a Gaussian function:

$$W(t) = \exp(-t^2/2\sigma^2). \tag{4.92}$$

The Gaussian function has the property of transforming, by the Fourier Transform, to another Gaussian function. On a decibel scale, its shape is that of an inverted parabola, with a characteristic which becomes successively steeper. Theoretically, the Gaussian function is defined between infinite time limits. For practical use, the function is truncated at three times the half-amplitude width, which is 7.06 times the standard deviation. As a consequence, sidelobes are established in the power spectrum but these are of the order of −44 dB down. The mainlobe noise bandwidth is wider than the previous windows, being $1.9/T$.

The final window, presented in Figure 4.17(f), is the Dolph–Chebyshev function, the discrete form of which is defined as

$$W(t) = \frac{(-1)^r \cos[N \cos^{-1}[\beta \cos(\pi r/N)]]}{\cos \mathrm{h}[n \cosh^{-1}(\beta)]}, \quad \text{for } 0 < r < N-1, \tag{4.93}$$

where r is an integer and N is the number of discrete samples of the window function.

$$\beta = \cos \mathrm{h}\left[\frac{1}{N} \cos \mathrm{h}^{-1}(10^{\alpha})\right],$$

and the inverse hyperbolic cosine is defined by

$$\cos \mathrm{h}^{-1} x = \begin{cases} \pi/2 - \tan^{-1}[x/\sqrt{(1-x^2)}], & \text{for } |x| < 1.0, \\ 1n[x + \sqrt{(x^2-1)}], & \text{for } |x| > 1.0. \end{cases}$$

This function provides the narrowest possible mainlobe width for a given specified sidelobe level, which is constant on a decibel scale. The sidelobe levels are controlled by the parameter α in Equation (4.93). With $\alpha = 4.0$ the sidelobes are at −80 dB (0.01%) with respect to the mainlobe.

4.4.4 Mainlobe Width Reduction

In obtaining low sidelobe levels to reduce spectral leakage, in all the window functions previously mentioned, there has been a sacrifice of mainlobe bandwidth. It is possible that in harmonic analysis, when evaluating, for example, the fundamental of a waveform, the resolution is such that the d.c. component and the second and third harmonics are included within the mainlobe. This causes considerable interference in the individual harmonic evaluations and restricts the identification of spectral leakage effects to higher-order frequencies and noise, outside the mainlobe.

However, with the ability to change the sidelobe level with the Dolph–Chebyshev window, an algorithm presents itself to effectively reduce the mainlobe bandwidth.

Consider the complex fundamental Fourier component of the waveform, multiplied by the Dolph–Chebyshev window function i.e. $W(r) \cdot x(r)^N{}_{r=1}$, obtained by the application of the DFT (using the FFT technique)

$$X_1 = W_0C_1 + W_1C_2 + W_2C_3 + W_{-2}C_{-1} + W_{-3}C_{-2} + W_{-4}C_{-3} + W_{-1}C_0 + \sum_{n=3}^{N/2} W_nC_{n+1} + \sum_{n=4}^{N/2-1} W_{-n+1}C_{-n}, \tag{4.94}$$

where W are the discrete window coefficients in the frequency domain and $C_n = C_{-n}$ are the complex periodic Fourier coefficients of harmonic order n.

By pre-processing the waveform, the d.c. component can be removed, thereby eliminating C_0 in Equation (4.94). In addition, since the last two terms of this equation include all the higher-order harmonics and noise, but are weighted with window coefficients of the order of 0.01%, they can also be neglected. Equation (4.94) can therefore be reduced to

$$X_1 = C_1(W_0 + W_{-2}) + C_2(W_1 + W_{-3}) + C_3(W_2 + W_{-4}). \tag{4.95}$$

Application of the DFT to the windowed discrete time-domain waveform yields a value for X_1. The window coefficients in the frequency domain are known by the defining Equation (4.91) and hence only the three harmonic terms C_1, C_2, C_3 are known.

The use of three windows, each with a different α parameter, and the application of the DFT three times to the same waveform, produces three simultaneous equations of the same form as Equation (4.95). The solution of these leads directly to the values of C_1, C_2 and C_3, i.e. the fundamental component and second and third harmonics of the original waveform.

As an illustration of the effectiveness of this algorithm, consider the function defined by

$$x(t) = C_1\cos(2\pi f_1t + \phi_1) + \sum_{n=2}^{7} C_n\cos(n2\pi f_1 + \phi_n), \tag{4.96}$$

where $C_1 = 1.0$ and $C_n = 0.2$ for $n = 2$ to 7 for which 32 samples were available.

With window filtering, the error introduced by the higher harmonics in the identification of C_1 was limited to 0.16%. However, when a non-periodic component of frequency $5.65f_1$ and magnitude 0.2 was introduced into $x(t)$, the error in identifying C_1 without window filtering was 2%. With the Dolph–Chebyshev window and using α values of 3.2, 3.5 and 3.8, the error was reduced to 0.36%.

4.4.5 Application to Inter harmonic Analysis [15]

If frequencies not harmonically related to the sampling period are present or the waveform is not periodic over the sampling interval, errors are encountered due to spectral leakage.

Section 4.4 has shown that the method normally used to minimise spectral leakage and obtain accurate magnitude and frequency information involves the use of windows.

Table 4.1 Frequency components of example system

Frequency	Magnitude
50	1.0
104	0.3
117	0.4
134	0.2
147	0.2
250	0.5

Windowing functions weight the waveform to be processed by the FFT in such a way as to taper the ends of the sample to near zero.

Figure 4.19 illustrates a waveform containing the six steady-state frequency components shown in Table 4.1. The resulting waveform is not periodic and appears even asymmetric depending on the observation interval.

Applying the Hanning window to the waveform of Figure 4.19 produces the waveform shown in Figure 4.20 and spectrum shown in Figure 4.21.

Even with the use of windowing functions, closely spaced inter harmonic frequencies are hard to determine due to the resolution of the FFT as determined by the original sampling period, which is 8 cycles of 50 Hz in this case. The use of the zero padding technique can result in a much more accurate determination of the actual inter harmonic

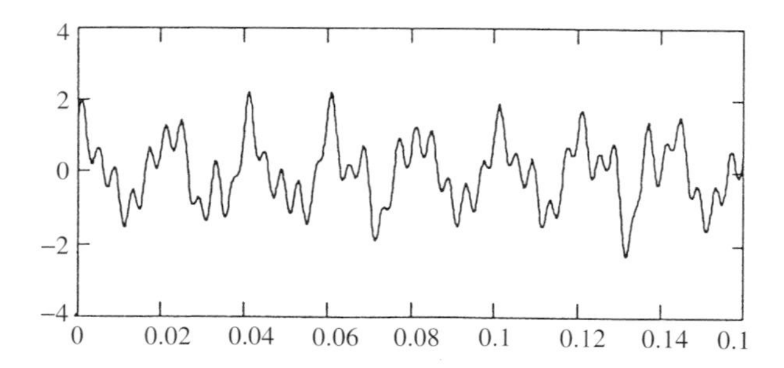

Figure 4.19 Waveform with harmonic and inter harmonic components

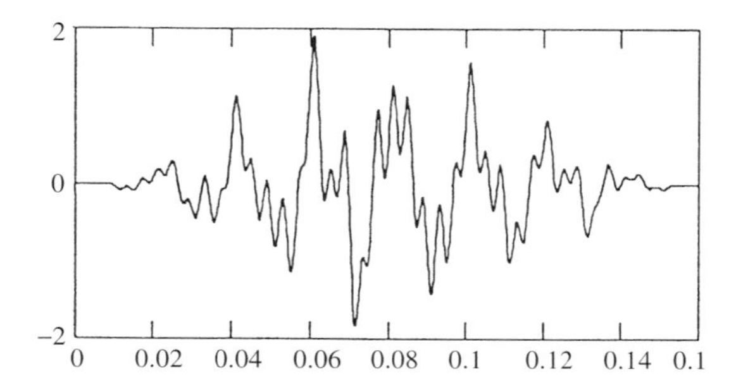

Figure 4.20 Result of hanning window

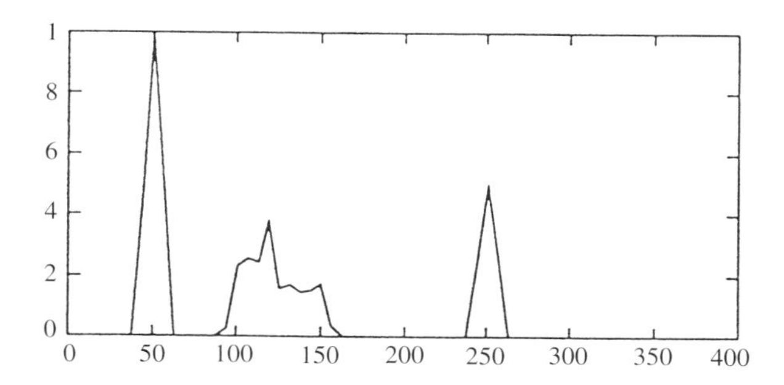

Figure 4.21 FFT spectrum of figure 4.19

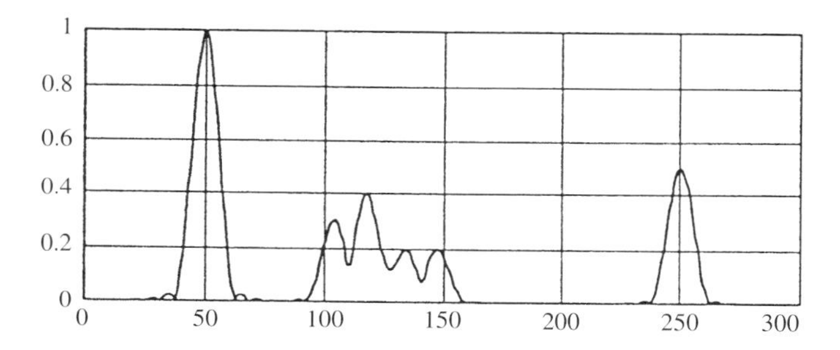

Figure 4.22 Result of FFT analysis of figure 4.21 with four-fold zero padding

frequency component magnitudes and frequencies. Figure 4.22 shows the results of applying a four-fold zero padding before performing the FFT.

Note that the resolution has been improved enough to accurately determine the magnitude and frequency of each component even though the sampled waveform was not periodic and some of the inter harmonic components are very close to each other.

4.5 Efficiency of FFT Algorithms

4.5.1 The Radix-2 FFT

The complex radix-2 [16] is the standard FFT version and is usually available in DSP libraries. A number of alternative algorithms, such as the higher radix, mixed radix and split radix, have been developed but the radix-2 is still widely used. It relies on a decomposition of the set of N inputs into successively smaller sets on which the DFT is computed until sets of length 2 remain. Figure 4.23 represents this approach which requires that the number of input points N is a power of two ($N = 2^{\nu}$). The decomposition in this way can then be performed ν times.

Each of the 2-point DFTs consists of a *butterfly* computation as depicted in Figure 4.24. It involves the two complex numbers a and b, where b is multiplied by the complex phase factor and then the product is added and subtracted from a. One butterfly hence requires one complex multiplication and two complex additions.

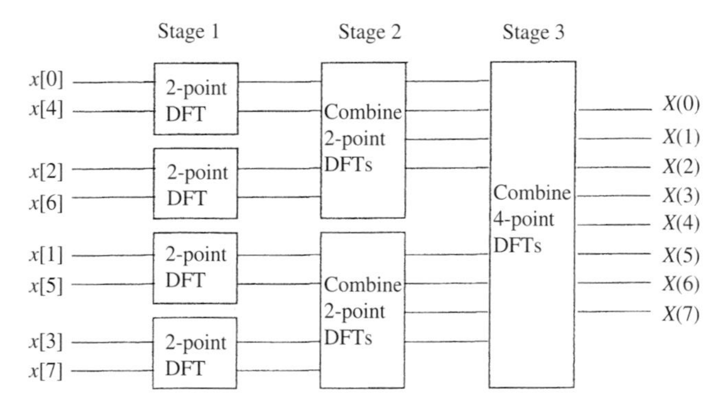

Figure 4.23 Decomposition of an 8-point DFT into 2-point DFTs by the radix-2 FFT algorithm

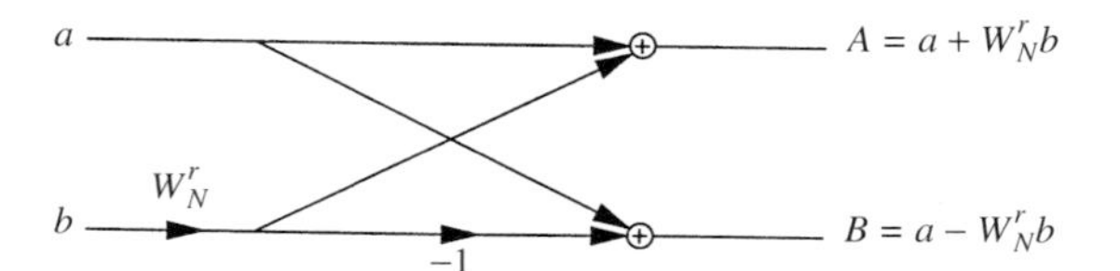

Figure 4.24 Butterfly computation with twiddle factor W_N^r

Each of the stages consists of $N/2$ butterflies. Thus, to compute the radix-2 FFT, $\nu N/2$ butterflies are required. The total number of real operations are $5\nu N$, consisting of $2\nu N$ multiplications and $3\nu N$ additions.

4.5.2 Mixed-Radix FFT

The sampling rate of measurement-equipment (and step-length for time-domain simulation) are not often chosen to result in 2^n sample points for one period of the fundamental frequencies of 50 Hz or 60 Hz. With the increase in computer power, the mixed-radix FFT is useful and practical for power system studies, giving flexibility in transform data size and sampling rate and eliminating the leakage problem that often occurs with radix-2 FFT.

FFT algorithms are not limited to the radix-2 family of algorithm only. An improvement of these algorithms is described in [17], where radix-2 and radix-4 routines are mixed to produce a more efficient FFT. Single-radix FFTs, such as the radix-3 and the radix-5 require 3, 9, 27, 81, 243, ... data points and 5, 25, 125, 625, ... data points, respectively. Algorithms based on routines higher than radix-10 require too many points for efficient computation.

More flexibility in the selection of data size is therefore provided by the mixed-radix FFT. For example, a 2000-point DFT can be achieved using a mixed-radix FFT to perform the DFT in $2 \times 2 \times 2 \times 2 \times 5 \times 5 \times 5$ or $2 \times 10 \times 10 \times 10$. Single-radix FFTs for radices 2, 3, 4, 5, 6, ... can be stacked in any order to accommodate the

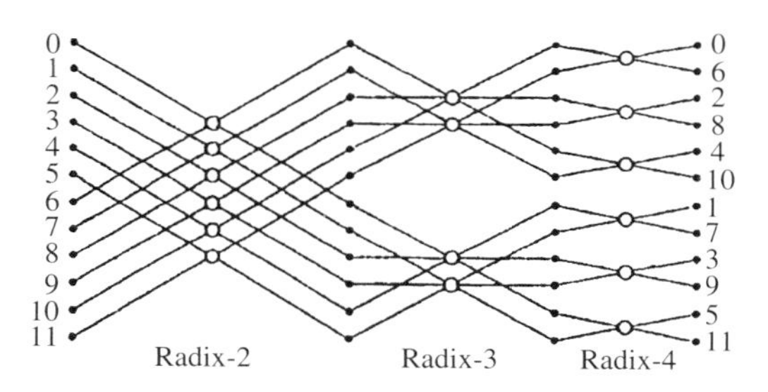

Figure 4.25 12-point $2 \times 3 \times 2$ FFT flow chart

desired number of data points. Figure 4.25 gives a simple flow chart showing a 12-point ($2 \times 3 \times 2$) FFT.

4.5.3 Real-valued FFTs

FFT algorithms are designed to perform complex multiplications and additions but the input sequence may be real. This can be exploited to compute the DFT of two real-valued sequences of length N by one N-point FFT. This is based on the linearity of the DFT and on the fact that the spectrum of a real-valued sequence has complex conjugate symmetry, i.e.

$$X(k) = X^*(N-k), \quad k = 1 \cdots \frac{N}{2} - 1, \tag{4.97}$$

while $X(k)$ and $X(N/2)$ are real. A complex-valued sequence $x[n]$ defined by two real-valued sequences $x_1[n]$ and $x_2[n]$ such that

$$x[n] = x_1[n] + jx_2[n] \tag{4.98}$$

has a DFT that, due to the linearity of the transform, may be expressed as

$$X(k) = X_1(k) + jX_2(k). \tag{4.99}$$

The DFTs of the original sequences are then given by:

$$X_1(k) = \frac{1}{2}[X(k) + X^*(N-k)],$$

$$X_2(k) = \frac{1}{j2}[X(k) + X^*(N-k)]. \tag{4.100}$$

The extra amount of computation needed to recover the DFTs according to Equation (4.100) is small. Since $X_1(k)$ and $X_2(k)$ represent the DFTs of real sequences, they must have complex conjugate symmetry (Equation (4.97)) and only the values for $k = 0 \ldots N/2$ need to be computed. For $k = 0$ or $k = N/2$, the result will be real. Thus, for two length N real-valued sequences, the total number of computations amounts to one N-point FFT plus $2N$-4 extra additions which essentially means the number of operations compared with the standard FFT is halved by this technique.

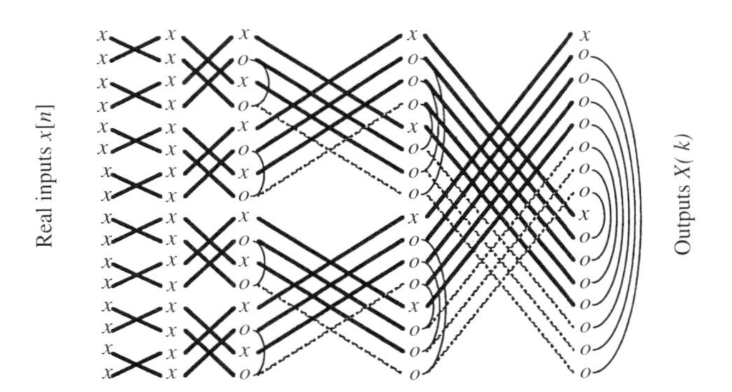

Figure 4.26 Radix-2 FFT of length 16 with real inputs. Real values are indicated by '*x*', complex ones by '*o*'. The complex values connected by arcs are conjugates of each other. The solid lines represent the butterflies that need to be computed. Reproduced from Sorensen et al., 1987

An even more efficient technique [18] is based on the fact that complex conjugate symmetry not only exists at the output of the FFT algorithm (coefficients $X(k)$) but also at every stage. This is shown in Figure 4.26 for a 16-point FFT with real-valued inputs. Thus, the complex conjugate values need not be calculated (saving five butterflies in Figure 4.26).

Although the concept can be applied to any kind of FFT algorithm, the *split-radix* FFT algorithm is recommended because it requires fewer operations than a radix-2 FFT, a radix-4 FFT or any higher-radix FFT. It is called *split-radix* because an N-point DFT is broken up into a length $N/2$ DFT over the inputs with even indices and two length $N/4$ DFTs over the inputs with odd indices. This scheme is then iterated through all stages of the transform. A corresponding real-valued inverse split-radix FFT exists (in this case the outputs being real-valued).

The computational complexity of the real-valued split-radix FFT is reduced to:

$$\#_{\text{mul}} = \frac{2}{3}\nu N - \frac{19}{9}N + 3 + \frac{(-1)^{\nu}}{9}$$

$$\#_{\text{add}} = \frac{4}{3}\nu N - \frac{17}{9}N + 3 - \frac{(-1)^{\nu}}{9}$$

$$\#_{\text{total}} = 2\nu N - 4N + 6 \tag{4.101}$$

where $2^{\nu} = N$.

4.5.4 Partial FFTs

FFT algorithms generally compute N frequency samples for a length N input sequence. If only a narrow band of the spectrum is of interest, i.e. fewer than the N outputs are needed, there are methods to reduce the complexity of the FFT accordingly.

One way is to compute the desired DFT points directly using digital filters, which resonate at the corresponding frequencies [19]. This method becomes computationally unattractive if the number of outputs is above $\nu/2$, as in the case of the flickermeter, where the range of at least 0–35 Hz has to be resolved in approximately ≤ 1 Hz steps by the DFT, which means that about 35 frequency samples are needed. To achieve the required resolution, N must be chosen to be 4096 (i.e. $\nu = 12$), for the 4.2 kHz sampling frequency.

Another well-known technique is FFT *pruning* [20] where the branches in the tree-like FFT flow graph (see Figure 4.26) that lead to unwanted outputs are removed. Instead of a full set of N points, a smaller subset of L points which must occur in a sequence can be calculated.

A more flexible and efficient method is *transform decomposition* (TD) [21] where S output points, which need not be in a sequence, are computed by decomposing the N-point DFT into QP-point DFTs ($N = PQ$). Subsequently, each of the P-point DFTs is computed and recombination (multiplication by the phase factors and summation) leads to the desired S output points. This process is shown in the block diagram of Figure 4.27 for an 8-point DFT. Any FFT algorithm may be used and therefore the real-valued split-radix algorithm described in the previous section is the best choice.

The overhead for the transform decomposition amounts to

$$\#_{\text{mul}} = 4QS$$
$$\#_{\text{add}} = 4QS - 2S$$
$$\overline{\#_{\text{total}} = 8QS - 2S} \tag{4.102}$$

To obtain the total number of operations for the TD, the computations for the Q length P split-radix RFFTs have to be added. These follow from Equation (4.101) where the operation counts of an N-point split-radix RFFT are given and the total becomes

$$\#_{\text{total}} = 2N[\log_2 P - 2 + (3 + 4S)/P)] - 2S. \tag{4.103}$$

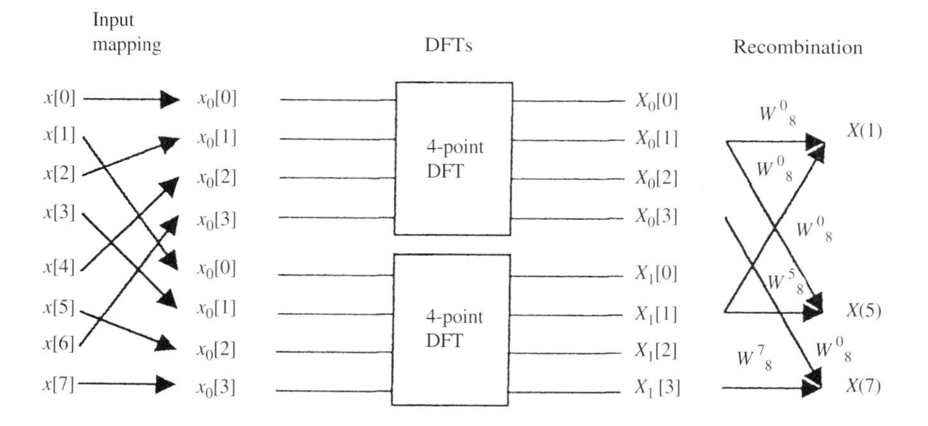

Figure 4.27 Transform decomposition of an 8-point DFT ($Q = 2$, $P = 4$). Only $S = 3$ output points are computed

Table 4.2 Number of operations (real multiplications and additions) required for the DFT computation by the discussed FFT algorithms. Equations (4.96), (4.101) and (4.103) have been evaluated for values of N and S relevant for the flickermeter. The last column indicates the savings of the transform decomposition over a standard FFT

Number of DFTs		Number of operations			Reduction [%]
Inputs N	Outputs S (0–5 Hz)	Standard FFT	Split-radix RFFT	TD + split-radix RFFT	
512	9	23 040	7174	4302	81.3
1024	17	51 200	16 390	10 430	79.6
2048	33	112 640	36 870	24 734	78.0
4096	65	245 760	81 926	57 438	76.6
8192	129	532 480	180 230	131 038	75.4

Table 4.2 lists the number of operations required by each of the FFT algorithms discussed in this section to compute the DFT for a number of inputs and outputs relevant to the flickermeter. The savings achieved by the TD are more than 75%.

4.6 Alternative Transforms

Traditionally, the Fourier Transform has almost exclusively been used in *Power* as well as most other engineering fields and, therefore, the book concentrates on assessment techniques based on Fourier analysis.

Three principal alternatives have been discussed at great length in recent literature with reference to potential power system applications. These are the Walsh, Hartley and Wavelet Transforms.

As in the Fourier case, the Walsh Transform [22] constitutes a set of orthogonal functions. It is extremely simple conceptually, since it involves only square wave components, but for accurate results it needs a large number of terms for the processing of power system waveforms. Moreover, it does not benefit from the differential-to-phasor transformation, and handling differentiation and integration, common operations in power systems, presents a problem as compared with Fourier's.

The Hartley Transform [22], also using the orthogonal principle, is expressed as

$$F(\nu) = \frac{1}{\sqrt{2\pi}} \int_{-\infty}^{\infty} f(t)\ \mathrm{cas}\ (\nu t)\,\mathrm{d}t, \tag{4.104}$$

where

$$\mathrm{cas}(\nu t) = \cos(\nu t) + \sin(\nu t)$$

and ν is identical to the ω of Fourier, i.e. rad/s.

Moreover, its inverse, i.e.

$$f(t) = \frac{1}{\sqrt{2\pi}} \int_{-\infty}^{\infty} F(\nu)\ \mathrm{cas}\ (\nu t)\,\mathrm{d}t, \tag{5.105}$$

has exactly the same form and this leads to simpler software. An important difference between the Fourier and Hartley Transforms is that the latter is all real and thus requires only one half of the memory for storage (i.e. one real quantity as compared with one complex quantity of Fourier's). Also, the convolution operation requires only one real multiplication as compared with four multiplications in the Fourier domain.

However, the Fourier Transform is widely spread throughout the power system field and permits a very convenient assessment of magnitude and phase information. The latter is not always required, however, and in such cases the efficiency of the Hartley Transform may be sufficiently attractive to be used as an alternative to the traditional philosophy.

Owing to the interest generated in the potential applications of wavelets, this transform is given special consideration in the next section.

4.6.1 The Wavelet Transform

The Wavelet Transform(WT), originally derived to process seismic signals, provides a fast and effective way of analysing non-stationary voltage and current waveforms. As in the Fourier case, the WT decomposes a signal into its frequency components. Unlike the Fourier transform, the wavelet can tailor the frequency resolution, a useful property in the characterisation of the source of a transient.

The ability of wavelets to focus on short time intervals for high-frequency components and long intervals for low-frequency components improves the analysis of signals with localised impulses and oscillations, particularly in the presence of a fundamental and low-order harmonic.

A wavelet is the product of an oscillatory function and a decay function. A *mother* wavelet is expressed as [23]

$$g(t) = \mathrm{e}^{-\alpha t^2}\mathrm{e}^{j\omega t} \quad (4.106).$$

An example of a *mother* wavelet is shown in Figure 4.28.

A variety of wavelets originating from a *mother* wavelet can be used to approximate any given function. These wavelets are derived by scaling and shifting (in time) *mother* wavelets as shown in Figure 4.29 and can be expressed as

$$g'(a, b, t) = \frac{1}{\sqrt{a}} g\left(\frac{t-b}{a}\right). \quad (4.107)$$

Therefore, the derived wavelets have the same number of oscillations as the mother wavelet. The one-dimensional signal has been transformed to the two-dimensional function of a (scale) and b (translation).

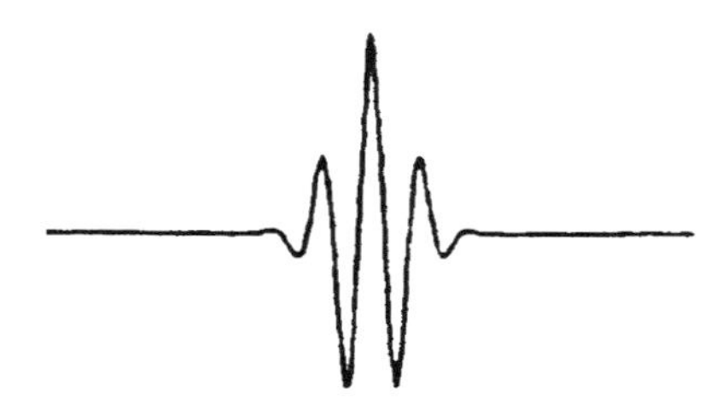

Figure 4.28 A sample mother wavelet

Figure 4.29 A sample daughter wavelet

The wavelet transform of a continuous signal *x(t)* is defined as

$$WT(a, b) = \frac{1}{\sqrt{a}} \int_{-\infty}^{\infty} f(t) g\left(\frac{t-b}{a}\right) \mathrm{d}t. \tag{4.108}$$

The time extent of the wavelet $g((t-b)/a)$ is expanded or contracted in time depending on whether $a > 1$ or $a < 1$. A value of $a > 1 (a < 1)$ expands (contracts) $g(t)$ in time and decreases (increases) the frequency of the oscillations in $g(t - b/a)$. Hence, as a is ranged over some interval, usually beginning with unity and increasing, the input is analysed by an increasingly dilated function that is becoming less and less focused in time.

The WT has a digitally implementable counterpart, the Discrete Wavelet Transform (DWT).

In discrete wavelet transformation, the scale and translation variables are discretised but not the independent variable of the original signal. It is to be noted that the two variables a and b are continuous in the continuous transform. However, in the reconstruction process, the independent variable will be broken down in small segments for the ease of computer implementations. A DWT gives a number of wavelet coefficients depending upon the integer number of the discretisation step in scale and translation, denoted by m and n, respectively. So any wavelet coefficient can be described by two integers, m and n. If a_o and b_o are the segmentation step sizes for the scale and translation, respectively, the scale and translation in terms of these parameters will be $a = a_o^m$ and $b = nb_o a_o^m$.

In terms of the new parameters a_o, b_o, m and n, Equation (4.107) becomes

$$g'(m, n, t) = \frac{1}{\sqrt{a_o^m}} g\left(\frac{t - nb_o a_o^m}{a_o^m}\right) \tag{4.109}$$

or

$$g'(m, n, t) = \frac{1}{\sqrt{a_o^m}} g(ta_o - nb_o) \tag{4.110}$$

and the discrete wavelet coefficients are given by

$$\mathrm{DWT}(m, n) = \int_{-\infty}^{\infty} \frac{1}{\sqrt{a_o^m}} f(t) g(a_o^{-m} t - nb_o) \,\mathrm{d}t. \tag{4.111}$$

Although the transformation is over continuous time, the wavelets' representation is discrete and the discrete wavelet coefficients represent the correlation between the original signal and wavelets for different combinations of m and n.

The inverse DWT is given by

$$f(t) = K \sum_{m=0}^{\infty} \sum_{n=0}^{\infty} W_g f(m, n) \frac{1}{\sqrt{a_o^m}} g(a_o^{-m} t - nb_o), \tag{4.112}$$

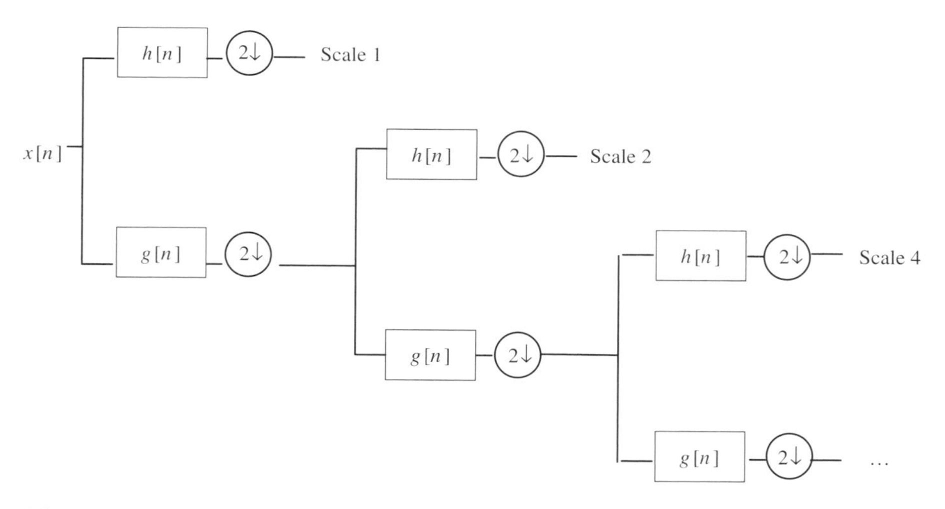

Figure 4.30 Multi-resolution signal decomposition implementation of wavelet analysis

where $K = (A + B)/2$, and A and B are the frame bounds (maximum values of a and b).

Wavelet analysis is normally implemented using multi-resolution signal decomposition (MSD). High-and low-pass equivalent filters, h and g, respectively, are formed from the analysing wavelet. The digital signal to be analysed is then decomposed (filtered) into smoothed and detailed versions at successive scales, as shown in Figure 4.30 where $(2 \downarrow)$ represents a down-sampling by half.

Scale 1 in Figure 4.30 contains information from the Nyquist frequency (half the sampling frequency) to one quarter the sampling frequency; Scale 2 contains information from one quarter to one eighth the sampling frequency, and so on.

Choice of Analysing Wavelets The decomposition can be halted at any scale, with the final smoothed output containing the information of all the remaining scales, i.e. Scales 8, 16, 32, ... if it is halted at Scale 4, one of the desirable properties of MSD.

The choice of mother wavelet is different for each problem at hand and can have a significant effect on the results obtained. Orthogonal wavelets ensure that the signal can be reconstructed from its transform coefficients [24]. Wavelets with symmetric filter coefficients generate a linear phase shift and some wavelets have better time localisation than others.

The wavelet family derived by Daubechies [25] covers the field of orthonormal wavelets. This family is very large and includes members ranging from highly localised to highly smooth. For short and fast transient disturbances, Daub4 and Daub6 wavelets are the best choice, while for slow transient disturbances, Daub8 and Daub10 are particularly good.

However, the selection of an appropriate mother wavelet without knowledge of the types of transient disturbances (which is always the case) is a formidable task. A more user-friendly solution [26] utilises one type of mother wavelet in the whole course of detection and localisation for all types of disturbances.

In doing so, higher-scale signal decomposition is needed. At the lowest scale, i.e. Scale 1, the mother wavelet is most localised in time and oscillates most rapidly within a very short period of time. As the wavelet goes to higher scales, the analysing wavelets become less localised in time and oscillate less due to the dilation nature of the WT analysis. As a result of higher-scale signal decomposition, fast and short transient disturbances will be detected at lower scales, whereas slow and long transient disturbances will be detected at higher scales.

Example of Application [26] Figure 4.31(a) shows a sequence of voltage disturbances. To remove the *noise* present in the waveform, squared wavelet transform coefficients (SWTC) are used at Scales $m = 1, 2, 3$ and 4, respectively (shown in Figures 4.31(b), (c), (d) and (e)); these are analysed using the Daub4 wavelet.

Figure 4.31(a) contains a very rapid oscillation disturbance (high frequency) before time 30 ms, and is followed by a slow oscillation disturbance (low frequency) after time 30 ms. The SWTCs at Scales 1, 2 and 3 catch these rapid oscillations, while Scale 4 catches the slow oscillating disturbance which occurred after time 30 ms. Note that the high SWTCs persist at the same temporal location over Scales 1, 2 and 4.

It must be pointed out that the same technique can be used to detect other forms of waveform distortion (like notches and harmonics) and other types of disturbance such as momentary interruptions, sags and surges.

However, rigorous uniqueness search criteria must be developed for each disturbance for the WT to be accepted as a reliable tool for the automatic classification of power quality disturbances.

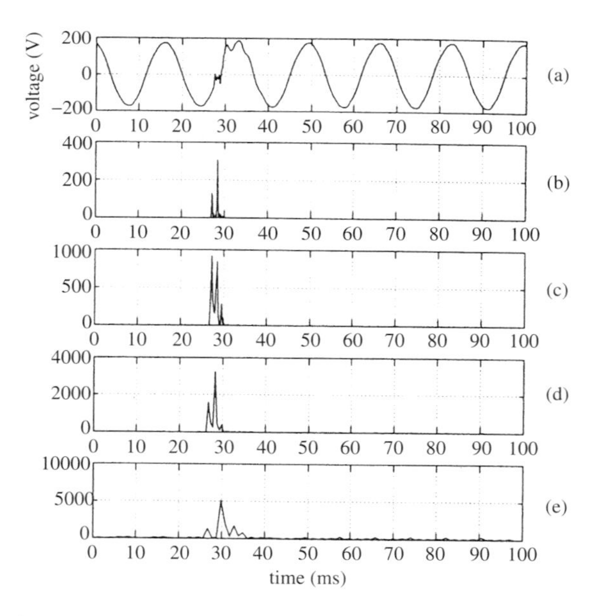

Figure 4.31 Wavefault disturbance detection using Daub4: (a) the voltage disturbance signal; (b), (c), (d) and (e) the SWTCs at scales 1, 2, 3 and 4, respectively Copyright © 1996 IEEE

4.6.2 Automation of Disturbance Recognition

Some proposals are being made to automate the process of disturbance recognition [27,28] to improve the speed, reliability and ease of data collection and storage. Such a scheme involves the three separate stages illustrated in Figure 4.32. These are a pre-processing stage to extract the disturbance information from the generated power signal; a main-processing stage to carry out pattern recognition on the disturbance data; and a post-processing stage to group the output data and form decisions on the possible nature and cause of the disturbance.

The WT is an obvious candidate to extract the disturbance information owing to its greater precision and speed over Fourier methods. A collection of standard libraries of wavelets can be developed to fit specific types of disturbance or transient.

Artificial neural networks can be used in the main processing stage to perform pattern recognition. The neural network can be trained to classify the preliminary information extracted in the pre-processing stage.

The most commonly used type of neural network for pattern recognition is the multilayered perceptron. This is constructed as shown in Figure 4.33 and is usually trained using the recursive error back-propagation algorithm or a modification thereof [29].

The output of the network is

$$y_k(p) = \varphi_k\left(\sum_j W_{kj}(p) \cdot Y_j(p)\right). \tag{4.113}$$

Finally, fuzzy logic [30,31] is well suited in the post-processing stage to make decisions on the disturbance category. It is simple and fast to compute.

Fuzzy rules must be obtained in order to take the information provided by the neural networks and produce a belief in each disturbance category. These take the form of fuzzy IF ... THEN rules and are based on human knowledge into the problem.

The output of the disturbance recognition system is produced as one or a list of disturbance categories with an associated degree of belief. A list of disturbance categories with a belief degree is necessary, since pattern recognition systems are inexact by nature. The system should, however, produce high belief degrees only for disturbance categories that are likely causes.

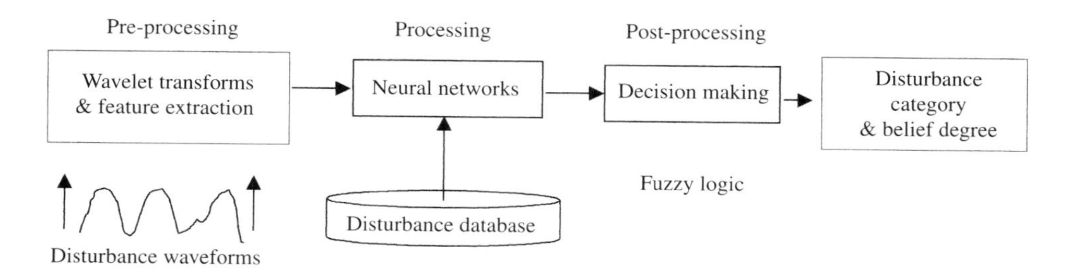

Figure 4.32 Block diagram of the automatic disturbance recognition system

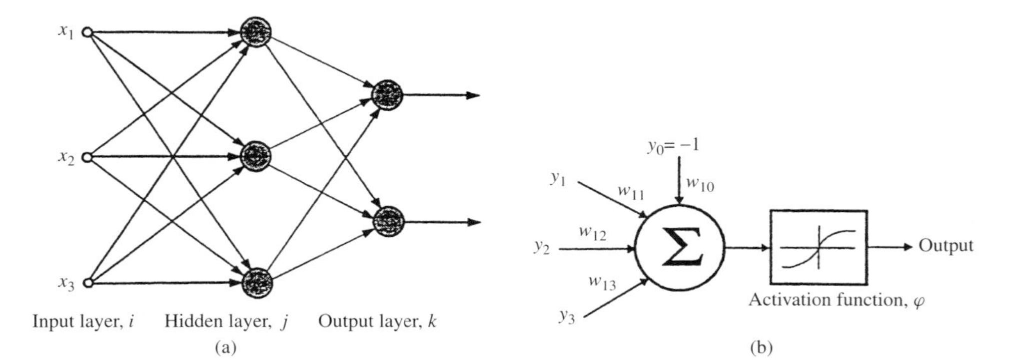

Figure 4.33 (a) Multi-layered perceptron; (b) configuration of an individual neuron

4.7 Summary

Estimation techniques based on the least square error and curve fitting have been shown to be very effective in the derivation of discrete frequency components such as the fundamental frequency voltages and currents.

As the most widely utilised signal processing tool, the Fourier Transform has been given special coverage in this chapter. A variety of techniques have been discussed to improve the accuracy and efficiency of the Fourier processes. The use of windows with non-stationary signals and their application to obtain inter harmonic information has been explained and illustrated with typical test examples.

Owing to its increasing acceptance in signal processing, the WT has also been described and its use in the processing of power system transient recordings has been illustrated.

Finally, neural networks and fuzzy logic techniques have been shown to enhance the post-processing of power system signals.

4.8 References

1. Soliman, S A Al-Kandari, A M El-Nagar, K and El-Hawary, M E, (1995). New dynamic filter based on least absolute value algorithm for on-line tracking of power system harmonics, *IEE, Proceedings of the Part C*, **142**(1), pp. 37–44.
2. Moo, C S Chang, Y N and Mok, P P, (1994). A digital measurement scheme for time varying transient harmonics, *IEEE/PES*, Summer Meeting, Paper 490–3, PWRD.
3. Sachdev, M S and Nagpal, M, (1991). A recursive least error squares algorithm for power system relaying and measurement applications, *IEEE Transactions on Power Delivery*, **6**(3), pp. 1008–1015.
4. Sidhu T S, and Sachdev M S, (1998). An iterative technique for fast and accurate measurement of power system frequency, *IEEE Transactions on Power Delivery*, **13**(1), pp. 109–115.
5. Terzija V V, Djuric M B, and Kovacevic B D, (1994). Voltage phasor and local system frequency estimation using Newton-type algorithm, *IEEE Transactions on Power Delivery*, **9**(3), pp. 1368–1374.

6. Fourier J B J, (1822). *Théorie Analytique de la Chaleur*,
7. Kreyszig E, (1967). *Advanced Engineering Mathematics*, 2nd edition, John Wiley and Sons.
8. Kuo F F, (1996). *Network Analysis and Synthesis*, John Wiley and Sons.
9. Brigham E O, (1974). *The Fast Fourier Transform*, Prentice-Hall.
10. Cooley J W, and Tukey J W, (1965). An algorithm for machine calculation of complex Fourier series, *Math Computation*, **19**, pp. 297–301.
11. Cochran W T, et al, (1967). What is the fast Fourier Transform. *IEEE, Proceedings of the*, **10**, pp. 1664–1677.
12. Bergland G D, (1969). A guided tour of the fast Fourier Transform, *IEE Spectrum*, pp. 41–42.
13. Bergland G D, (1968). A fast Fourier Transform algorithm for real-values series. *Numerical Analysis*, **11**(10), pp. 703–710.
14. Harris F J, (1978). On the use of windows for harmonic analysis with the discrete Fourier transform, *IEEE Proceedings of the*, **66**, pp. 51–84.
15. IEEE Task Force on Inter harmonics and CIGRE WG 36.05/CIRED 2-CCO2.
16. Lu I D and Lee P, (1994). Use of mixed radix FFT in electric power system studies, *IEEE Transactions on Power Delivery*, **9**(3), pp. 1276–1280.
17. Sorensen H V, Heiderman M T and Burrus C S, (1986). On computing the split-radix FFT, *IEEE Transactions on Acoustics, Speech and Signal Processing*, **ASSP-34**(1), pp. 152–156.
18. Sorensen H V, Jones D L, Heiderman M T and Burrus C S, (1987). Real-valued FFT algorithms, *IEEE Transactions on Acoustics, Speech and Signal Processing*, **35**(6), pp. 849–864.
19. Soertzel G, (1958). An algorithm for the evaluation of finite trigonometric series, *American Mathematics Monthly*, **65**(1), pp. 34–35.
20. Markel J D, (1971). FFT pruning, *IEEE Transactions on Audio and Electroacoustics*, **19**(4), pp. 305–311.
21. Sorensen H V and Burrus C S, (1993). Efficient computation of the DFT with only a subset of input or output points, *IEEE Transactions on Signal Processing*, **41**(3), pp. 1184–1200.
22. Heydt, G T, (1991). *Electric Power Quality*, Stars in a Circle Publications, West LaFayette (USA).
23. Ribeiro, P F Haque T, Pillay P and Bhattacharjee A, (1994). Application of wavelets to determine motor drive performance during power systems switching transients, *Power Quality Assessment*, Amsterdam.
24. Chui C K, (1992). *An Introduction To Wavelets*, Academic Press, pp. 6–18.
25. Daubechies I, (1988). Orthonormal bases of compactly supported wavelets, *Communications in Pure and Applied Mathematics*, ,**41**, pp. 909–996.
26. Santoso S Powers E J, Grady W M and Hofmann P, (1996). Power quality assessment via wavelet transform analysis, *IEEE Transactions on Power Delivery*, **11**(2), pp. 924–930.
27. Ringrose M and Negnevitsky M, (1998). Automated disturbance recognition in power systems, *Australasian Universities Power Engineering Conference (AUPEC'98)*, Hobart, pp. 593–597.
28. Ribeiro P F and Celio R, (1994). Advanced techniques for voltage quality analysis: unnecessary sophistication or indispensable tools, Paper A–206, *Power Quality Assessment*, Amsterdam.
29. Haykin, S, (1994). *Neural Networks; A Comprehensive Foundation*, Macmillan, pp. 138–229.
30. Zadeh, L, (1965). Fuzzy sets, *Information and Control*, **8**(3), pp. 338–354.
31. Tanaka, K, (1997). *An Introduction to Fuzzy Logic for Practical Applications*, Springer.

5

POWER QUALITY MONITORING

5.1 Introduction

Power quality monitoring involves the capturing and processing of voltage and current signals at various points of the power system. The signals to be captured are normally of high voltage and current levels and thus require large transformation ratios before they can be processed by the instruments.

The first part of this chapter describes the characteristics of conventional and special types of current and voltage transformers for use in power quality assessment. The transducers are normally placed in outdoor switchyards and the transformed low-level signals have to travel through a hostile electromagnetic environment before they reach the control rooms; thus the transmission of data in that environment also needs to be considered. Once these signals reach the control room the whole science of signal processing becomes available for the derivation of power quality information. Chapter 4 has already described the main waveform processing techniques available with reference to the power system signals. The implementation of these techniques in modern digital instrumentation and various examples of their application constitute the main part of the chapter.

5.2 Transducers

The function of a current or voltage transformer is to provide a replica of power system current or voltage, at a level compatible with the operation of the instrumentation, in circumstances where direct connection is not possible.

Whilst the behaviour of the conventional current and voltage transformers at a fundamental frequency is well understood and defined, the behaviour at higher frequencies has not been as fully examined. With the need to measure power system harmonic content, their performance in transforming current and voltage signals containing harmonic components is essential to the measurement process.

In line with the accuracy requirements suggested for instrumentation, the IEC 61000-4-7 standard indicates that the errors of voltage and current transformers shall not exceed 5% (related to the measured value) in magnitude and 5° in phase angle.

5.2.1 Current Transformers

The most common type of current transformer is the toroidally wound transformer with a ferromagnetic core. This has, by virtue of its construction, low values of primary and secondary leakage inductance and primary winding resistance. Under normal operating conditions, the transformer primary current will be substantially less than that required for saturation of the core, and operation will be on the nominally linear portion of the magnetisation characteristic.

The frequency response of current transformers is effectively determined by the capacitance present in the transformer and its relationship with the transformer inductance. This capacitance may be present as interturn, interwinding or stray capacitance. Tests have shown [1,2] that, while this capacitance can have a significant effect on the high-frequency response, the effect on frequencies to the 50th harmonic is negligible.

In addition to harmonic frequencies, it is also possible that the primary current will contain a d.c. component. If present, this d.c. component will not be transformed but will cause the core flux of the transformer to become offset. A similar condition could arise from remnant flux present in the transformer core as a result of switching.

For this reason, where the presence of a d.c. component is suspected, or remanence a possibility, a current transformer with an air-gap in the core could be used. This air-gap reduces the effect of the d.c. component by increasing core reluctance and enables linearity to be maintained. Because the current transformer burden tends to increase with frequency, the associated power factor reduces with increasing frequency and the transformer will produce a higher harmonic output voltage than it would for a purely resistive load. The resulting increased magnetising current will cause further error.

For measurements of harmonic currents in the frequency range up to 10 kHz, the normal current transformers that are used for switchgear metering and relaying have accuracies of better than 3%. If the current transformer burden is inductive, then there will be a small phase shift in the current. Clamp-on current transformers are also available to give an output signal that can be fed directly into an instrument.

The following practical recommendations are worth observing whenever possible:

1 If the current transformer is a multi-secondary type, the highest ratio should be used. Higher ratios require lower magnetising current and tend to be more accurate.

2 The current transformer burden should be of very low impedance to reduce the required current transformer voltage and, consequently, the magnetising current.

3 The burden power-factor should be maximised to prevent its impedance from rising with frequency and causing increased magnetising current errors.

4 Whenever possible, it is suggested that the secondary of the measuring current transformer is short-circuited and the secondary current monitored with a precision clamp-on current transformer.

Unconventional Types of Current Transformer [3] Various alternatives to the conventional current transformer have been investigated, some of which are already finding a place in power system monitoring. Among them are:

- *Search coils* The magnetic field in the proximity of a conductor or coil carries information on the components of the current which generates the field. The amplitude of the induced harmonic voltage in a search coil is proportional to the effective coil area, number of turns, the amplitude of the harmonic magnetic field perpendicular to the coil surface and the frequency of the harmonics.

  In such measurements, the measured magnetic field can arise from the contributions of more than one source. The magnetic field is inversely proportional to the distance from the source. Where it is possible to place the search coil at a small distance d from the conductor, while other conductors are located at distances larger than $20d$, the measurements of values in the chosen conductor are not substantially changed by the fields of the other conductors.

- *Rogowski coils* These devices are coils wound on flexible plastic mandrels, and they can be used as clamp-on devices. They have no metallic core, so problems of core saturation are avoided in the presence of very large currents, such as the 60 kA to 100 kA in the feed to an arc furnace or in the presence of direct current.

- *Passive systems* In a passive system, a transmitted signal is modulated by a transducer mounted at the conductor. No power source is required at the conductor.

  Optical systems use the Faraday magneto-optic effect by which the plane of polarisation of a beam of linear polarised light is rotated by a magnetic field along its axis.

  Designs for Faraday-effect current transformers use either the open path or the closed path optical system [4].

  Microwave systems make use of gyromagnetic materials to modulate a microwave carrier by a magnetic field. The form of modulation is controlled by the arrangement of the gyromagnetic material and the form of polarisation of the microwave signal.

- *Active systems* An active system uses a conductor-mounted transducer to provide a modulating signal for a carrier generated at the conductor. Transmission of the carrier to the receiving station is then achieved via a radio or fibre-optic link. The power for the transmitter is usually line derived using a magnetic current transformer together with some battery back-up.

- *Hall effect transducers* The Hall effect is used in a variety of probes and transducers covering a range of current levels. For current transformer applications, a major problem is that of maintaining calibration over long periods.

5.2.2 Voltage Transformers

Only on low voltage systems can the analyser be connected directly to the terminals where the voltage components must be determined. On medium and high voltage systems, means of voltage transformation are required.

Magnetic voltage transformers, of extensive use for medium voltage levels, are designed to operate at fundamental frequency. Harmonic frequency resonance between winding inductances and capacitances can cause large ratio and phase errors. For voltages to about 11 kV, and harmonics of frequencies under 5 kHz, the accuracy of most

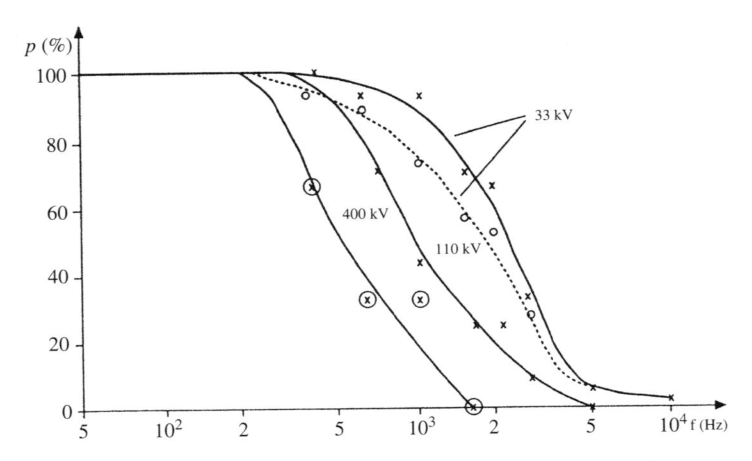

Figure 5.1 Percentage of voltage transformers, the transfer ratio of which has a maximum deviation (from the nominal value) of less than 5% or 5° up to the frequency f. ——— minimum error (5%); - - - - - maximum error (5%)

potential transformers is within 3%, which is satisfactory, the response being dependent upon the burden used with the transformer [5].

At higher voltage levels, the transformer tends to exhibit resonances at lower frequencies, since the internal capacitance and inductance values vary with insulation requirements and construction. The precise response for a particular unit will be a function of its construction [6].

Figure 5.1 contains test results carried out in over 40 voltage transformers at levels between 6 kV and 400 kV. The figure indicates the percentage of transformers that maintained the required precision (i.e. 5% and 5°) throughout the frequency range. The main conclusions are as follows.

- At medium voltage, all the transformers perform adequately up to 1 kHz, while only 60% of them manage to cover the whole harmonic spectrum. The figures reduce further, to 700 Hz and 50% respectively, when the phase precision level requirement is included.
- At high voltage, the transformers' response deteriorates quickly for frequencies above 500 Hz unless special designs are introduced.
- The conventional voltage transformers of magnetic type do not provide accurate information for harmonic orders above the 5th.

Capacitive Voltage Transformer The capacitive voltage transformer (CVT) combines a capacitive potential divider with a magnetic voltage transformer, as shown in Figure 5.2. This combination enables the insulation requirements of the magnetic unit to be reduced with an associated saving in cost.

The additional capacitance provided by the capacitive divider will influence the frequency response of the CVT producing resonant frequencies as low as 200 Hz, which makes them unsuitable for harmonic measurements.

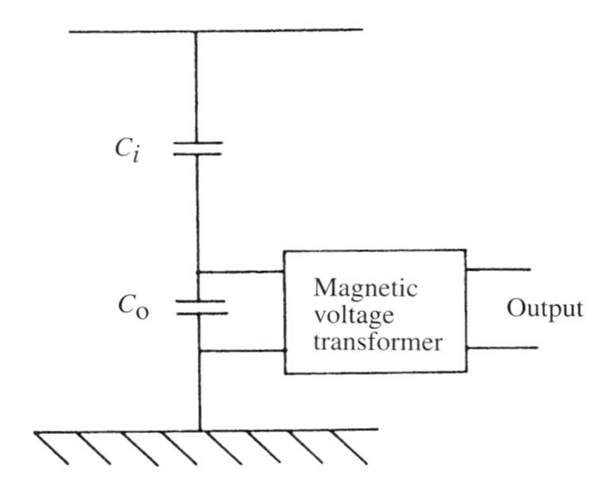

Figure 5.2 Capacitive voltage transformer (CVT)

The form of frequency response obtained is also dependent upon the magnitude of the fundamental component and its relationship to any transition point in the magnetisation characteristic of the transformer steel.

Capacitive Dividers For harmonics measurements, either a purpose-built divider could be assembled or, alternatively, use could be made of the divider unit of a capacitive voltage transformer, with the magnetic unit disconnected, or the loss tangent tap on an insulating bushing (Figure 5.3).

When subject to an impulse, such as might arise from local switching, the capacitive divider is subject to *ringing* due to the interaction between the divider capacitors and their internal inductances. This can lead to high common mode voltages, particularly in areas of high earth impedances. To minimise ringing, the capacitors forming the

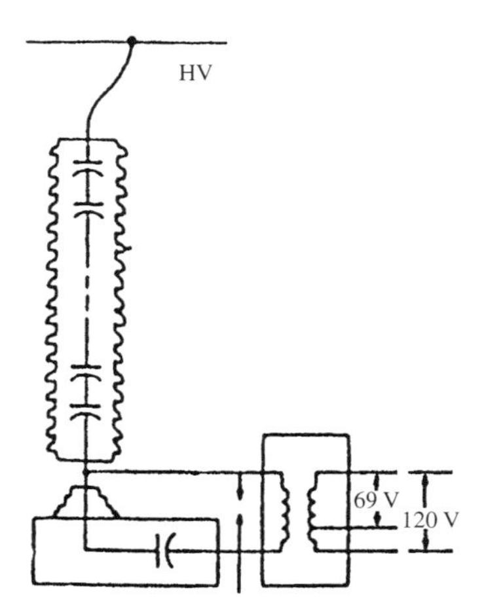

Figure 5.3 Capacitive voltage divider using a bushing tap

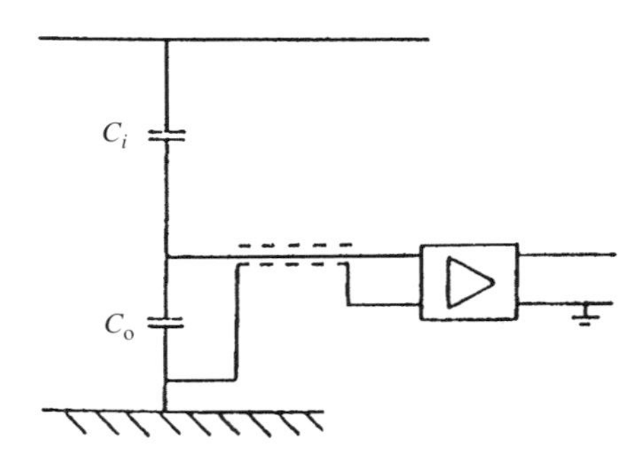

Figure 5.4 Capacitive voltage divider incorporating an amplifying circuit

divider circuit should have a low inductance and the low voltage capacitor should be screened.

In recent years, a number of amplifier-based capacitive divider systems have been developed [7,8]. Although intended primarily for use with high-speed protection schemes, they also have an obvious application in harmonic measurements.

The basic arrangement for a single phase unit is shown in Figure 5.4. High-input impedance instrumentation amplifiers must be included in such measurements. For best results, the input amplifier should either be battery operated or use a suitably shielded and isolated supply. The leads from the low voltage capacitors to the input amplifier should be as short as possible. In general, short leads from the amplifier to the analyser will greatly reduce the angle error when measuring phase angles. These devices have a limit on the burden that they can supply without saturation, hence the requirement for a high-impedance amplifier.

Unconventional Voltage Transformers [3] Electro-optical and electrogyration effects can be used to measure voltage in a manner akin to the Faraday effect used for current measurement.

The electro-optical effect causes linearly polarised light passing through the material to become elliptically polarised. As the two mutually perpendicular components propagate in the crystal at different velocities, they have a differences in phase as they emerge from the material. This phase difference will be proportional to the path length in the material and to either the field (the Pockels effect) or the square of the field (the Kerr effect). By measuring this phase difference, the electric field strength, and hence the voltage, can be obtained.

If a linearly polarised beam is propagated through an electrogyrational material in an electric field, the effect is to rotate the plane of polarisation in a manner analogous to that occurring in magneto-optic materials. The electrogyration effect can, therefore, be used to measure voltage in a manner similar to the Faraday-effect current transformers.

5.3 Power Quality Instrumentation

Some commercial instruments have been specifically designed for power systems use (e.g. harmonic analysers) while others are of more general use (signal analysers). The main difference between these two categories is the need to follow the variations in fundamental frequency in harmonic analysis.

Portable instruments are of small size and lightweight, easy to set up and use in the field. The transducers (clip-on type) and interconnecting cable normally are part of the unit. To reduce cost, portable instruments are normally restricted to one or two channels. Battery operating is essential for usage flexibility without dependence on external wires and power supplies, particularly since the instrument has to be close to the measurement point due to the length of the clip-on cable.

Glitch capture and recording minimum and maximum levels are normally present to capture transient events such as motor in-rush current or spikes and surges. Generally, these instruments are not automated and therefore they need a person controlling their operation, although some now have a logging feature. The operator must control the transducer locations, what quantity is displayed, the storing of data to memory or down-loading to PC. Generally, the logging feature allows periodic down-loading of the instrument readings to a PC through an RS323 interface. The capabilities of these instruments are fixed to the features originally designed into the unit and cannot be changed. Therefore, upgrades are achieved by buying a newer model.

For permanent or semi-permanent monitoring, the instruments require many channels since they are intended to operate without human intervention and, generally, the channels are not designed to be moved from one location to another. Transducers do not come as part of these instruments, since it is assumed that the CTs or VTs already existing in the system will be used. Owing to the high cost of such hardware, the functionality is designed into the software to permit upgrades. Since some of the measuring points are likely to be in outdoor switchyards, the cables and transducers must be designed to withstand all weather conditions and operate satisfactorily in a hostile electromagnetic environment.

These instruments operate unattended over long periods and software is thus required to automate the data collection, processing and storage as well as deciding on what events to capture and store. Such instrumentation is normally required at a multitude of sites and therefore synchronisation and the ability to control them all from a central location is an important feature. Owing to the enormous amount of data being acquired by the instrumentation, *smart* algorithms are required to decide what is to be kept and what is not. Moreover, the enormous amount of data requires innovative ways to display it to make it meaningful. Statistical analysis of the data is one approach, as shown in Figure 5.5. Another is to superimpose the events on the CBEMA curve described in section 2.3.3, as shown in Figure 5.6.

The processing of the waveforms can be carried out in analogue or digital form, although the latter has practically displaced the analogue-type analysers.

The purpose, system requirements and implementation of digital instrumentation are discussed in the following sections.

5.3.1 System Requirements

At first glance, a power quality monitoring system resembles a general-purpose data acquisition system, where analogue input signals are merely converted into digital formats before being processed and stored for analysis at a later date. However, due to

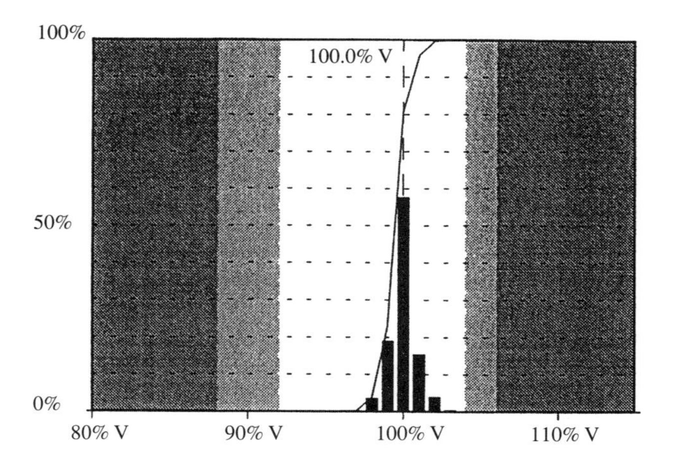

Figure 5.5 Steady-state voltage histogram

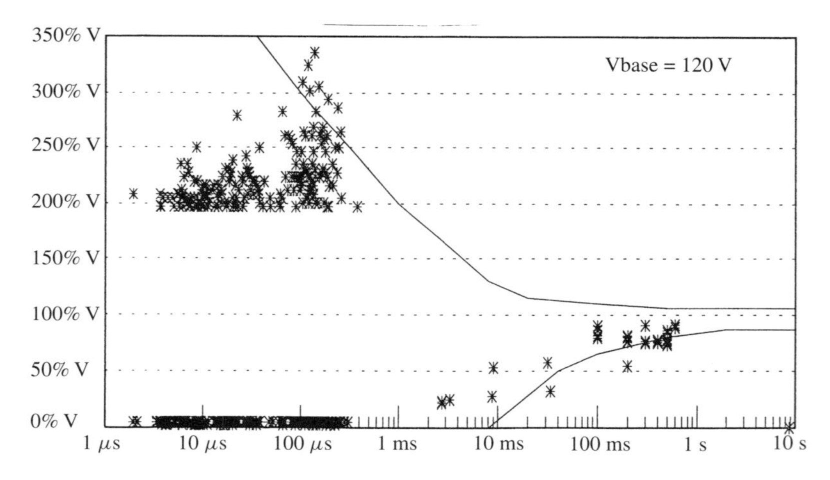

Figure 5.6 Power quality envelope

the many aspects of power quality disturbances, many general-purpose data acquisition systems are not capable of undertaking the measurements or are simply not suitable for such measurements. The family of power quality disturbances ranges from low-speed low-level variations to high-frequency transients with large amplitude changes. These contrasting characteristics place very different demands on the monitoring system, which are not usually considered by general-purpose data acquisition systems. Hence, it is often necessary to construct a monitoring system specifically for measuring power-quality-related disturbances, taking into consideration the wide-ranging requirements created by the different characteristics [9].

Other than this characteristic difference, the requirements placed on the monitoring system also depend on the purposes of the measurement assignment. Some of the possible purposes are:

- monitoring existing values of disturbances for checking against recommended or admissible limits,
- testing a piece of equipment that generates or causes disturbances in order to ensure its compliance to certain standards or guidelines,
- diagnosing or trouble-shooting a situation in which the performance of a particular piece of equipment is unacceptable to the utility or to the user,
- observing existing background levels and tracking the trends with time for any hourly, daily, weekly, monthly or seasonal patterns.
- verifying simulation studies or techniques and fine-tuning the modelling of the devices and the system under analysis.
- determining the driving point impedance at a given location. This impedance is useful for gauging the system capability to withstand power quality disturbances.

Other than these possible purposes, there may be other reasons for undertaking monitoring and they will affect the requirements on the monitoring system. Monitoring background levels will require the system to be operated continuously over a long period of time and the measured data will have to be stored. The stored data may comprise the average, maximum and minimum values over predetermined intervals so as to provide an overall picture of the phenomenon. On the other hand, testing a piece of equipment simply requires certain snapshots under certain operating modes. The captured data most likely will consist of several cycles of the time-domain waveforms, which can be further processed to extract the necessary information. These differing purposes will affect the level of complexity required of the monitoring system.

A monitoring system can be broadly divided into three subsystems of input signal conditioning and acquisition subsystem; digital processing and storage subsystem; and the user interface subsystem as shown in Figure 5.7. The roles played by each subsystem are briefly illustrated in the following subsections and the requirements pertaining to each subsystem are also described.

Input Signal Conditioning and Acquisition Subsystem This subsystem forms the basis of the modern monitoring system. Its function is the conversion of analogue signals into digital formats. This digitisation simplifies the design of analogue circuitry and provides greater flexibility for altering the algorithms to be used for processing the data samples. The basic parameters of this conversion process are the sampling rate, the input dynamic range and the resolution.

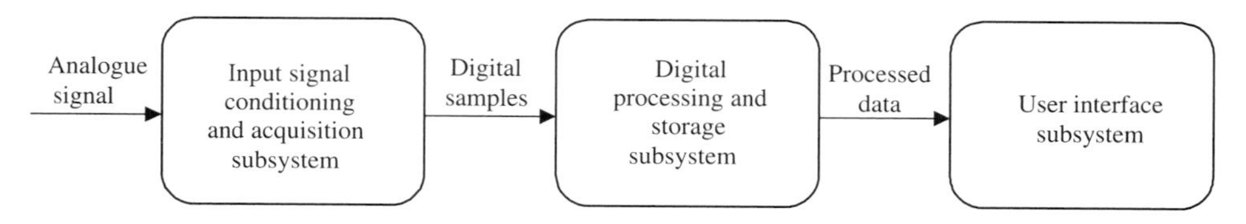

Figure 5.7 Major subsystems making up a monitoring system

- *Sampling rate*: the corresponding sampling rate required depends on the bandwidth of the measured signals. For capturing a transient waveform, sampling rates in the MHz range are necessary, while harmonic analysis or r.m.s. voltage variation analysis requires only sampling rates in the kHz range.
- *Input dynamic range*: the required dynamic range depends on the magnitude of the disturbance to be measured. High overvoltages will require a wide dynamic range in order to measure the actual magnitude changes.
- *Resolution*: together with dynamic range, the resolution ensures that the desired signal information can be extracted from the samples. Measurement of low-level voltage variations or high-frequency harmonics demands high resolution so that a small disturbance can be detected.

If a monitoring system needs to measure signals of different characteristics, it is important that the number of binary bits used in the conversion process can be tailored to provide the appropriate dynamic range and resolution for capturing such disturbances. Other than these basic requirements, the following are some of the other factors that may become important under certain situations or under certain monitoring purposes.

- *Anti-aliasing technique*: the required amount of anti-aliasing filtering is determined by the bandwidth of the signal to be measured. The complexity (i.e. number of orders) of the anti-aliasing analogue filter can be reduced by moving a certain amount of filtering into the digital processors. Multirate digital signal processing techniques has been used to measure harmonic levels with a simplified analogue filter design [10].
- *EMI susceptibility*: power system environments are extremely noisy electromagnetic environments. Hence, the monitoring equipment must have relatively high immunity to such interference.
- *Number of channels and modularity*: almost all monitorings require more than one input channel and hence the monitoring system must be able to provide multiple channels in a modular manner.
- *Synchronisation and timing*: when multiple channels are used, data measured from different channels must be able to be referenced (in time) to each other. Hence, the sampling processes on all channels must be synchronised. Moreover, if it is necessary to carry out simultaneous measurements at different geographical locations such as in system-wide harmonic monitoring, the samplings at one monitoring unit must be synchronised to those at other units. This requirement is also necessary even when only snapshots are taken but the phase information is needed.
- *Automatic ranging*: a single analogue input channel is sometimes used for monitoring different kind of signals of different magnitudes. Hence, an automatic ranging feature will enable full use of the dynamic range provided by the system. A monitored current can vary significantly between light and heavy loading conditions, autoranging will ensure that the full dynamic range is fully utilised under all conditions.

- *Automatic calibration*: the properties of semiconductor devices like analogue filters and A/D are known to drift over time and under different operating environments. Hence, an automatic calibration procedure will help to minimise such errors. Routine automatic calibration in on-going monitoring will ensure that changes in temperature and humidity over different time of day or over different seasons will not adversely affect the measurements.

Digital Processing and Storage Subsystem Digital samples are transferred to this subsystem for processing and recording. This subsystem can simply be a data logger or a powerful parallel processing computer system. Generally, it is the design of this subsystem that determines whether continuous real-time data acquisition is possible or only snapshots can be captured and stored for offline processing. The following requirements are important if the system is to achieve continuous real-time data acquisition.

- *System architecture*: the architecture and bus system adopted for the monitoring system can have a major impact on the system scope and flexibility to handle the large quantities of data that are acquired in a multi-channel measurement. The classical centralised architecture has a higher data bandwidth constraint than the more modern distributed architecture. These designs are further illustrated in subsequent sections.
- *Memory architecture*: the memory size and structure being employed at various stages of the system depends on the amount of data to be transferred and the mode of data transfer. This design is critical if continuous data acquisition with no breaks in the acquired data is to be achieved.
- *Processing capability*: the ability to process captured samples continuously in real time will help to reduce the amount of data to be stored. Along with system architecture and memory architecture, the processing capability determines whether it is feasible to process data continuously in real time. Such real-time processing will be able to compress the amount of data need to be stored since only the relevant information is retained rather than the entire raw samples.
- *Modularity and multiple processing*: the type, quantity and distribution of digital processors in the system must be made modular so as to complement the number of input channels and the level of data processing intensity required by each channel. This requires the system to be easily reconfigured or expanded in order to meet the number of input channels and the data processing requirement.
- *Storage capacity*: this must be sufficient to record the necessary information for the entire period of measurement. This includes both fixed storage media as well as removable storage devices. In the event of monitoring over a long period of time, stored data can be transferred into removable media so as to free the fixed storage space for further recording.
- *Customisation of processing algorithms*: analysing different disturbances usually calls for different processing algorithms. Hence, the system must be able to execute different algorithms when used for different types of monitoring. Sometimes, different algorithms are required to be used simultaneously in a common processing system.

- *Real-time operating system*: a proper system hardware design must be complemented with sound system software design. Owing to the possible large volume of data to be transferred through the system, real-time operating systems with guaranteed latency are necessary to ensure continuous and smooth transfer of data through the system and storage or distribution for display and analysis purposes.
- *Industry standard*: industry standards should be used for hardware and software wherever possible and practical. This avoids heavy dependence on certain proprietary products, which may become obsolete without any substitute or upgrade option. Moreover, industry standard hardware and software tend to be less costly but at the same time are more reliable with better customer services than proprietary products.

User Interface Subsystem The main purpose of the user interface is to provide users with access to the measured data, either through on-screen displays or in hard copy forms. It also provides the users with the ability to control and configure the monitoring system. The basic requirement on this subsystem is to hide the complicated details of the monitoring system from the users. It should aim to make it easy for the user to access the data, and to make control of the operation of the system transparent to the user.

- *Graphical user interface*: the graphical user interface forms the essential *face*, representing the monitoring system, to the user. The interface should reflect the current configuration of the system and should be intuitively easy and friendly to use.
- *Presentation of measured data*: different disturbances are processed differently for different attributes and hence the processed result will vary. Therefore, different displays, either graphically or in text form, are required to present the different information. Certain amounts of continuous real-time displays are required in order to provide immediate information whenever it is called for. Multiple displays should be able to be invoked simultaneously for showing the characteristics of different disturbances.
- *Integrated control*: through the graphical user interface, the user must have access to the controlling functions of the monitoring system. This integrated control enables the user to configure the system according to the monitoring purposes and to control its operation during the course of measurement.
- *Remote monitoring and control*: this feature provides the user with the ability to monitor and control the operation of the measurement system from a remote site. This is particularly necessary if the main processing system is situated in certain out-of-bounds or hard-to-access locations. This control can be extended to include the controlling and monitoring of several geographically separated processing systems simultaneously from a single user interface.
- *Interfacing with commercial reporting packages*: the graphical user interface, particularly the displays for presenting the measured data, must be able to interface with other commercial word-processing, spread-sheet or database packages. This is to enable the measured result to be imported into these packages for report writing purposes.

- *Retrieval and analysis of stored data*: easy and transparent access must be provided to the stored data. The graphical user interface should include the facility to retrieve the stored data and a certain amount of further processing be provided to the user for analysis purposes. The stored data should also be able to be imported into commercial packages for further processing, analysis or simply for display and reporting purposes.

5.3.2 System Structure Designs

Power system quality monitoring by means of data acquisition systems places specific demands on the system and processing architectures used. These architectures include both the system hardware and software structures. Existing power quality monitors are usually designed for certain specific applications and according to certain standards. Any deviation from such pre-defined usage generally calls for significant amounts of additional effort and capital investment. Recent advances in computer hardware and software technology have made it possible to implement flexible multi-channel real-time data acquisition systems for power quality assessment [11,12].

System Hardware Design Power system quality assessment inherently requires precise and continuous knowledge of the voltage and current waveforms over a wide bandwidth and with good resolution, usually at multiple nodes of the power network. Harmonic measurement, for example, demands a minimum sampling rate of 6 kHz if the 50th harmonic of a 60 Hz fundamental waveform is to be correctly captured. Furthermore, in order to capture the very low magnitude higher-order harmonics in the presence of the large fundamental component, analogue to digital conversion resolutions of better than 16 bits are called for. In the case of current waveforms, which vary dynamically depending on the load, front-end gain or attenuation signal conditioning may also be required. By contrast, transient wave-front capture may require 1 MHz sampling rates with lower level analogue-to-digital conversion resolutions of the order of 8–10 bits. Most existing power quality monitoring instruments are only capable of acquiring snapshots of the system waveform. The validity of the acquired data relies heavily on the assumption that the system is operating in a true or quasi-steady state. Few, if any, existing instruments are readily equipped with the capability of synchronising the acquisition of data samples across multiple channels and between multiple instruments and/or nodes on a power network. Consequently, the steady-state assumption is again taken as implicit when snapshots, gathered at different parts of a power system network, are used alongside each other in order to make simultaneous power quality assessment of the system.

The need to know the precise state of the system at all times is particularly important when endeavouring to locate the source of power disturbances or distortions. Locating the source of harmonic distortions, for example, requires good magnitude and phase measurements, often at more than one location on the network and preferably synchronised. In terms of policing power quality it is essential that the monitoring system is capable of acquiring complete and unquestionable knowledge of all the relevant voltage and current waveforms. This is particularly important where there may be legal consequences.

Many existing data acquisition systems used in power systems such as the Supervisory Control And Data Acquisition (SCADA) systems inherit the centralised processing architecture of the 1980s as shown in Figure 5.8. Despite the advances in computer hardware and software technology, the centralised processing structure places a constraint on the data processing capability of the system. From the power quality monitoring perspective, this limited real-time processing capacity results in off-line post-processing of the acquired data to derive the necessary information. The lack of on-line analysis processing capability means that large volumes of raw data have to be acquired and stored. Consequentially, the limited system throughput, bandwidth and storage volumes only allow the system to record snapshots.

The configuration described above relies on the CTs' and VTs' outputs being routed to a central location, typically in the control or metering room. Although this configuration is normally sufficient for relay operation or metering purposes, the limited bandwidth and EMI susceptibility of the long analogue communication links create serious concerns over the integrity of the measurements. In general, the designs of metering systems are optimised for precise operation at the system nominal fundamental frequency while protection systems may give some cognisance to fast transients.

The functions of the local acquisition modules are usually confined to some signal conditioning and possibly A/D conversion, leaving the data processing to be implemented in a centralised system. Although the centralised system may possess distributed processor architecture with industrial standard high-speed bus systems, extensive use of the bus system to shuffle large volumes of raw data through the system is usually required. Ultimately, this not only limits the data acquisition to snapshots, but also places constraints on the number of data channels, with little opportunity for expansion.

In order to overcome the above shortcomings due primarily to insufficient online processing capability, some form of distributed processing architecture is required. Figure 5.9 shows a possible configuration, but many variants of it may suffice. There are two key features in this architecture compared with the conventional one of Figure 5.8. The first is the shift of A/D conversion from the central location to the switchyard close to the transducers. This enables the use of a digital communication link between the

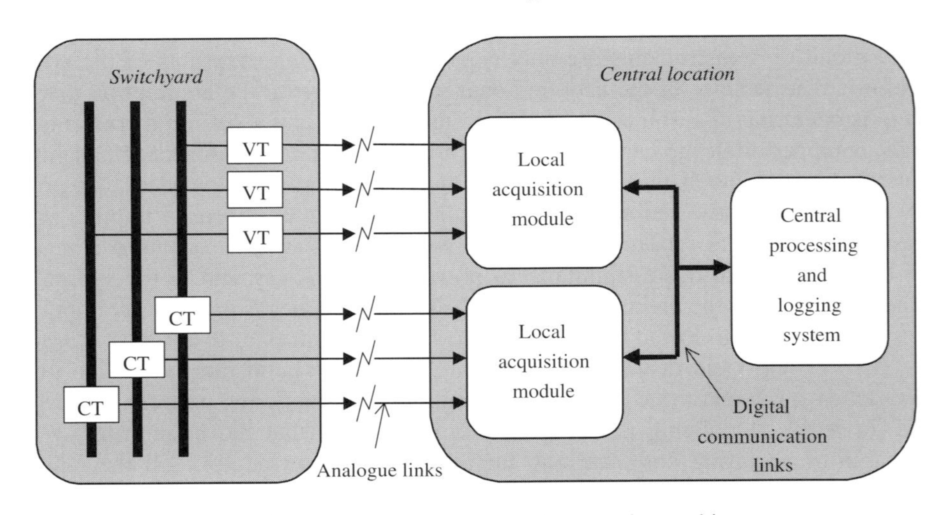

Figure 5.8 Conventional centralised processing architecture

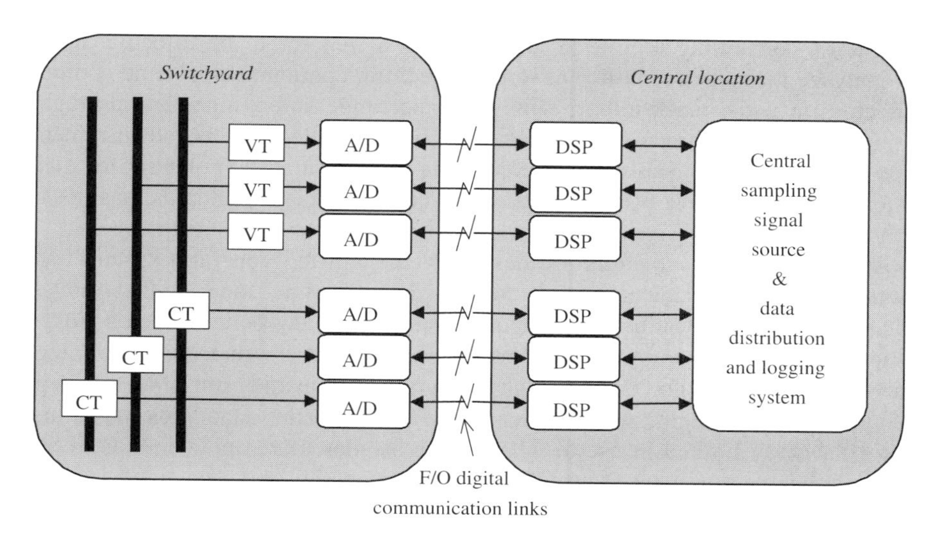

Figure 5.9 A possible distributed processing architecture

switchyard and the central system, improving the bandwidth, dynamic range and noise immunity of the acquired signals. The fibre optic link also reduces the system's susceptibility to electromagnetic noise, which can be significant in a switchyard environment. The second key element is the provision of a DSP (or CPU) for undertaking the data processing for each individual channel. By dedicating a single DSP/CPU to every data channel, computationally intensive manipulations can be implemented on-line and thereby reduce the traffic through the system.

A centralised source of sampling signals provides the opportunity to synchronise the samplings across all data channels. The real challenge in this configuration is a mechanism for routing the sampling pulses from the central location to the front end A/D in the switchyard. An efficient technique is needed so that all the data acquisitions can be synchronised. Furthermore, the modular structure provides flexibility for future expansions. New A/Ds and DSPs can be gradually added to the system in more cost-effective and manageable stages.

The DSP usually communicates with the central data collection system through multi-processing bus architectures. This enables the DSP modules to be designed as plug-in cards facilitating flexible expansion. The bandwidth of the bus system needs to be carefully considered even though the on-line data processing and filtering capability has reduced the amount of data transferred through the system.

There are clearly many variants and extensions of the proposed system and, indeed, one of the prime attributes of the system is the ease with which it can be extended and expanded, as well as the ease with which its performance can be upgraded. The adoption of standard busses, packaging, interfaces and software wherever possible makes it a relatively simple and efficient task to adapt the designed components and subassemblies to new configurations and system designs.

System Software Design Power quality monitoring such as harmonics and voltage flicker often requires long-term assessment over several days or weeks. Hand-held or portable meters, albeit providing some indications of the extent of the problem, are incapable of performing long-term analysis. For a monitor to be effective in long-term analysis, it requires a sufficient storage medium, continuous real-time computing architecture, a wide bandwidth, a wide dynamic range and a high data throughput. Although these requirements have generally been fulfilled by the advancement in computer technology and distributed processing architecture, the potential for making a system flexible, capable of being adapted to new issues and standards, has not been greatly explored.

Existing commercial monitors are designed either in too general a system configuration format or as dedicated tools for certain pre-defined and limited ranges of applications. A general-purpose data acquisition system may be constructed from off-the-shelf hardware and software modules but much effort is still required to integrate them. On the other hand, even though instrument companies are able to integrate systems tailored to specific needs or to certain power quality standards, these instruments are usually limited in use to where these standards are applied. Moreover, the quality of the measurement is usually pre-determined by the design of the instrument and it is often not possible to shape the system to accommodate specific or new measurement requirements. In general, the limitations of the measuring instruments have had a major impact on power quality standards rather than power quality standards being solely set by the effects of poor quality power on the consumer — i.e. the instruments have tended to dictate the standards by their own limitations.

Secondly, due to the dedicated nature of existing designs, it is difficult to adapt them for new measuring techniques or algorithms. At the moment, researchers have to construct the prototype system complete with all the rudimentary hardware and software. While this process is becoming easier with the growing supply of standard hardware and software components, it should be possible to design a flexible system fully equipped with the basic hardware and software which allows new customised applications to be incorporated into the system with minimal effort.

A flexible multi-purpose data acquisition system should ideally consist of a unit capable of implementing user-definable algorithms, while allowing the use of specialised display and user interfaces. This flexibility has been achieved to the degree that the technology permits by extending the functionality of existing monitors and by using standard, expandable data acquisition modules. A flexible system comes in the form of a general interface module interconnecting two sets of user definable modules as shown by Figure 5.10. The interface enables customised applications to be implemented at the front end near the source of the data, while at the same time enabling specialised control, set-up and display interfaces to be utilised at the user end. This functionality has to be achieved without requiring the user to implement the complicated and tedious interfaces between the two sets. The interface module undertakes the transfer of control and set-up instructions from the users to the front ends, and of the data in the opposite direction. Moreover, a customised application is more likely to be associated with certain custom-designed control, set-up and display interfaces. Such transparent links have to be included in the design of the interface module so that different applications and user interfaces can be used concurrently in a single unit.

The design of this general interface module resembles that of a computer operating system and hence has been referred to as a *virtual operating system*. An operating system is customarily defined as the interface between the human user and

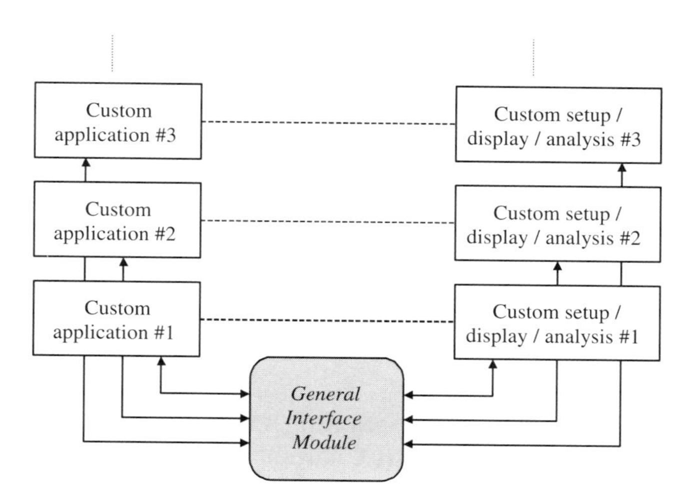

Figure 5.10 Philosophy behind a flexible system

the computing hardware. Its main function is to provide a user-friendly method for instructing the computer to perform a certain task. It can be regarded as made up of three separate parts comprising a user interface, a command interpreter and the implementation of the instruction. A command or instruction is issued via the user interface and is decoded by the command interpreter. The decoded instruction tells the operating system what task is to be conducted. This command can be an operating system command or an instruction to activate a particular application. The operating system then proceeds to carry out the instruction and, if necessary, it can fetch the program binary from secondary storage devices.

The virtual operating system closely resembles the operating system terminology outlined above. Figure 5.11 shows that system software can be divided to incorporate a virtual operating system into its framework. The software is divided into an application and a virtual operating system. The virtual operating system provides the application with interfaces to the custom hardware devices and to the computer operating systems. This arrangement eliminates the need for the user to learn about the hardware design and the different kinds of operating systems used in the system. It allows users to introduce their algorithms or applications into the system with minimum

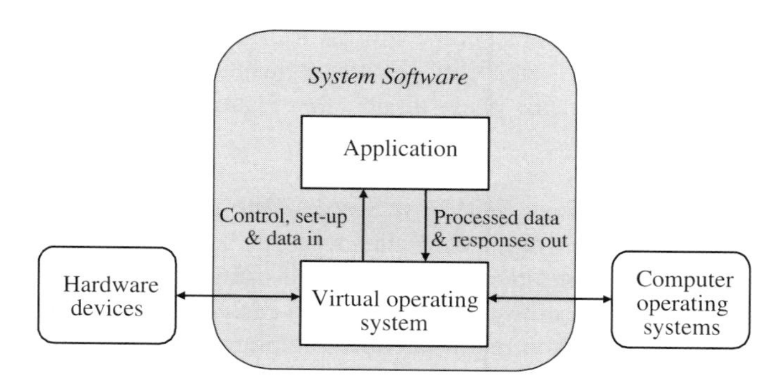

Figure 5.11 Virtual operating system

effort. The virtual operating system transmits the control and set-up commands, as well as the data acquired from the hardware devices, to the application. In return it receives the responses to the control and set-up commands, and the processed data. Through this virtual layer or system, the user-definable application is able to access the hardware devices and to make use of the services provided by the computer operating systems. The virtual operating system also undertakes the necessary conversion between formats used in different hardware devices and in different computer operating systems. In short, the virtual operating system acts as a communicator and translator for the application. This application can equally be a processing algorithm or a specialised display with the virtual operating system being made up of the environment housing the display.

5.3.3 Synchronisation of Sampling Processes Across Multiple Input Channels

Assessing power system quality always requires monitoring several voltage or current channels concurrently. Even when the monitoring is undertaken at a single node within a power system network, there are at least three voltage phases and three current phases to be measured. In certain cases, multiple input channels are multi-plexed across a single sampling unit or A/D converter. This converter is typically driven by one specific timer/counter, and hence the delays between the samplings of each subsequent multi-plexed channel are known and are usually constant. Therefore, there is no need to provide synchronisation across the multiple channels since they are all driven by a common timer/counter, and the constant delay between them can easily be compensated. However, most monitorings require a large number of input channels, which cannot be handled by a single A/D converter. Since each converter is driven by its own timer/counter with its own clocking signal sources, their samplings will not coincide in time with each other. Hence, these non-synchronised samplings create an ambiguity over the timing of the samples acquired from different input channels. In order to remove such ambiguity, the sampling processes of all A/D converters, whether they reside in a single unit or in separate units, have to be properly synchronised.

Synchronisation of sampling across multiple channels requires the timing and sampling signals to be efficiently distributed to every corner of the system wherever there are sampling processes. This difficulty, and hence the high cost involved, has deterred vendors from including this feature in the majority of data acquisition systems. The importance of this synchronisation feature varies, and in most cases it may only be an advantage to have this capability. However, when exact timing information is required such as when measuring phase angles, this feature becomes a necessity rather than a nice-to-have feature.

Synchronisation of Samplings Within a Single Monitoring System In a single-point measurement and if all input data channels can be accommodated by a single monitoring system, synchronisation of samplings is simply ensuring that all sample-and-holds and/or A/D conversions are coherent with each other. In such cases, the most obvious solution is to have a single source of sampling or timing signals, i.e. using a single timer/counter to drive all sampling units and A/D converters in the system. This requires an efficient signalling architecture to distribute the sampling or timing signals from the centralised source to all corners of the monitoring systems wherever samplings are carried out. High-speed computer buses have been used successfully to distribute these signals to the front ends in specialised power quality monitoring systems [13] and in general-purpose data acquisition systems [14].

Synchronisation of Samplings Across Geographically Separated Systems In system-wide monitoring where multiple monitoring systems are located at different geographical locations, or when multiple systems are required to accommodate a large number of input channels in single-point measurements, synchronisation of samplings involves the synchronisation of the centralised signal generating unit in each of the systems. This means that some signals must be fed into each of the systems and these signals must be in synchronism. Recent advances in the Global Positioning System (GPS) have provided accurate timing signals, which are maintained in synch with each other irrespective of where the receivers are located on the globe. These advances have also been employed in the phasor measurement units used for the relaying and control of power systems [15,16]. Using this technology, each monitoring system is equipped with a GPS receiver and the accurate and synchronised timing signals are used as the reference for generating sampling signals for the entire unit. Hence, by synchronising the centralised generation of the sampling signals and by ensuring that all the input channels are driven by these sampling signals, synchronisation of multiple channels across separate systems is achieved.

5.3.4 Data Transmission

A high proportion of power system harmonic measurements will be made using instrumentation sited remotely from the transformers providing, replicas of the system current and voltage. Some means of communication must, therefore, be provided between the transformers and the instruments.

This communication link may pass wholly or partially through a high voltage switchyard, in which case particular attention must be given to the effects of electrostatic and electromagnetic interference as well as the necessary screening.

The provision of increased noise immunity may require the use of other forms of data transmission such as current loop systems, modulated data or digitally encoded data.

Shielded conductors (coaxial or triaxial cables) are essential for accurate results but proper grounding and shielding procedures should be followed to reduce the pick-up of parasitic voltages. Moreover, coaxial cable is only suitable for relatively short leads.

Where high common-mode voltages can occur, the communications link may be required to provide insulation up to several kilovolts to protect both users and equipment. This may require the use of isolation amplifiers or, where higher levels of isolation are required, fibre-optic links.

Information may be transmitted either as an analogue signal for direct connection of the instrument, or in a modulated or encoded form using both analogue and digital data systems. If direct analogue transmission is used, then a system of sufficiently high signal-to-noise ratio is obviously required. For certain harmonic measurements, a dynamic range of the order of 70 dB may well be required and, hence, the achievable signal-to-noise ratio must be in excess of this figure.

5.4 Harmonic Monitoring

As shown in Figure 5.12, A/D converters change the analogue signals into digital form as required by digital instruments; these signals are then processed by digital filters or the FFT.

Some digital analysers still use the digital filtering method which, in principle, is similar to analogue filtering. Before starting a series of measurements, the range of frequencies to be observed must be defined and this information selects the required digital filters. At the same time, the bandwidth is varied to optimise the capture of all the selected harmonics in the presence of a large fundamental frequency signal. All the recent instruments, however, use the FFT (described in Chapter 4), a very fast method of analysis that permits capturing several signals simultaneously via multi-channel instruments.

Generally, digital instruments use microprocessors for the processing of the signals and co-ordination of their functions. Some instruments include a PC with a data acquisition card that collects the voltage and current signals from the transducers; the PC contains a microprocessor that calculates the harmonic levels, a hard disk for data storage and a screen for the visual display of the results.

The main components of an FFT-based instrument are illustrated in Figure 5.13. First, the signal is subjected to low pass filtering to eliminate all the frequencies above the spectrum of interest; this is normally limited to orders below the 50th. Once filtered, the analogue signal is sampled, converted to digital and stored. The FFT is then applied to the 2^i samples included in a period T_w multiple of the fundamental wave period, i.e. $T_w = NT_1$, the sampling frequency being $f_s = 2^i/(NT_1)$.

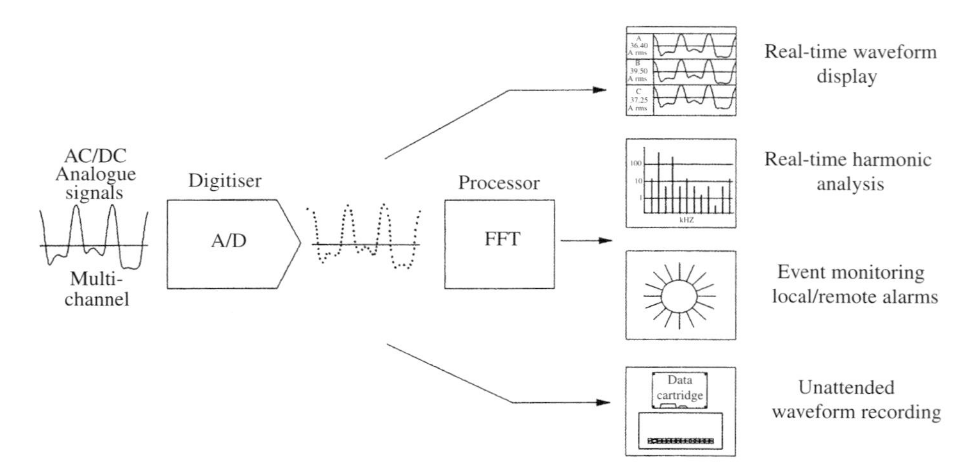

Figure 5.12 Components of a multi-wave monitoring system (from Electric Research and Management Inc.)

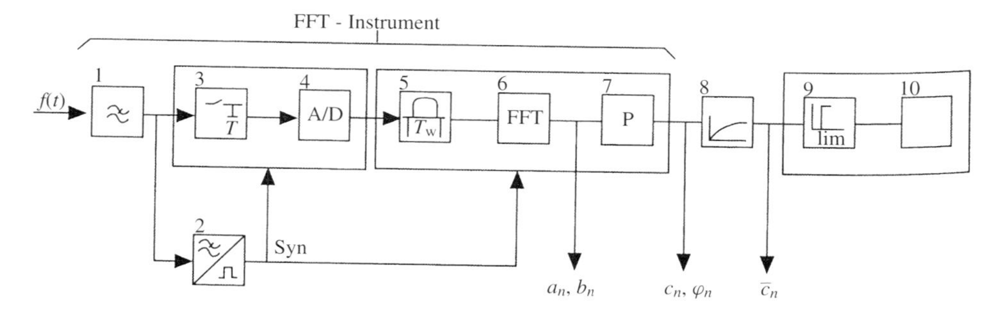

Figure 5.13 An FFT: based instrument, (1) Anti-aliasing low pass filter; (2) Synchronisation; (3) Sample and hold; (4) Analogue/digital converter; (5) Shape of window unit; (6) FFT processor; (7) Arithmetic processor; (8) Unit for evaluation of transitory harmonics; (9) Programmable classifier; (10) Counter and storage unit

To reduce spectral leakage, the samples contained in T_w are sometimes multiplied by a window function as described in Chapter 4.

The FFT process derives the Fourier coefficients a_k and b_k of frequencies $f_x = k(1/T_w)$ for $k = 0, 1, 2, \ldots, 2^{i-1}$ and with adequate synchronisation the nth harmonic of the fundamental frequency is given by $n = k/N$.

Finally, an arithmetic processor calculates the harmonic amplitudes

$$C_n = \sqrt{a_n^2 + b_n^2} \tag{5.1}$$

and phases

$$\varphi_n = \arctan\left(\frac{b_n}{a_n}\right). \tag{5.2}$$

5.4.1 Sampling for the FFT

The sampled signal $\hat{h}(t)$ of Equation (I.1) must be truncated to a finite number of samples for machine computation. This is achieved by multiplying the sampling signal, illustrated in Figure 5.14(a), by the rectangular truncation function of Figure 5.14(b) to yield the finite sequence of N samples illustrated in Figure 5.14(c). A sinewave is used to illustrate this process in both the time and frequency domains.

The Fourier transform of the rectangular truncation function is the sin *f/f* function, also illustrated in Figure 5.14(b). The Fourier Transform of the sampled truncated signal of Figure 5.14(c) is obtained by convolving the sampled signal's Fourier Transform with the sin *f/f* function of Figure 5.14(b). If the periodic signal $h(t)$ is represented by a Fourier series expansion and is band-limited to N harmonics, the sampled truncated signal's Fourier transform becomes

$$H'(f) = T_0 f_s \sum_{n=0}^{N} \alpha_n \frac{\sin(\pi T_0[f - n f_0])}{\pi T_0[f - n f_0]}, \tag{5.3}$$

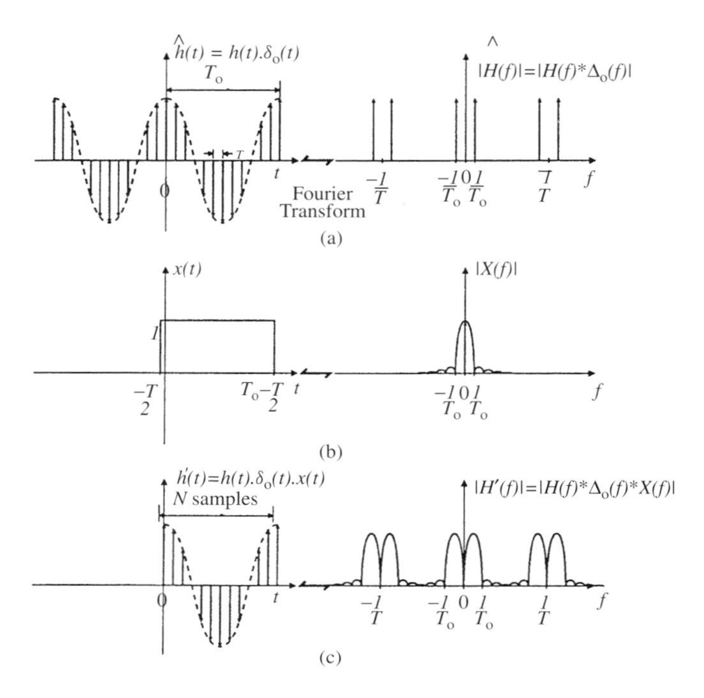

Figure 5.14 Truncation of a periodic signal and the resultant Fourier Transform

where α_n are the complex coefficients of the Fourier series, and f_0 is the fundamental frequency of the periodic signal *h(t)*. Figure 5.15(a) depicts $H'(f)$, which is essentially the summation of sin *f/f* functions centred on each harmonic frequency. The width of the sin *f/f* functions is dependent on the truncation interval width of Figure 5.14(b). If the truncation interval is equal to an integer multiple of the period of *h(t)* (i.e. $T_0 = n/f_0$, where *n* is an integer) the sin *f/f* function centred on each harmonic will be maximum at the harmonic frequency and zero at all adjacent harmonics.

When the truncation interval is not equal to an integer multiple of the fundamental period ($T_0 f_0 \neq n$, where *n* is an integer), then the sin *f/f* function at each harmonic will interfere with adjacent harmonics producing spectral leakage. The interference of a harmonic adjacent to another is given by

$$A(\text{dB}) = 20\log_{10}\left|\frac{\sin(\pi T_0 f_0)}{\pi T_0 f_0}\right|, \tag{5.4}$$

where f_0 is the fundamental frequency and T_0 is the truncation interval length.

If a voltage or current signal is sampled at a constant frequency, f_s, such that the truncation interval corresponds to exactly one period of the 50 Hz or 60 Hz fundamental frequency, no interference from adjacent harmonics will occur. However, fluctuations in the fundamental frequency will effectively change the truncation interval, causing spectral leakage. The attenuation of adjacent harmonics is shown in Figure 5.15(b) for a minimal 50 Hz fundamental. New Zealand legislation requires that this attenuation be greater than 40 dB [17]. It also states that harmonic voltage and current measurements shall be made when the system frequency is within 0.5%

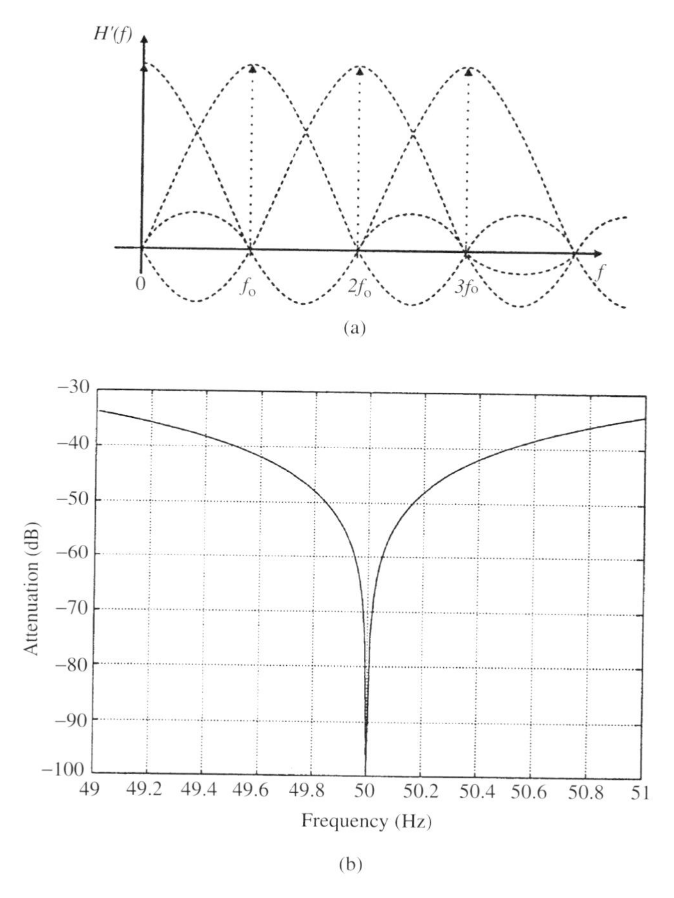

Figure 5.15 (a) the spectrum resulting from the rectangular windowing of a sampled signal; (b) the attenuation of a harmonic at an adjacent harmonic when the rectangular window is exactly equal to the period of the nominal 50 Hz system frequency but with the system frequency changing around the nominal 50 Hz

above or below the standard of 50 Hz. Evaluating the attenuation for the two limits of 49.75 Hz and 50.25 Hz gives 46 dB attenuation at each limit. Hence, it is feasible to sample at a constant frequency for harmonic monitoring to comply with New Zealand legislation. Ideally, the harmonic monitor should be equipped with a device [18] that produces a sampling signal of frequency f_s locked to the fundamental frequency f_0 such that $f_0 T_0$ is very close to unity, so as to monitor harmonics even when the system frequency is well outside the limits set by the legislation. Because spectral leakage is reduced by sampling coherently with the fundamental, windowing of the time-domain data prior to FFT computation is not necessary.

The FFT algorithm [19] is used to compute efficiently the DFT of voltage and current signals and, as discussed previously, the transform is computed over exactly one cycle of the fundamental. In this way, the system works on a cycle-by-cycle basis, treating each cycle separately and producing harmonic results for each cycle. Harmonics up to the 50th are required, and in order to realise this, as well as to satisfy the sampling theorem, the sampling frequency must satisfy the relation

$$f_s > 2 \times 50 f_0, \tag{5.5}$$

where f_0 is the fundamental frequency. This corresponds to exactly 100 samples per cycle. The FFT algorithm requires a record length of $N = 2^{\gamma}$ samples, where γ is an integer [19]. The lowest value of N satisfying Equation (5.5) is 128, meaning that 128-point FFTs are required to resolve up to the 50th harmonic, giving a required sampling frequency of 128 f_0 and 128 samples per cycle of the fundamental.

5.4.2 Anti-Aliasing Filtering

Sampling at a frequency of 128 f_0, which is equivalent to 2.56 f_{50} where f_{50} is the frequency of the 50th harmonic, leaves a band between f_{50} and $f_s/2$ of unwanted harmonics in the FFT results, illustrated in Figure 5.16(a). Because these results are not required, it does not matter if they are distorted by an anti-aliasing filter and by aliasing. Hence, this band is used to implement an anti-aliasing filter with a roll-off from $f_c = f_{50}$ sufficiently steep to give 98 dB attenuation at $f_s - f_c$, illustrated in Figure 5.16(b). Aliasing will occur in this band, but because harmonic analysis is performed outside of it, it is of no consequence. The use of 16-bit converters gives a theoretical r.m.s. signal-to-quantisation noise ratio of 98 dB.

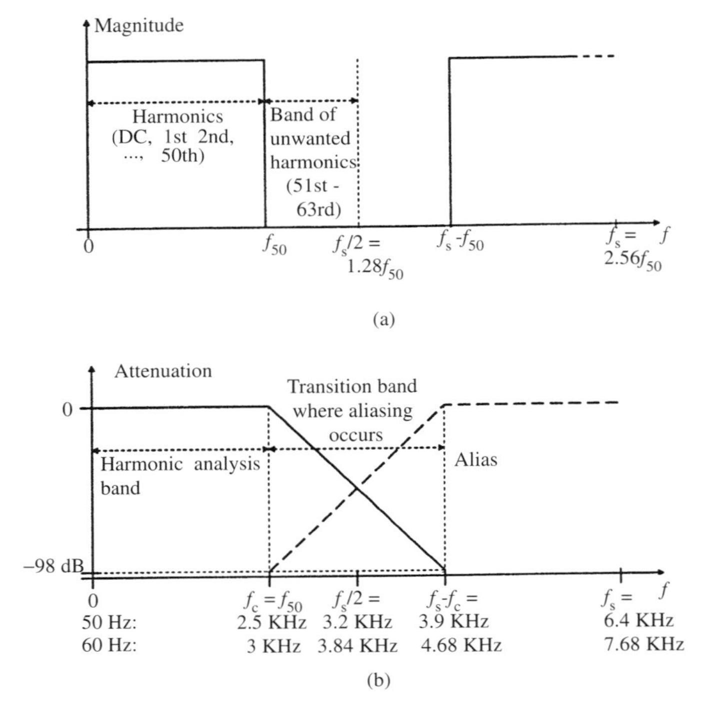

Figure 5.16 (a) spectrum showing the band of unwanted harmonics resolved by the FFT; (b) the anti-aliasing filter response implemented inside this band

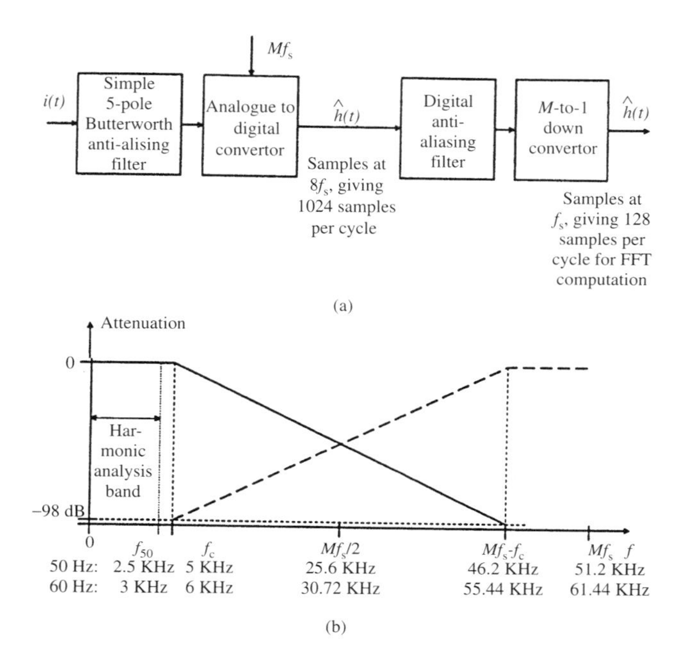

Figure 5.17 (a) oversampling of a signal with anti-aliasing filtering performed by a digital filter; (b) the required filter response of an anti-aliasing filter preceding an oversampling ADC. Details of 50 Hz and 60 Hz power systems are shown below for an oversampling factor of $M = 8$

The anti-aliasing filter required for the steep roll-off depicted in Figure 5.16(b) can be realised by using a digital FIR (Finite Impulse Response) filter, implemented in DSPs. The analogue voltage and current signals are sampled at a higher frequency than that required to give 128 samples per cycle. This reduces the required roll-off slope of the anti-aliasing filter preceding the ADC (illustrated in Figure 5.17(b)), making its design much simpler. It can also improve the effective Signal-to-Noise Ratio (SNR) of the sampled signal if the signal bandwidth is held constant [20]. This is known as *oversampling*, and is illustrated in Figure 5.17(a), with the required anti-aliasing filter response preceding the converter shown in Figure 5.17(b). An oversampling rate of $M = 8$, results in a sampling frequency of 1024 f_0. This enables a simple 5-pole filter to be used as the front-end anti-aliasing filter. The main features of the FIR filter suited to the application are its stability, phase linearity and simple design and implementation. Appendix I illustrates how a FIR filter can be designed to achieve the above specifications.

5.4.3 The FFT Implementation

The voltage and current signals are sampled coherently with the fundamental frequency, providing a finite set of samples in a form ready for analysis of spectral content by the DFT, using an FFT algorithm. They are also filtered sufficiently to ensure that aliasing will not distort spectral information, and decimated to a rate high enough

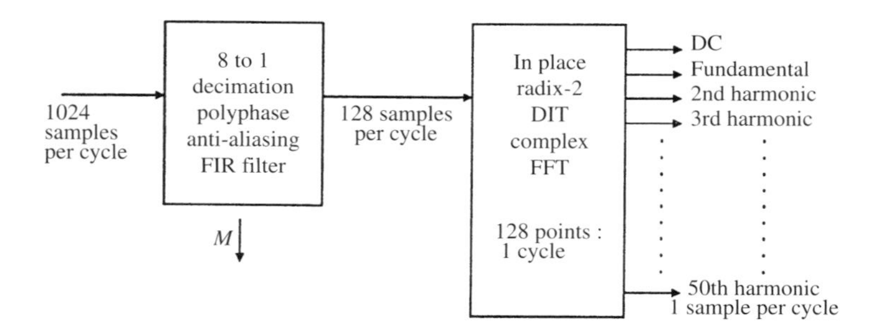

Figure 5.18 Computation of harmonics

to resolve up to the 50th harmonic but not so high that unnecessary computation is performed in taking the FFT.

Figure 5.18 illustrates that the 1024 samples taken per cycle are FIR filtered and decimated down to 128 samples. A radix-2 complex FFT is then used to compute the magnitude and phase of each of the 50 harmonics.

An FFT computation can be undertaken for each fundamental cycle, producing a set of harmonics every cycle, or several cycles of samples can be joined together before using a longer FFT to achieve better frequency resolution.

The frequency resolution of the FFT is given by the reciprocal of the record length

$$\Delta f = \frac{1}{T_0}. \tag{5.6}$$

Hence, to resolve harmonic values, separated by 50 Hz or 60 Hz, the record length must be one period of the fundamental frequency (20 ms or 16.7 ms). An FFT output bin is not, however, an impulse function centred on a particular harmonic. Instead, it has a non-zero response to frequencies between harmonics. This means that signals present in the signal to be transformed, that are not harmonics, will contribute to particular harmonic outputs from the FFT, making them erroneous. The severity of this problem can be reduced by taking the FFT over longer time spans, thereby resolving interharmonic frequencies which would otherwise contribute to harmonic outputs. This does, however, require more data processing. In harmonic analysis this is seen as a disadvantage because only harmonics are required–as opposed to a spectrum analyser which is required to have a very good resolution.

It is possible to achieve the same effect of taking the FFT over several periods by averaging several periods to one and computing the FFT of that, provided that sampling is synchronous with the fundamental. This preserves the bandwidth of the FFT with only a small processing overhead. It must be noted that if subharmonics are required, averaging is not suitable. Instead, the FFT must be taken over multiple cycles to produce the submultiples of the 50 Hz or 60 Hz fundamentals.

5.4.4 (Quasi) Steady-State Harmonic Measurements

In terms of the quantity of information to be processed, harmonic analysers should perform continuous or discontinuous analysis, the choice between these depending on the characteristics of the signals to be measured, which can be of the following types:

- Quasi-stationary harmonics
- Fluctuating harmonics
- Intermittent harmonics
- Interharmonics.

Only in the case of quasi-stationary harmonics is it possible to use instruments of discontinuous analysis. Examples of this type are the harmonic currents produced by well-defined loads, such as TV receivers and PCs.

In other cases, the content of the signal must be analysed in real time; for example, the current harmonics caused by electrical drives during speed changes, and the supply voltage to rolling mills.

5.4.5 Interharmonics Measurement

Quasi-stationary interharmonics of fixed frequencies may be measured with the harmonic instruments specified earlier. Both time and frequency domains may be used if the instruments can be synchronised or adjusted to any frequency.

However, frequency-domain methods are not appropriate for the measurement of fluctuating or rapidly changing interharmonic frequencies, since the synchronisation of the measuring instrument to those fast-changing frequencies is difficult. Time-domain instruments must be used instead, since they provide measuring results at all frequencies $f_m = mT_w$, T_w being the width of time window and $m = 0, 1, 2, \ldots$. A window width of 0.16 s provides frequency lines all 6.25 Hz; this is recommended [21] as a compromise between the desired bandwidth and the ability to follow rapidly changing amplitudes and/or frequencies.

If only a few interharmonic lines are predominant, then they can be interpreted as vectors rotating with the difference $\Delta f = f_{\text{int}} - f_m$ in the complex plane, f_{int} being the actual interharmonic frequency and f_m being the measured centre frequency. The actual frequency f_{int} may be recalculated, and its attenuation by the transfer function of the FFT may be corrected by software handling.

5.4.6 Phase-Angle Measurement

The measurement of phase angles between harmonic currents and voltages is required to:

- evaluate the harmonic flows throughout the power system,
- determine the location of harmonic sources,
- combine the harmonic currents from different disturbing loads if connected to the same node,
- evaluate the active harmonic power at the point of common coupling.

Digital instrumentation normally provides phase angles. With FFT instruments, the phase-lag may be calculated from the coefficients a_n and b_n. Instrumentation not synchronised to the fundamental may be used provided that the samples are taken simultaneously on both channels.

The measurement of absolute phase angles, i.e. related to the fundamental component, needs a precise synchronisation, preferably to the zero crossing of the fundamental system voltage. With absolute phase angle assessment it is possible to compare the measurements at different nodes of the same system or at nodes of different systems.

The maximum measurement error of absolute harmonic phase angles shall not exceed $\pm(5^\circ \text{or } n \times 1^\circ)$, whichever is the greater for an accurate assessment of the direction of active harmonic power.

5.5 Transients Monitoring

The use of input digital buffer storage by modern instrumentation permits capturing and analysing transient data. The main reason for measuring transients is to compare their values with those used to verify the immunity of electrical and electronic equipment in the power system. In immunity tests, the reproducibility of transients is ensured by standardising the artificial mains network or line impedance stabilisation network. On the basis of this principle, EPRI's project RP2542-1 [22], set up to measure short transients in surge arresters, connected a known non-linear impedance to the measuring point and recorded the currents and voltages at the terminals of that impedance.

The interpretation of short-transient measurements with reference to the immunity test procedures is described in detail in a document by Hydro-Quebec [23]. The report describes the circuitry and techniques used in the source voltage assessment (with details of the window widths and sampling rates required) as well as the transient overvoltage envelope assessment.

Transients are usually detected by their voltage rise or voltage amplitude thresholds. However, the presence of a very high harmonic may be identified as a transient. The following technique can be used to differentiate transients from high-order harmonic distortion.

Each sampled point $A_\nu(i)$, in Figure 5.19, is first squared and then subtracted from a corresponding point $A_{\nu-1}(i)$ on the previous cycle to derive the variation in squared amplitude VSA_i of the point i in the half-cycle interval ν, i.e. $VSA_i = A_\nu^2(i) - A_{\nu-1}^2(i)$.

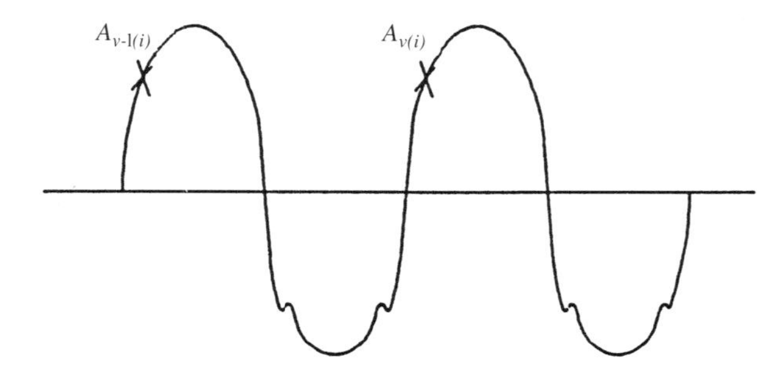

Figure 5.19 Location of points $A_\nu(i)$ and $A_{\nu-1}(i)$ used in example

The root mean square of the absolute variation in *AVSA*s of the half-cycle interval ν gives the root mean absolute variation in squared amplitude $MAVSA_\nu$.

$$MAVSA_\nu = \sqrt{\frac{\sum_{i=1}^{n} |A_\nu^2(i) - A_{\nu-1}^2(i)|}{n}}, \tag{5.7}$$

where n is the number of voltages sampled during each half-cycle, which should be at least 30 to detect most low-frequency transients. The variable ν indicates the positive or negative half-cycle interval and $\nu - 1$ the respective previous half-cycle interval.

The waveform contains a transient when the absolute difference between the $MAVSA_\nu$ of the positive or negative half-cycle exceeds the respective $MAVSA_{\nu-1}$ of the previous cycle by a specified level (17% in the EPRI report).

5.5.1 Wavelet Detection

As explained in Chapter 4 (section 4.6.1), the task of sorting and diagnosing transient events can be improved by the use of the Wavelet Transform (WT) as a pre-processing tool. The WT is particularly appropriate for power system transients since it partitions the frequency spectrum according to identifiable frequencies.

Referring back to Chapter 4, the output of Figure 4.30 is a multi-signal trace from each high pass filter $h[n]$, corresponding to a particular scale parameter, a_o^m = Scale 2^m. The traces labelled Scale 1, Scale 2, Scale 4, ... in Figure 4.30 correspond to the filter outputs $h[n]$, $h[n/2]$, $h[n/4]$, and the highest frequency impulse components appear in Scale 1.

The magnitudes of the continuous wavelet coefficients are calculated taking a few frequency steps, i.e. 50 Hz and higher frequencies, as shown by the frequency profiles in Figure 5.20(b) to (e) for the switching transient of Figure 5.20(a) [24]. Clearly, the magnitude of the 50 Hz signal is constant throughout the recording, whereas high frequency components change suddenly.

In order to detect a voltage transient, each peak detected in the high-frequency profile is compared with a normal value. Therefore, a part of the signal, assumed to be disturbance-free, is taken as reference and leads to 50 Hz, 350 Hz, 650 Hz or 1500 Hz benchmark values. The next part of the signal is then analysed and each peak exceeding the benchmark values is treated as a major signal discontinuity.

The duration and the number of peaks exceeding benchmark values differentiate transient events from voltage sags. Sharp and short peaks match simple voltage sags, whereas long peaks correspond to high-frequency transients.

When a transient is detected from the analysed signal, the time and frequency location of the main transient events are computed. A time–frequency plane is created and each frequency is automatically measured from the time-frequency plane. The frequency profile gives the maximum local value of the highest of the transients. For instance, Figure 5.21 is the time-frequency plane of the transient shown in Figure 5.20(a). The transient is detected to occur at 0.113 s. Figure 5.22 represents the corresponding frequency profile extracted from Figure 5.21 at that instant of time and indicates a 1500 Hz for the transient.

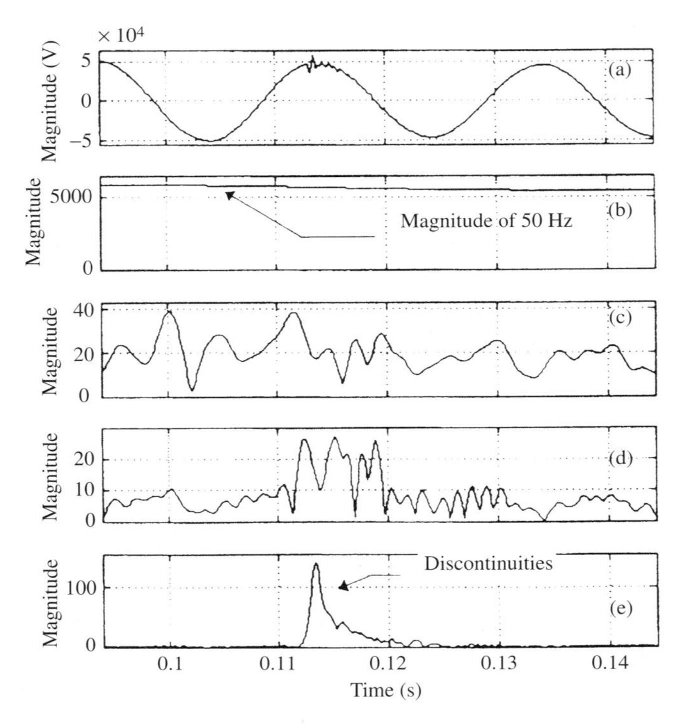

Figure 5.20 (a) example of switching transient; (b) profile at 50 Hz; (c) profile at 350 Hz; (d) profile at 650 Hz; (e) profile at 1500 Hz Copyright © 1998 IEEE

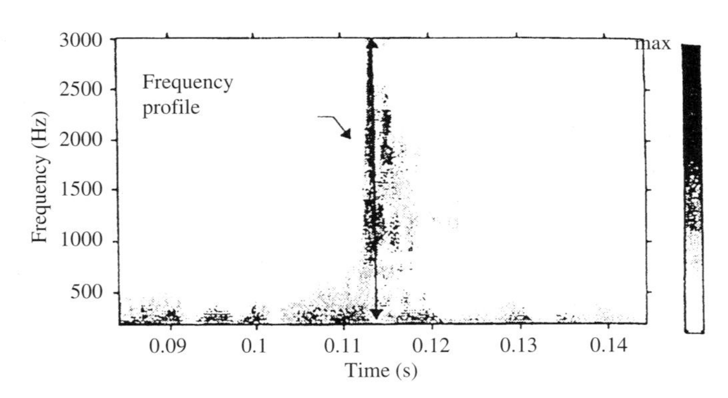

Figure 5.21 Time–frequency plane of transient from Figure 5.20(a) Copyright © 1998 IEEE

Conventional detection of transients is based on the derivative of the voltage signal. The maximum value of the derivative during the first 20 ms (for 50 Hz frequency supply) of the signal is taken as reference, and a transient is detected if a value overshoots this limit.

These comparisons give an experimental validation of the wavelet-based detection and characterisation algorithm. Comparisons have been made for over 1570 examples of actual records in [24].

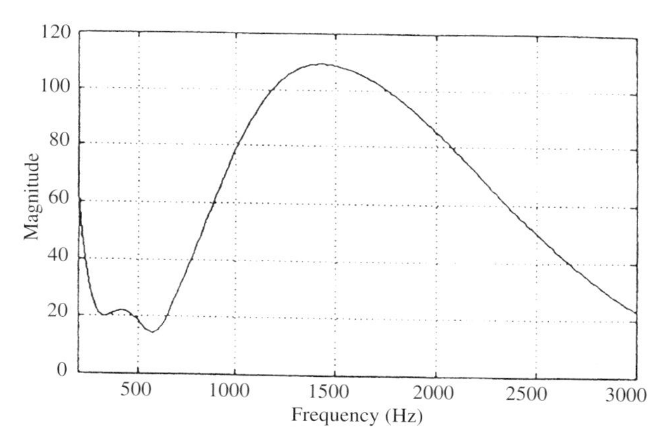

Figure 5.22 Profile frequency from Figure 5.20 at 0.113 s Copyright © 1998 IEEE

5.6 Event Recording

Probably the main power quality concerns of customers are voltage dips and cuts in number and duration. Earlier processing of these events gave only imperfect information to customers. Lack of reporting, inaccuracies in network mapping and changes in the operating schemes combined to reduce monitoring reliability. This encouraged the development and deployment of economic and reliable devices for general use throughout the customer network. An example of the new technology is the *Indicateur de Qualité de fourniture* (IQF) concept, used by Electricité de France [25].

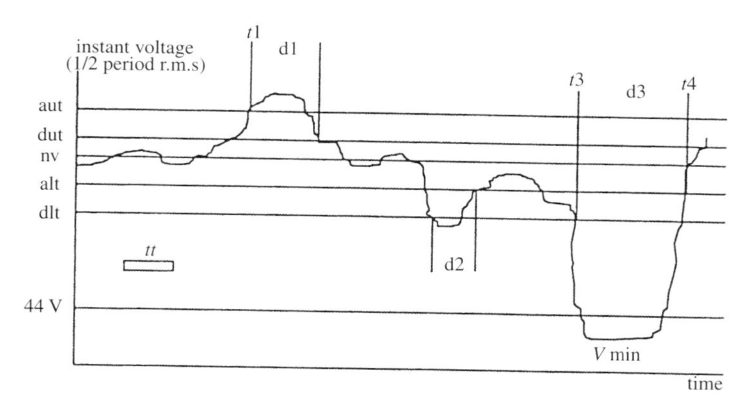

LEGEND

aut : ascending upper threshold
dut : descending upper threshold
alt : ascending lower threshold
dlt : descending lower threshold
nv : nominal value
tt : time threshold

d1 > tt event recorded as overvoltage at time t1 with duratoin t1 and rms value V1

d2 > tt not recorded

d3 > tt with Vmin < 44 V event recorded as outage at time t3 with duration t4-t3

Figure 5.23 An illustration of the definition of events Copyright © 1994 PQA

Such an approach, coupled to the introduction of a guarantee of supply quality in customers' contracts, has greatly improved power quality awareness, and given confidence to customers on its implementation.

The IQF instrument operates in two modes, as illustrated in Figure 5.23. In normal mode, the *instant r.m.s. value* defined as 10 ms ($= \frac{1}{2}$ period 50 Hz) r.m.s. value of the voltage is permanently compared with thresholds. When the voltage exceeds the upper ascending threshold, a timer is activated until the voltage goes under the upper descending threshold. The elapsed time is then compared with a time threshold. If the threshold is trespassed, an event is declared and recorded with time, duration, and mean value of the *instant r.m.s. value* computed on this duration. A similar process is defined for undervoltage. When the voltages goes below, say, 10% of the nominal, the device switches to a *power shortage mode* and the tracking of the voltage is interrupted. This event is recorded after power recovery, with time of occurrence and duration, with r.m.s. value. The voltage is derived from a resistor divider, and an ST9030 microprocessor performs the necessary processing 90 times every half-cycle.

5.7 Flicker Monitoring

5.7.1 The IEC Flickermeter

The two main parts of the ICE flickermeter [26] are shown in Figure 5.24. Voltage flicker is quantified on the basis of human irritation and thus the first part employs models of an incandescent filament lamp and of the human eye–brain system. From the monitored voltage, an indication of the momentarily experienced flicker (as 50% of the population would perceive it) is obtained which has been termed the *instantaneous flicker* level (IFL). In the second part, the generally unpredictable IFL is evaluated statistically to provide better consistency between the measurements. The statistical evaluation determines the distribution of magnitudes of the IFL signal and then characterises it by a single figure, the *short-term flicker severity index* (P_{st}). *Short-term* refers to periods of 10 minutes which the P_{st} indices represent. A *long-term flicker severity index* (P_{lt}) is also derived characterising the voltage flicker of the preceding two hours. The basis of the statistical evaluation has been described in Chapter 2.

Figure 5.25 shows the component blocks of the flickermeter to produce the IFL. The purpose of the voltage adapter is scaling the a.c. input voltage to its average r.m.s. value. In this way, slow changes due to voltage regulation will still be tracked. Furthermore, it allows the connection of the flickermeter to any voltage level. Squaring

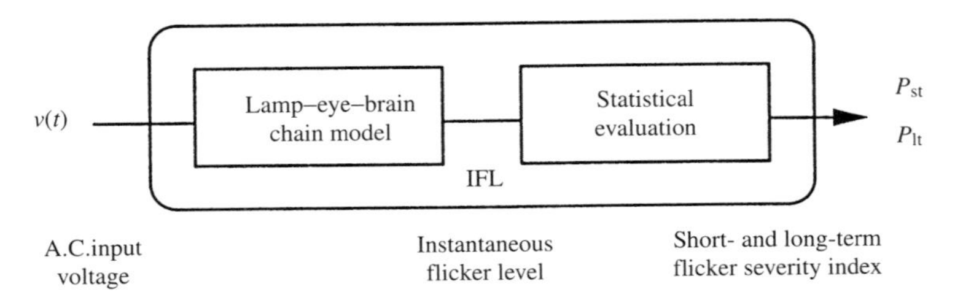

Figure 5.24 Working principle of the IEC flickermeter

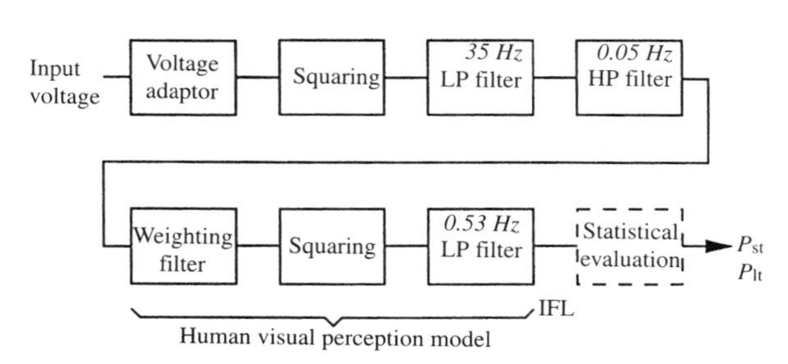

Figure 5.25 Block diagram of the analogue IEC flickermeter. Filter 3 dB frequencies are indicated in italics

of the voltage adapter output yields the normalised instantaneous power supplied to the lamp.

The following low pass filter must have an attenuation of around 90 dB at twice the mains frequency and low distortion in the passband. A 6th order Butterworth filter with a cut-off frequency of 35 Hz has been used for this purpose; for a 10% ripple ratio, which is assumed to be the maximum fluctuation occurring in a power system, the highest perceptible flicker frequency lies just under 35 Hz.

The high pass filter is designed to remove d.c. such that the filter output represents the low frequency normalised fluctuations of the instantaneous lamp power.

A tungsten coil filament lamp with a rating of 60 W/230 V is used as the reference lamp, whose low pass characteristic has been incorporated in the weighting filter representing the response of the eye to brightness fluctuations.

The weighting filter magnitude response of the IEC proposal is represented by the solid curve in Figure 5.26. The dotted characteristic is to be used instead for flicker measurements in 120 V systems [27].

The remaining blocks of the human perception model are the squaring block and a first-order low pass filter. Together they form a non-linear *variance estimator* for the weighting filter output. Thus, the resulting IFL is always positive.

A digital derivation of the IFL is described next.

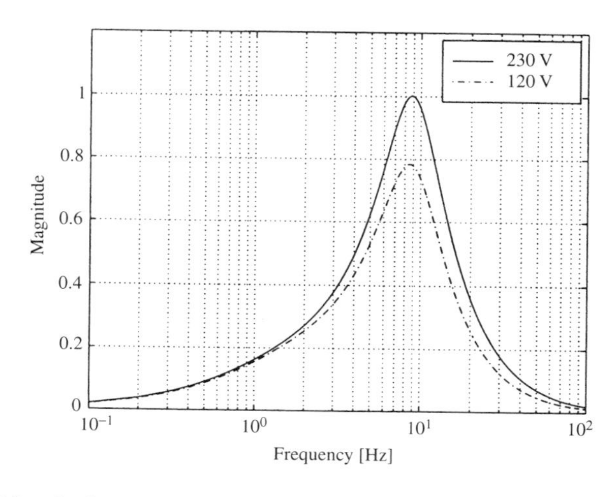

Figure 5.26 Magnitude responses of the weighting filters for 120 V and 230 V systems

5.7.2 Digital, Time-Domain-Based Flickermeter [28]

The digital flickermeter design, except for the statistical evaluation block, is based on the analogue technique. However, the DSP implementation requires a fully digital realisation.

The s-transfer functions for all filters have been specified by the IEC. The corresponding z-transfer functions can be obtained by standard analogue to digital filter conversion algorithms or designed directly to specification.

Figure 5.27 shows a block diagram of a digital flickermeter implemented in the Continuous Harmonic Analysis in Real Time (CHART) monitoring system described in Appendix II.

Sampling Frequency The digital flickermeter employs multi-rate processing and the use of the lowest possible sample rate at each stage ensures maximum computational efficiency.

Flicker is considered to be an amplitude modulation of the a.c. system voltage exhibiting the frequency spectrum of a carrier with sidebands. Under this assumption and for a maximum flicker fusion frequency of 35 Hz, the interesting frequency band for flicker measurements extends up to 85 Hz for a 50 Hz a.c. system (95 Hz for a 60 Hz a.c. system). The input bandwidth of the flickermeter should, however, not be restricted to this small range because there are other phenomena that have to be taken into account.

Interharmonics also produce light flicker [29]. Beating of certain components present in the frequency spectrum of the voltage can lead to voltage amplitude fluctuations causing perceptible light flicker. For example, a voltage with a spectrum consisting of the a.c. system fundamental, a 5th harmonic, and a 260 Hz component will contain

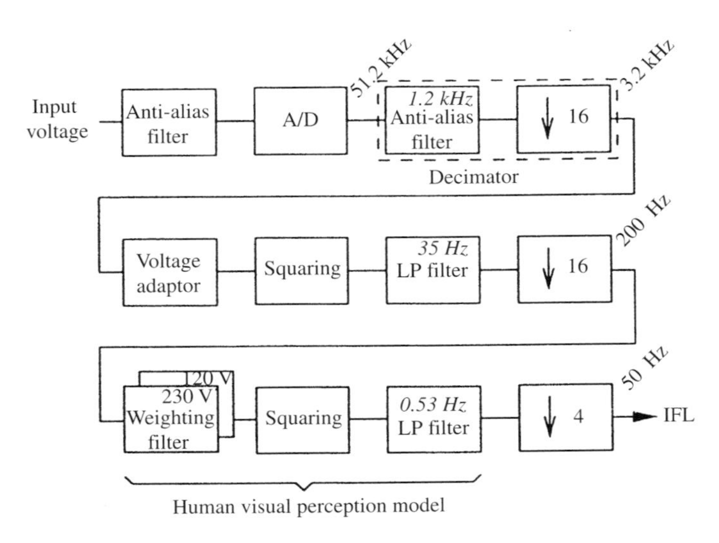

Figure 5.27 Block diagram of the digital flickermeter (statistical evaluation excluded). Sample rates and filter 3 dB frequencies are indicated in italics

a 10 Hz beat frequency in its envelope relevant for flicker perception, although the spectrum has no components at frequencies below or close to the fundamental.

Spectral components at non-harmonic frequencies can result from harmonic cross-modulation in the process of a.c./d.c. static conversion. In practice, the amplitudes are very small (below 1%) so that a single interharmonic is unlikely to cause perceptible flicker. The interesting frequency range is, therefore, limited to approximately 20 Hz–80 Hz if only the beating of single interharmonics with the fundamental is considered. However, several interharmonics and harmonics might result in noticeable flicker because the individual flicker levels are additive. Consequently, flickermeters should have a sufficiently high input bandwidth.

The bandwidth of the digital flickermeter described is 1.2 kHz, limited by a 255th-order linear phase FIR anti-aliasing filter. The stopband extends from 1.6 kHz allowing for a sample rate reduction to 3.2 kHz achieved by a decimation factor of 16 from the sample rate 51.2 kHz at which the analogue-to-digital converter of CHART operates. The FIR filtering is embedded in the down sampler and filter coefficient symmetry exploited to achieve an efficient decimator.

The first (LP) filter of the IEC flickermeter has an attenuation of at least 90 dB at 100 Hz and above to attenuate twice the fundamental frequency component resulting from squaring. Therefore, the sample rate can be reduced further after the filter. The remaining filters operate at a sample rate of 200 Hz.

The final LP filter reduces the signal bandwidth again (first-order filter, $f_{\text{cut-off}} = 0.53$ Hz) so that the sample rate can be lowered to minimise computations involved in the statistical evaluation of the IFL. The minimum rate at which samples should be supplied to the statistical evaluation block is 50 Hz according to [30].

Digital Weighting Filter Design The magnitude response of the digital weighting filter influences the overall characteristic of the meter strongly and, therefore, the digital filter should match the IEC specification closely.

The bilinear, impulse–invariant, and covariance–invariant transforms were considered for the analogue-to-digital filter conversion. Figure 5.28 shows the maximum

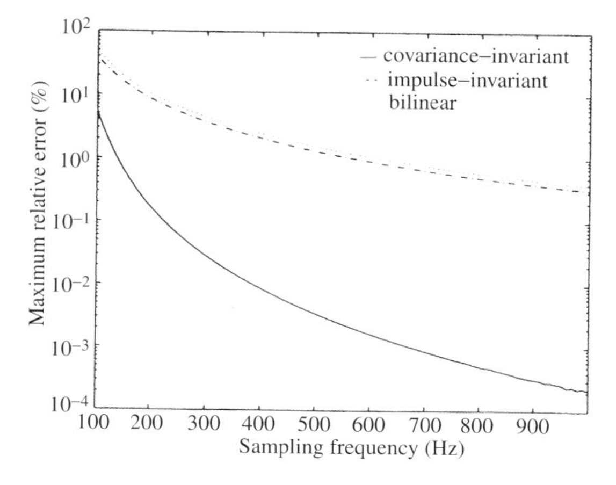

Figure 5.28 Magnitude errors of digital filter designs compared with the analogue filter vs sampling frequency

Table 5.1 Coefficients of weighting filter for 230 V systems (sample rate: 200 Hz)

Order	Numerator coeff.	Denominator coeff.
z^0	$+9.487215e-02$	$+1.0$
z^{-1}	$-1.582865e-01$	$-3.167151e+00$
z^{-3}	$+4.023729e-02$	$+3.752054e+00$
z^{-3}	$+2.317702e-02$	$-1.958255e+00$
z^{-4}		$+3.747149e-01$

relative errors in magnitude for digital weighting filter designs referenced to the IEC analogue filter specification. The filters were designed for a range of sample rates and the maximum occurring error in the interesting frequency range (0.5 Hz–35 Hz) was recorded each time. The superior magnitude response fit of the covariance–invariant filter can be confirmed [31]. For the chosen sample rate of 200 Hz, only the covariance–invariant filter gives an acceptable accuracy.

The d.c. rejection of the digital filter given directly by the covariant–invariant transform has been improved by shifting the relevant filter zero onto the unit circle (i.e. from $0.9989 + j0$ to $1.0 + j0$). This eliminates the need for a separate high-power (HP) filter preceding the weighting filter. The determined filter coefficients are given in Table 5.1.

As already mentioned, the original filter had been specified for 230 V lamps preventing the use of the flickermeter in 120 V systems. A digital flickermeter has the flexibility to switch between different weighting filter characteristics as indicated in Figure 5.27. Table 5.2 gives the coefficients for the digital filter corresponding to the 120 V system's adapted weighting filter of [27]. Measurements using the appropriate filter will be consistent and allow for the comparison of results.

Compliance Tests To verify the compliance of the time-domain-based flickermeter with the IEC standard, the standardised tests were conducted for a MATLAB simulated flickermeter and the CHART implementation.

Figure 5.29 represents the maximum IFL for sinusoidal and rectangular flicker over the range of frequencies and magnitudes specified by the IEC. All values lie well within the ±10% tolerance band around one unit of perceptibility. This is equivalent to the criterion of the IEC standard which demands that for unity output each test signal amplitude must lie within ±5% of the reference value.

Table 5.2 Coefficients of weighting filter for 120 V systems (sample rate: 200 Hz)

Order	Numerator coeff.	Denominator coeff.
z^0	$+6.568031e-02$	$+1.0$
z^{-1}	$-1.081596e-01$	$-3.236145e+00$
z^{-3}	$+2.660934e-02$	$+3.93687e+00$
z^{-3}	$+1.586997e-02$	$-2.127023e+00$
z^{-4}		$+4.275761e-01$

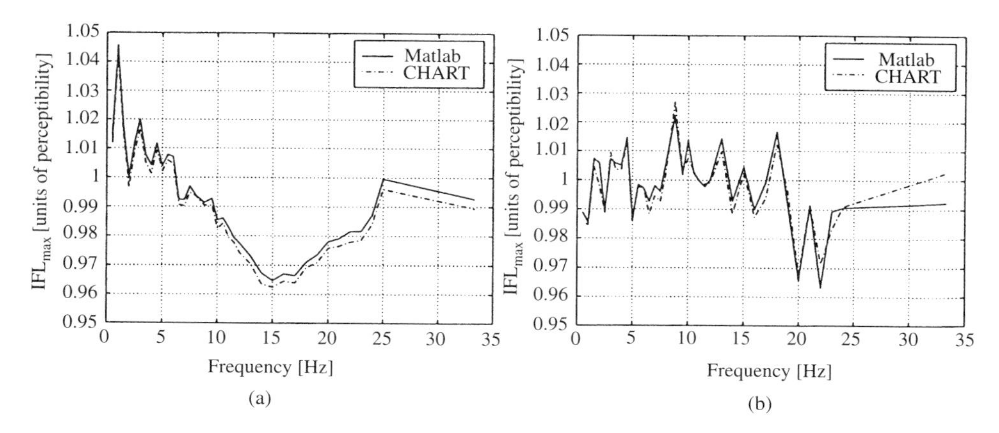

Figure 5.29 Compliance test results for the simulated flickermeter (MATLAB) and the CHART implementation (a) sinusoidal test signals; (b) rectangular test signals

5.7.3 Digital, f-Domain-Based Flickermeter

For compliance with the IEC flickermeter specification, an f-domain-based meter must incorporate the frequency-domain equivalents of all function blocks. Processing in the f-domain is beneficial when realising the filters, allowing for arbitrarily shaped frequency responses and straightforward filter design. However, replacing the squaring operations by convolutions, which require much more computation, is not desirable.

Figure 5.30 shows the block diagram of the developed design. The transformations via FFT and inverse FFT (IFFT) take place just before and after the weighting block. All other function blocks remain the same as for the time-domain-based flickermeter (Figure 5.27) because f-domain processing would decrease the efficiency with little or no gain in performance. A partial FFT algorithm which calculates only the frequency samples in the band 0.05–35 Hz saves computations and conveniently realises the ideal bandpass filtering of the relative brightness changes. The *transform decomposition*

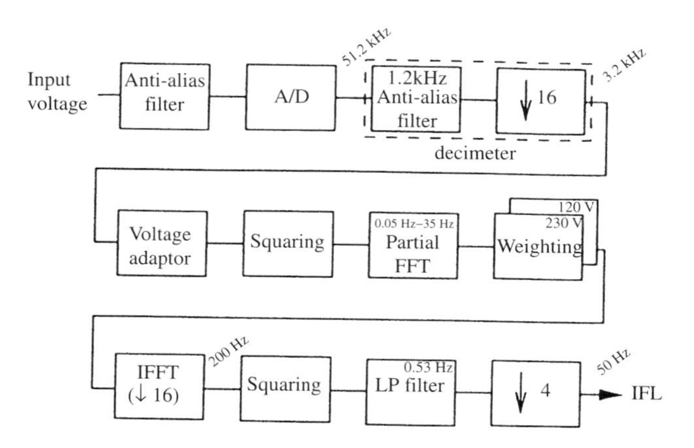

Figure 5.30 Block diagram of the digital, frequency-domain-based flickermeter

method devised by Sorensen [32] is 75% more efficient than a complex radix-2 FFT (for the present number of FFT input and output points and real inputs).

The number of frequency samples given by the FFT is small depending on the frequency resolution (e.g. 35 for a resolution of 1 Hz). The complex weights according to the frequency response of the analogue weighting filter can, therefore, be taken from a look-up table and then applied to the corresponding spectral component.

Once again, the band-limited spectrum allows for a sample rate reduction, which is best achieved by performing an IFFT of reduced length. If the FFT has N inputs, then an IFFT with N/D outputs reproduces the signal sampled at a rate reduced by D. Since, for the chosen decimation factor of 16, the number of non-zero inputs to the IFFT is still far less than the number of outputs (due to oversampling), using the *transform decomposition*, as well as exploiting Hermetian symmetry, saves computations.

The use of rectangular windows and data sampled synchronously with the a.c. system fundamental eliminates spectral leakage.

Compliance Tests The f-domain flickermeter was simulated using MATLAB running under UNIX. An implementation on CHART's DSPs was not undertaken,

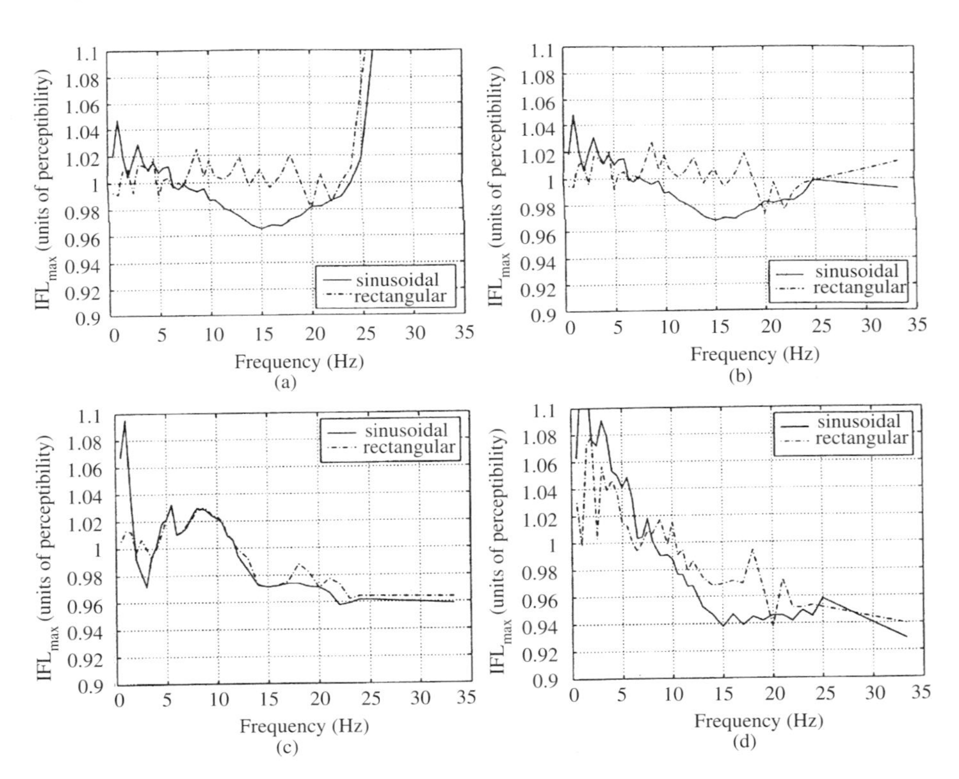

Figure 5.31 Compliance test results for the f-domain flickermeter. Hamming analysis windows and rectangular synthesis window used. $N = N_f = 8192$: (a) $N_h = 4096$, $L = 1024$ (75% overlap); (b) $N_h = 4096$, $L = 1024$ (75% overlap), demodulation low pass included; (c) $N_h = 6720$, $L = 4096$ (39% overlap), demodulation low pass included; (d) $N_h = 6720$, $L = 3360$ (50% overlap), demodulation low pass included

since both systems are fully digital and only very small differences, due to the single precision number representation on the DSP, are expected.

Figure 5.31 presents the test results using 8192-point FFTs, various analysis window lengths N_h, and configurations. L is a power of two in cases (a)–(c), so that the reconstruction ripple from the synthesis windows is zero. Compliance with the IEC standard, which currently only comprises the test signals in the 0 Hz–25 Hz frequency range, is proven for those configurations, since the curves lie within the tolerance band extending from 0.9–1.1 units of perceptibility.

However, with the proposed inclusion of a test signal at 33.33 Hz to achieve better consistency between different flickermeter implementations, the results in Figure 5.31(a) would lead to a failure of the test. The reason is that the demodulation low-pass filter has been omitted in this implementation (see Figure 5.30) and thus the maximum IFL, obtained for 33.33 Hz, exceeds the upper limit. This problem is eliminated by combining the response of the weighting and the demodulation low-pass filter in the DFT filter bank; this does not increase the complexity of the system. The result is represented in Figure 5.31(b), now verifying accordance with the standard over the whole range of test frequencies.

Figures 5.31(c) and 5.31(d) show $\mathrm{IFL}_{\max}$ curves for a greater analysis window length $N_h = 6720$, leading to a computationally more efficient system.

5.7.4 Arc-Furnace Flicker Measurement

To test the flickermeter for more realistic measurements, arc-furnace voltage recordings were used as input. Arc-furnace flicker has varying spectral content which is of particular interest when testing an FFT-based application, otherwise problems due to spectral leakage may not be recognised.

The obtained IFL signals were compared with the output from the time-domain implementation using the Integral Absolute Error (IAE) as a figure of merit. Figure 5.32 represents the resulting output signals and their relative differences. From the magnitude of the differences, it is clear that $N_h = 6720$ provides no acceptable reconstruction, while in Figure 5.32(a) there is no appreciable difference between the two IFL curves. The errors are small reaching only a few percent. This manifests itself in the IAEs which evaluate to 0.078% and 2.97%, respectively.

Although the overlap used in Figure 5.32(a) is only 50% of the theoretical value, the results for an overlap of 75% show no appreciable difference.

5.7.5 Steady-State Flicker Estimation

In the general case, voltage fluctuations leading to light flicker are non-deterministic and thus the flicker severity indices are based on a statistical evaluation of a time series of IFL. Flicker severity evaluation thus inherently requires the time-domain waveform of the IFL.

Monitoring the IFL is also a non-linear process (including two squaring operations). This means that the IFLs at different nodes of a power system cannot be related to each other through the (linear) frequency characteristics of the network in the way this is done for harmonics. This also applies to the flicker severity indices P_{st} and

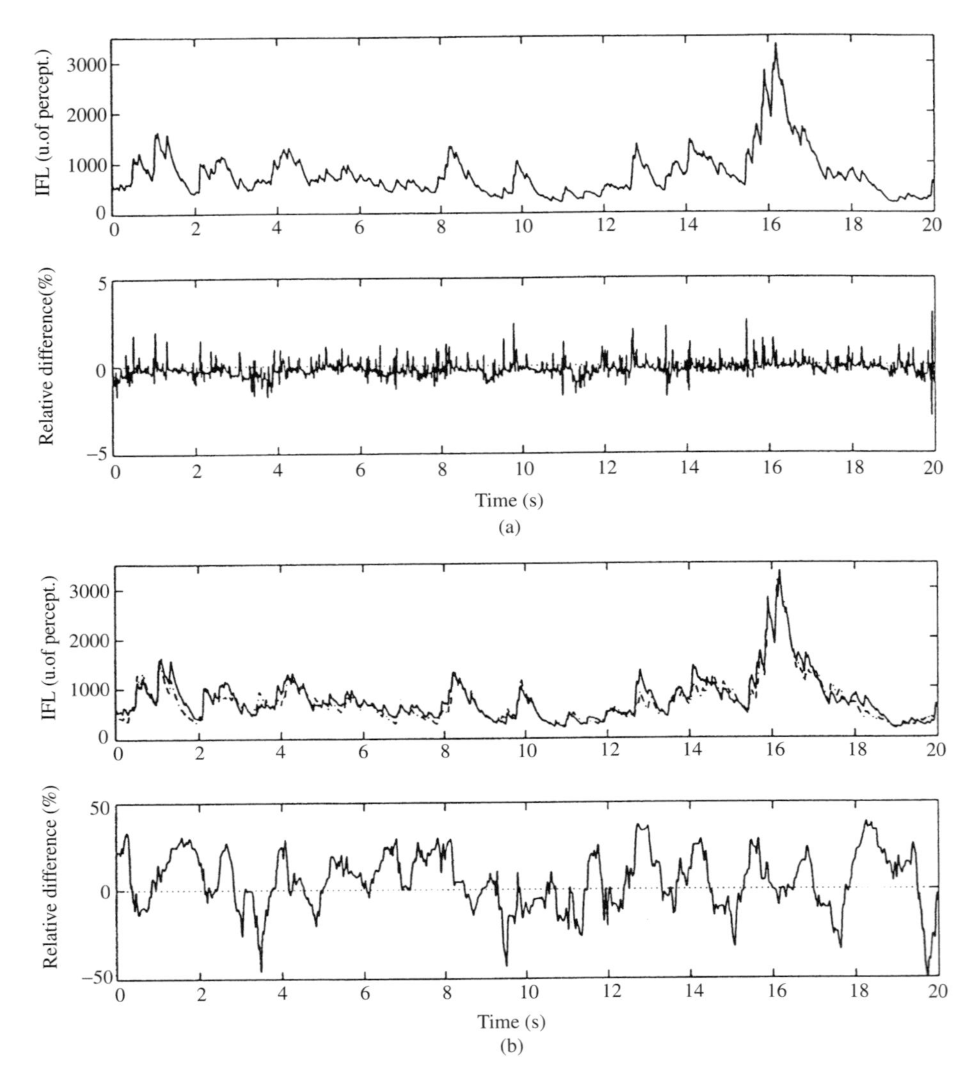

Figure 5.32 IFL resulting from arc-furnace flicker. The solid lines show the output from the time-domain flickermeter, the dash-dotted ones indicate the IFL from the f-domain approach. (a) and (b) correspond to the configurations that yield the $\mathrm{IFL}_{\max}$ curves in Figures 5.31(c) and (d) $N = N_f = 8192$: (a) IAE $= 0.078\%$, $N_h = 4096$, $L = 2048$ (50% overlap); (b) IAE $= 2.97\%$, $N_h = 6720$, $L = 4096$ (39% overlap)

P_{lt}, since the former is derived from the IFL time series and the latter is based on the former. Hence, flicker severity at a desired bus cannot directly be computed from P_{st} values measured at other buses. Instead, use must be made of the computed time series of the voltage at the desired location, from which the IFL and P_{st} can then be derived.

Consequently, f-domain simulation is only suited to computing the flicker severity for a steady-state disturbance, i.e. periodic flicker.

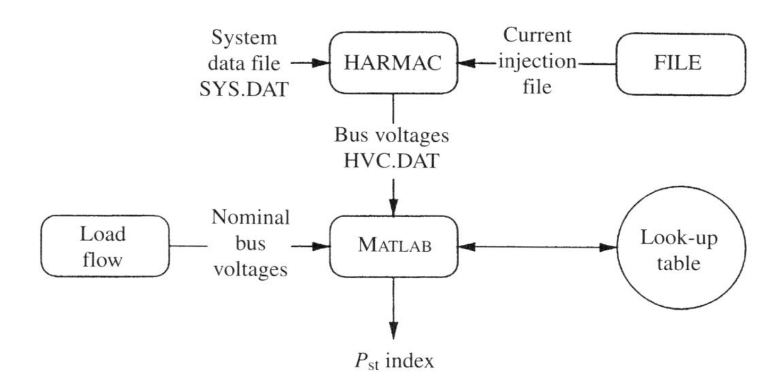

Figure 5.33 Using HARMAC's output to estimate the P_{st} index

For periodic flicker, the IEC flickermeter relates flicker severity and flicker magnitude linearly. This linear relationship can be exploited to estimate the flicker severity index throughout a power system using frequency-domain techniques. In this case, the voltage spectra occurring at all buses can be calculated by the harmonic penetration program.

Figure 5.33 illustrates the process of estimating P_{st} from spectra calculated by the harmonic penetration program (HARMAC), described in Chapter 6. In the simulation process, the current injections can be entered manually or read from file.

The bus voltages are found through superposition of the modulation voltages that are excited by the current injection (HVC.DAT) and the fundamental frequency voltages obtained from a load-flow program.

In the next step, the flicker magnitudes have to be determined which can then be related to their P_{st} values using the appropriate proportionality factor based on the type and frequency of the flicker waveform (taken from a look-up table). To illustrate

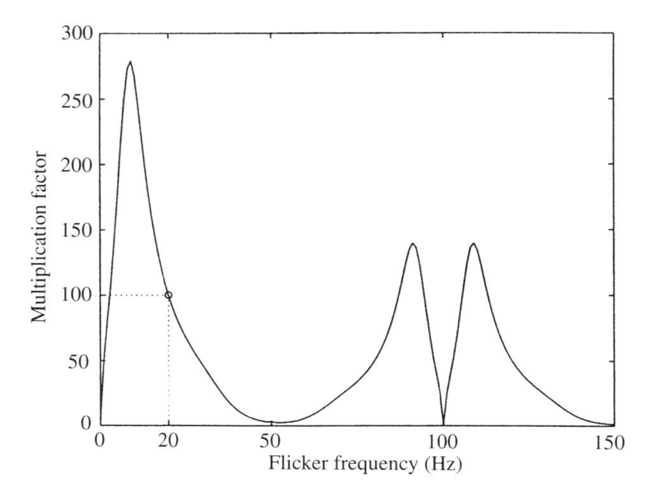

Figure 5.34 Proportionality factors relating modulation magnitude and flicker severity for sinusoidal flicker

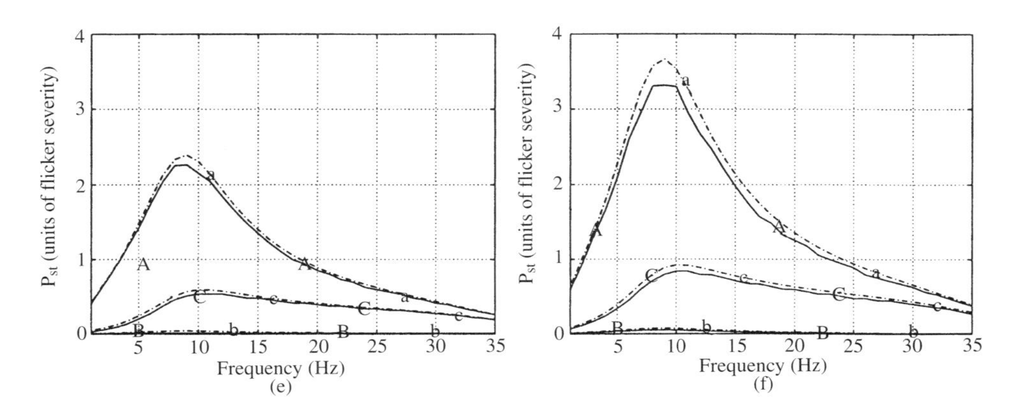

Figure 5.35 Comparison of P_{st} indices resulting from a single-phase current injection (i.e. all sequences) at phase A of the Tiwai bus. PSCAD results are shown as solid lines (phases A, B, C), HARMAC results as dash-dotted lines (phases a, b, c): (e) Roxburgh 11 kV; (f) Invercargill 33 kV

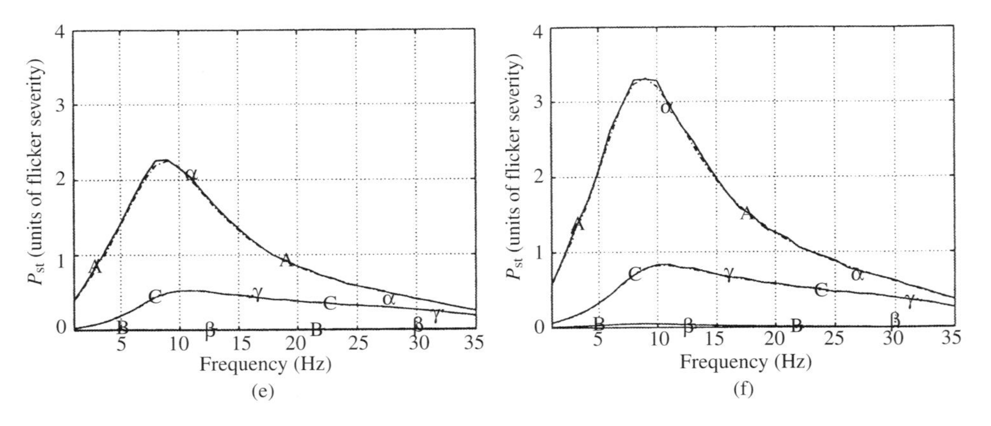

Figure 5.36 Comparison of P_{st} indices given by the PSCAD flickermeters to those obtained directly from PSCAD voltages using a single-phase current injection at Tiwai as an example. The flickermeter results are shown as solid lines (phases A, B, C), the dash-dotted lines represent the P_{st}s derived from the FFTed voltages (phases α, β, γ): (e) Roxburgh 11 kV; (f) Invercargill 33 kV

this further, the factors for sinusoidal flicker are shown in Figure 5.34. For example, a modulation magnitude of 3% occurring at 20 Hz would yield a P_{st} of 3 units of severity.

To illustrate the steady-state flicker estimation using frequency-domain (PSCAD) [33] simulation, the lower south island of the New Zealand test system (shown in Figure 7.8 of Chapter 7) was used with the aluminium smelter at Tiwai representing the flicker source. Figure 5.35 compares the P_{st} values obtained for six of the busbars obtained using frequency-domain simulation and time-domain simulation. The comparison is very good with the small discrepancies being due to the different degrees of frequency dependency modelled. This is verified by applying an FFT to the time-domain (PSCAD) [33] simulation results and then processing it with the frequency-domain flickermeter, and comparing this with the time-domain flickermeter results. This comparison is shown in Figure 5.36.

5.8 Assessment of Voltage and Current Unbalance

In a three-phase system voltage (current), unbalance is manifested by inequality in the magnitude values of the phase voltages and/or differences in the phase angles between consecutive phases.

An unbalanced three-phase voltage or current system can be replaced by three *symmetrical sequence* components of the same frequency, referred to as the positive, negative and zero sequence sets.

The main reasons for using symmetrical components in harmonics assessment are as follows.

- Positive and negative sequence impedances differ from zero sequence impedances for nearly all loads and network equipment (lines, cables, transformers); therefore, to assess harmonic voltages caused by injected currents a separate treatment of the system is necessary.
- Zero sequence currents and voltages are not transferred by commonly used transformer connections and delta-connected loads are not affected by zero sequence voltages.
- The characteristic harmonic currents generated by a.c./d.c. converters are only of positive or negative sequence.

The Unbalance Factor The symmetrical components may be derived from synchronised measured three-phase values using the modal decomposition matrix or they can be measured directly with special instruments.

The resulting positive (U_1) and negative (U_2) phasors are expressed as

$$U_1 = U_1\cos(\phi_1) + jU_1\sin(\phi_1), \tag{5.8}$$

$$U_2 = U_2\cos(\phi_2) + jU_2\sin(\phi_2). \tag{5.9}$$

Once the symmetrical components are derived from the phase voltages, the unbalance factor (i_u) is obtained using the ratio of negative to positive sequence, i.e.

$$i_\mathrm{u} = \frac{|U_2|}{|U_1|} = \frac{\sqrt{(U_2\cos(\phi_2))^2 + (U_2\sin(\phi_2))^2}}{\sqrt{(U_1\cos(\phi_1))^2 + (U_1\sin(\phi_1))^2}}. \tag{5.10}$$

Similarly, the degree of zero sequence voltage unbalance can be determined by the ratio of the zero sequence component to the positive sequence component.

The zero sequence voltages mainly result from the zero sequence currents of unbalanced loads flowing in the network. It can affect three-phase equipment connected line-to-neutral, but does not affect the majority of three-phase equipment connected line-to-line.

The voltage unbalance factors must be measured at the fundamental frequency (50 Hz or 60 Hz). If not, the contribution of the zero sequence component, such as the 3rd harmonic voltage, and/or the negative sequence component, such as the 5th harmonic voltage, can increase the measured unbalance factor and consequently introduce an error because this contribution does not cause the same effects as the fundamental frequency unbalance on equipment.

5.9 Examples of Application

This section reports on field measurements carried out in the New Zealand power system to illustrate the use and capability of modern digital instrumentation.

5.9.1 South Island (NZ) Synchronised Tests

Figure 5.37 shows the 220 kV network of New Zealand's South Island system. Between the Islington and North Makawera substations, there are a number of hydro stations and an HV d.c. scheme. The distribution network at Islington had reported a substantial amount of 5th harmonic distortion caused by the presence of a large number of industrial sites connected to this bus. The 5th harmonic current is largely absorbed by the compensation capacitors at the Islington substation. However, the capacitors are usually removed from service during light load conditions, causing the 5th harmonic current to flow into the 220 kV transmission system. On the other hand, at North Makawera, the opposite effect had been observed, with the 5th harmonic current flowing from the 220 kV transmission network into the distribution system. The main consumer of electrical power fed by the North Makawera substation is an aluminium smelter at the Tiwai bus. A recently installed 5th harmonic filter at Tiwai was frequently overloaded and the source of the 5th harmonic current distortion was traced to the 220 kV transmission system.

The first objective of the test was to decide whether there was any connection between the 5th harmonic problems at the Islington and North Makawera substations. If the two problems were related, then the response of the 5th harmonic distortion at North Makawera to the switching of capacitors at Islington was to be identified. This required that the measurements undertaken at both sites should be synchronised as accurately as possible. The 5th harmonic problems at both substations were known to vary with the daily operation of the South Island system. Therefore, the monitoring systems had to gather harmonic information over a fairly long period, covering a variety of operating conditions. Two CHART units (described in Appendix II) were installed at the North Makawera and Islington substations to monitor the currents flowing between the 220 kV transmission system and the substations. Details of the CHART instrumentation are given in Appendix II and their geographical location is illustrated in Figure 5.38.

At Islington, the two transmission lines considered to have the lowest impedance between the substation and the generating stations around Benmore are the Islington–Livingston and Islington–Timaru–Twizel lines. The three phases on these two circuits were therefore measured. Besides measurements on the 220 kV system, the

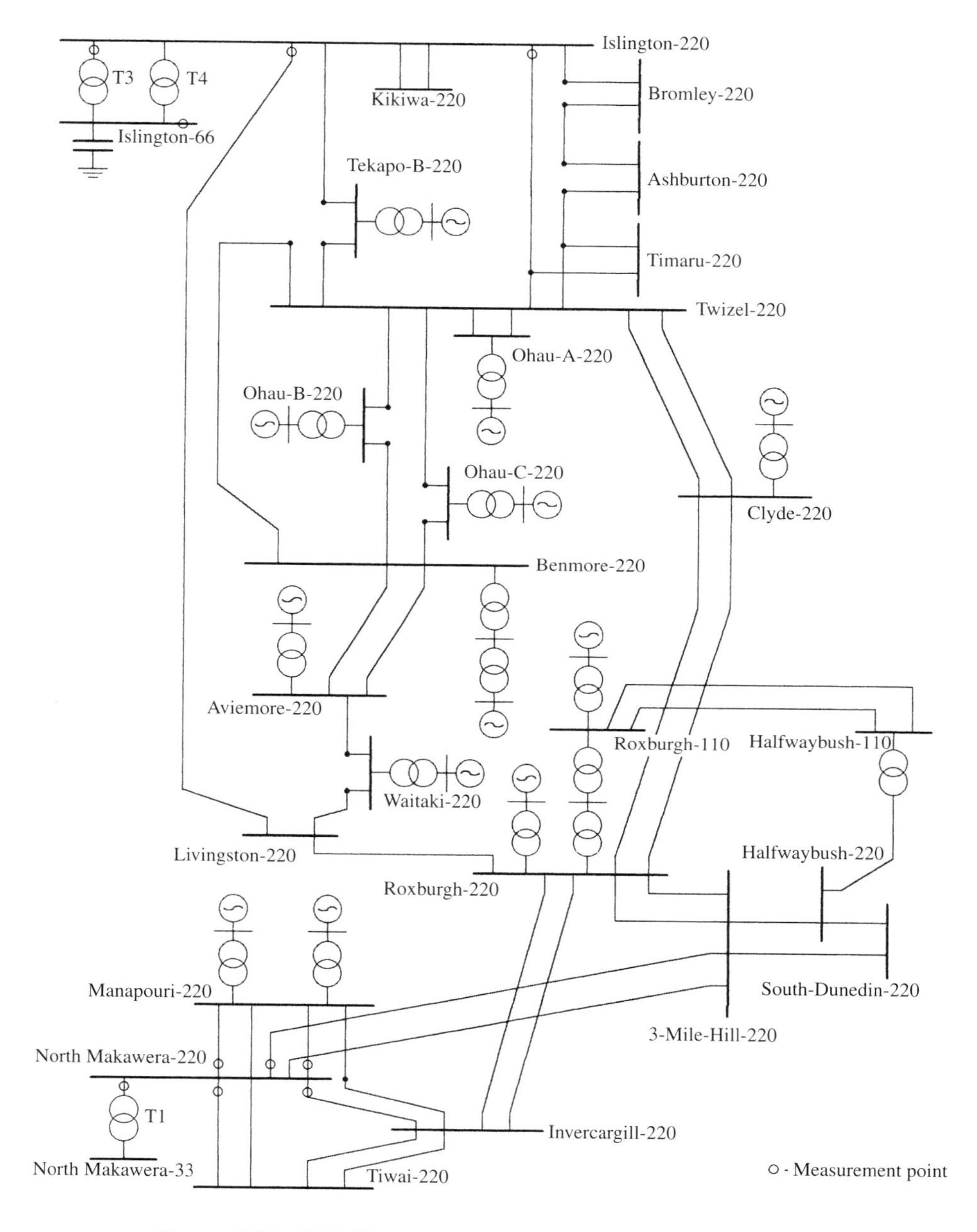

Figure 5.37 CHART measurement points in the synchronised test

voltage distortion on the 66 kV bus was also monitored. The compensation capacitors, connected to the 66 kV distribution circuit at Islington, are known to contain a substantial amount of 5th harmonic distortion. The CHART unit was also set up to record the current of one of the transformers (T3) interconnecting the 66 kV distribution network and the 220 kV transmission network.

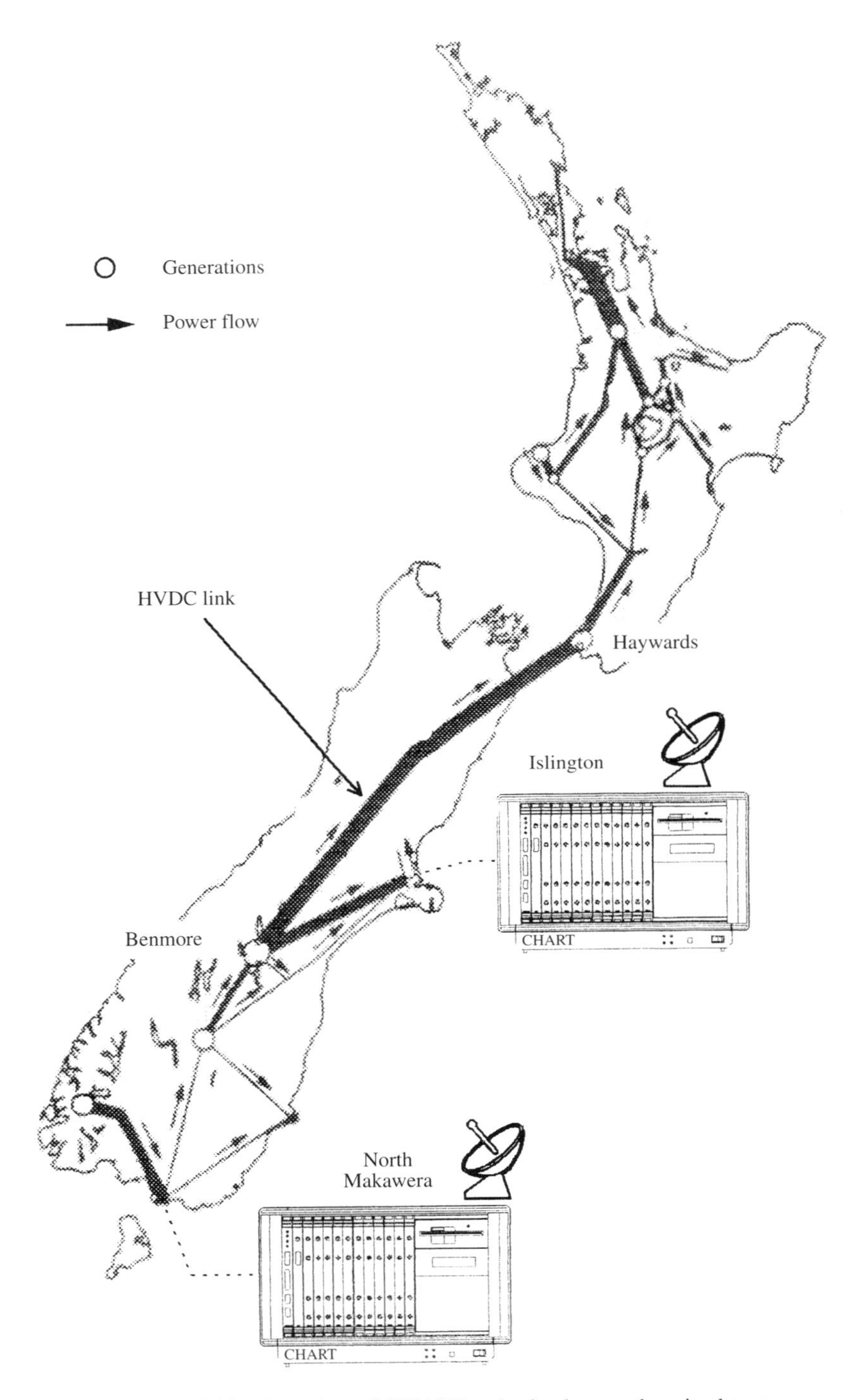

Figure 5.38 Location of CHART units in the synchronised test

At North Makawera, the main flow of power is between the generation at Manapouri and an aluminium smelter at Tiwai. The CHART unit was set up to monitor the current between these two places and the North Makawera substation. The connections from North Makawera to the rest of the 220 kV network were also monitored by measuring the current between North Makawera and 3 Mile Hill, and between North Makawera and Invercargill. Moreover, the current of transformer T1 feeding the 33 kV distribution network at North Makawera was monitored to determine whether there was any 5th harmonic source within the local load. The voltage distortion on the 33 kV distribution busbar was also recorded. However, due to the limited number of channels available on this particular CHART unit (12 channels), it was necessary to forgo some of the phases at several measurement points.

The main requirement of this test was to record the harmonic distortion at both substations simultaneously. The data samples were time-stamped to enable them to be matched in time. It was decided to average the harmonics over one second and to compute the mean, maximum and minimum harmonics over a minute throughout the entire measurement. These requirements call for special design of the DSM and DAPM applications in the CHART system described in Appendix II.

Figure 5.39 shows the operation of the DSM in this test. The Sample Rate Multiplier (SRM) of the DSM is initialised to generate 1024 sampling pulses per fundamental cycle. The mains frequencies at both substations are used as the reference for the fundamental frequency. The sampling pulses are gated at the P2 Interface and the DSM application decides when the pulses are to be dispatched onto the P2 time-stamping bus. The DSM application monitors the precision 1 pps (pulse per second) and the date and time information received from the GPS system. Through the parameter set-up process of the CHART virtual operating system, the DSM application is programmed to release the sampling pulses at a user-configurable time. The application can also be configured with a predefined stop sampling time. Upon the arrival of every 1 pps, which is also used to register the new date and time received from the GPS system, the DSM application compares the new time against the pre-defined start-sampling and stop-sampling time. If the new time falls between them, then the gate in the P2 interface is opened, releasing the sampling pulses onto the P2 bus. This method of making use of the precision 1 pps to initiate and terminate the sampling process ensures that the data are acquired at the same instant in both substations.

The DAPM application is programmed to operate on a cycle-by-cycle basis with 1024 samples designated as one cycle. Figure 5.40 shows the data processing carried out by the DAPM application in the synchronised test. This application operates over every 3000 cycles corresponding to one minute if the fundamental frequency remains constant at 50 Hz.

The 1024 samples per cycle of raw time-domain data are passed through a FIR filter with 8:1 decimation to reduce the number of samples to 128. Every 10 filtered cycles are averaged and the harmonic information is obtained by applying the FFT to the averaged cycle. This process results in 300 sets of harmonic packets and every five of them are again averaged to produce one-second harmonic averages. The mean, maximum and minimum harmonics over a minute are then computed from the 60 packets of one-second averages. The five averaged time-domain cycles used to compute the FFT are dispatched from the DAPM application alongside the one-second harmonic averages. This provides the option of analysing the time-domain waveform if there are doubts

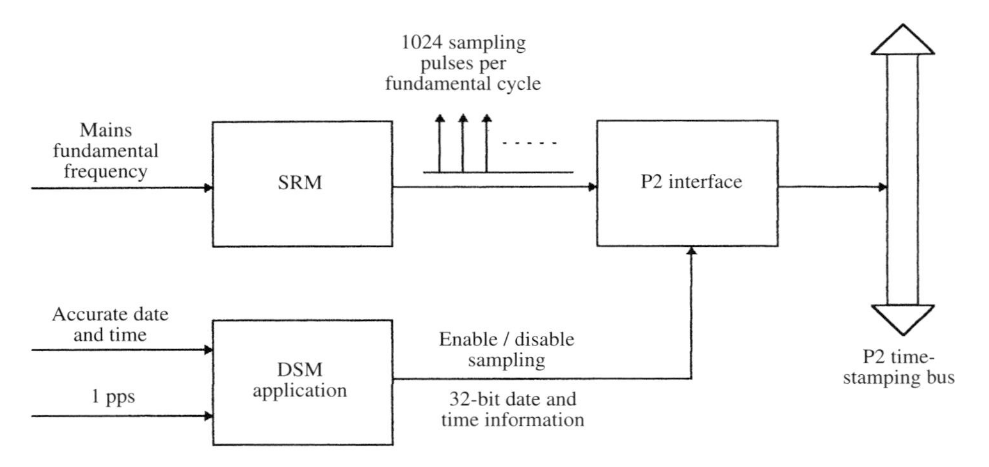

Figure 5.39 Operation of the Digital Services Module DSM in the synchronised test

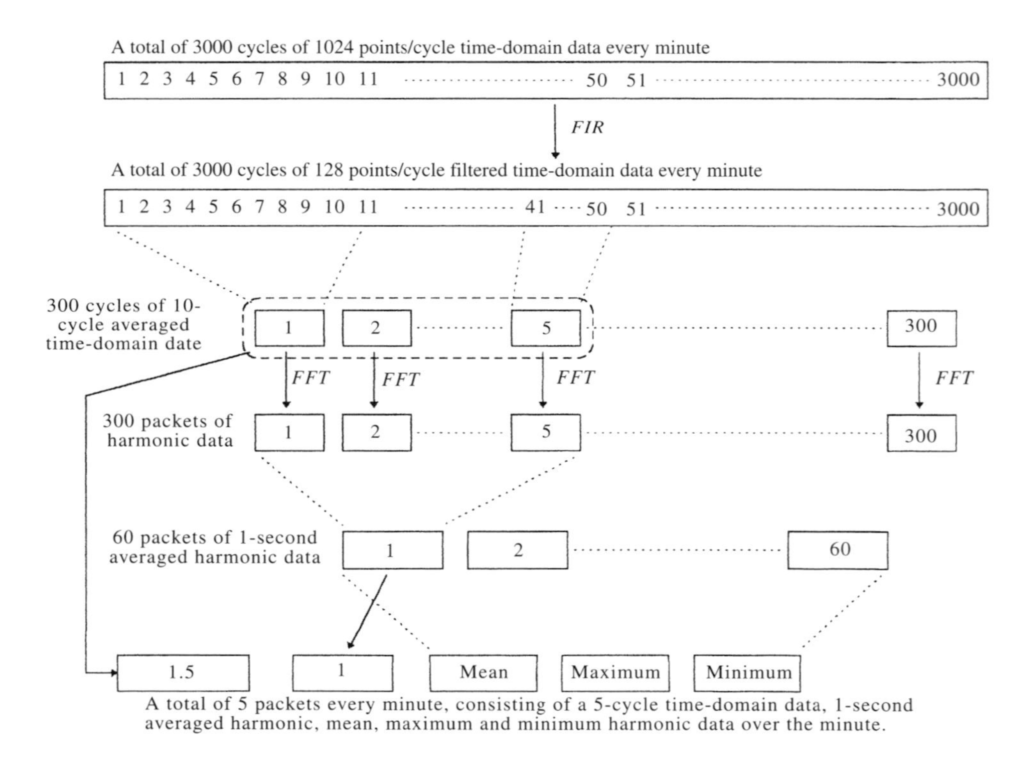

Figure 5.40 Data processing carried out by the DAPM in the synchronised test

about the computed harmonic information. The HUB processor of the CHART system was set up to store these five forms of data over the entire test.

Summary of Test Results The test was carried out over a period of 13 h with the two CHART units initiated simultaneously through the synchronisation mechanism built into the DSM application. The measurement period was chosen to coincide with the switching-out of the capacitors at the Islington substation which takes place during light load conditions. Although the one-second harmonic averages were computed every second, it was decided that the mean, maximum and minimum harmonics over a minute were sufficient and, hence, the one-second averages and the corresponding five-cycle averaged waveform are only recorded once every minute.

The sampling process is initiated by the 1 pps received from the GPS system. During the course of the measurement, the DSM locks the sampling pulses to the system mains frequency. Therefore, if the mains frequencies at both sites remain fairly similar to each other, the sampling processes in both CHART units are synchronised. The measured fundamental frequencies at Islington and North Makawera are shown in Figure 5.41. The fundamental frequencies are identical at both sites with similar deviations throughout the measurements. This ensured that the sampling processes were synchronised between the two CHART units.

A selection of acquired data is shown in Figure 5.42. A number of steps are observed on the 66 kV bus voltage which may be caused by the switching of the compensation capacitors. However, only three of these show concurrent changes in the 5th harmonic current flowing out of the Islington substation. The three cases are highlighted in the figure, together with the possible capacitor switching instants. The currents in the two transmission lines, Islington–Livingston and Islington–Timaru–Twizel, show an increase in the 5th harmonic current when the capacitors are switched out. A decrease in the 5th harmonic current is also observed when the capacitor is switched back into service towards the end of the test.

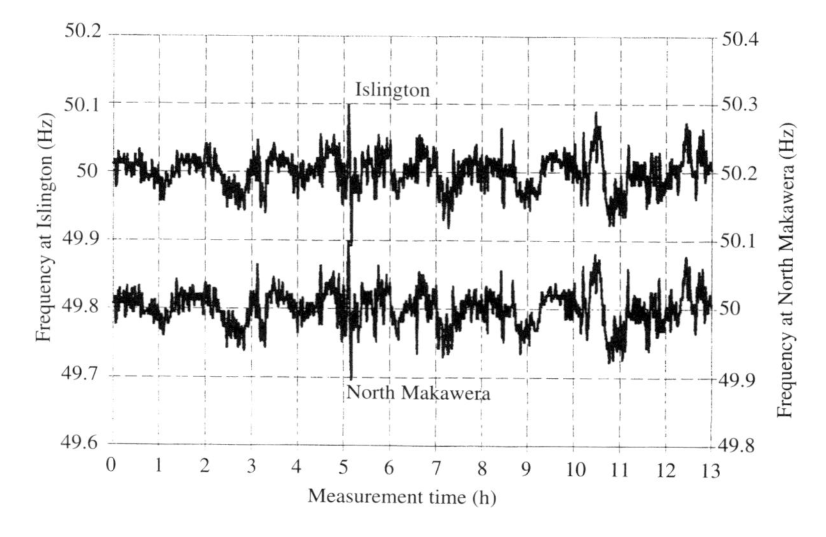

Figure 5.41 Fundamental frequency measured at Islington and North Makawera

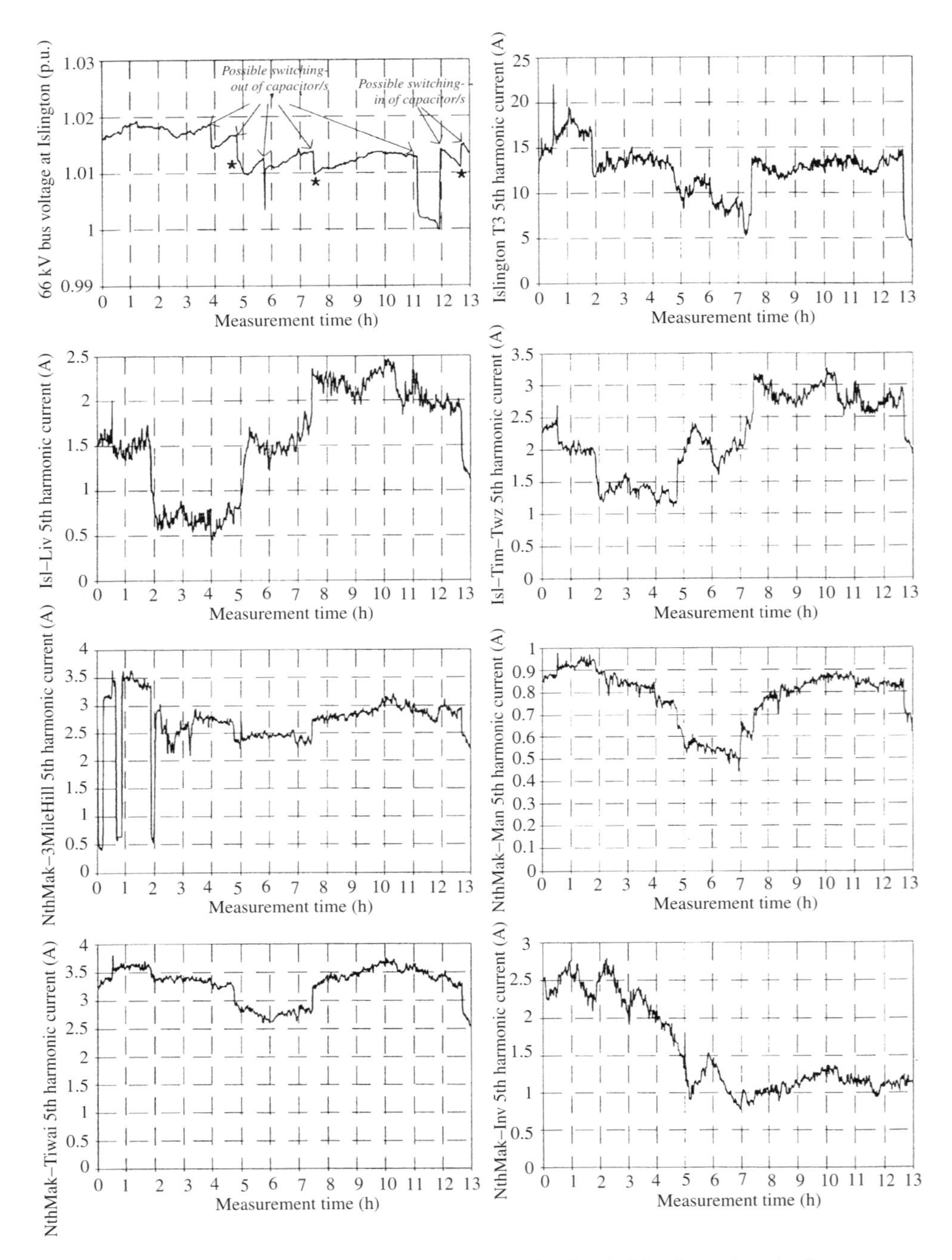

Figure 5.42 Selection of data acquired in the South Island synchronised test

The variations in the 5th harmonic current at North Makawera do not always correspond with the changes at Islington. Among the three aforementioned instants when there are changes to the 5th harmonic current flowing out of Islington, only the last two show corresponding changes in the distortion at North Makawera. At the second highlighted switching just after the 7th hour, when one of the capacitors is removed from service, the sudden increase in the 5th harmonic currents in the two outgoing lines from Islington coincides with similar increases in the lines between 3-Mile Hill, North Makawera and Tiwai. A similar coincidence is observed when one of the capacitors is put back into service near the end of the test. The decrease in the 5th harmonic current flowing out from the Islington 220 kV system coincides with the decreases in the 5th harmonic distortion around North Makawera.

These observations indicate that under certain operating conditions the switching of capacitors at the Islington substation can affect the 5th harmonic distortions at North Makawera. However, more detailed analyses of the system, in particular during the period when the measurement was carried out, will be required to finalise the above findings.

5.9.2 Group-Connected HV d.c. Converter Test

During a maintenance period at the Benmore HV d.c. converter station, the opportunity arose to test the possibility of operating the converter plant in the group connection mode, i.e. islanded from the South Island a.c. network and without filters on it's a.c. side, thus subjecting the generators to greater harmonic distortions than under normal operation conditions. Moreover, the fundamental frequency of the islanded network could deviate from the nominal 50 Hz depending on the d.c. load and the amount of generation.

Besides capturing the waveforms needed for the validation of computer models, this test provided the opportunity to illustrate some of the capabilities of the CHART instrumentation. These include the ability to perform alternative data processing tasks, the transparent handling of the different data format results from the tests, the ability to track the varying fundamental frequency in the islanded a.c. system and its use to ensure coherent sampling.

Figure 5.43 shows the islanded system at the Benmore converter station. Two CHART units were used to provide a total of 24 data channels. The measurements include the voltage at the 16 kV bus, generator currents, converter transformer currents, Pole 1B d.c. line voltage and current, and the firing angle of the Pole 1B converter. The main task of the test system was to capture the time-domain waveforms under different d.c. current and a.c. generation configurations. It was decided to capture two different forms of time-domain data, a small number of samples per fundamental cycle over a longer period for steady-state analysis and a handful of cycles of data with a much higher sampling rate for resolving the switching instants of the converter. Therefore, the instrument was programmed to gather the following two sets of time-domain data: 256 cycles of 128 sample points per fundamental cycle and 8 cycles with 1024 points per cycle. In addition to these two sets of data, their harmonic contents were also computed to be used for online real-time display and for post-processing the data after the test.

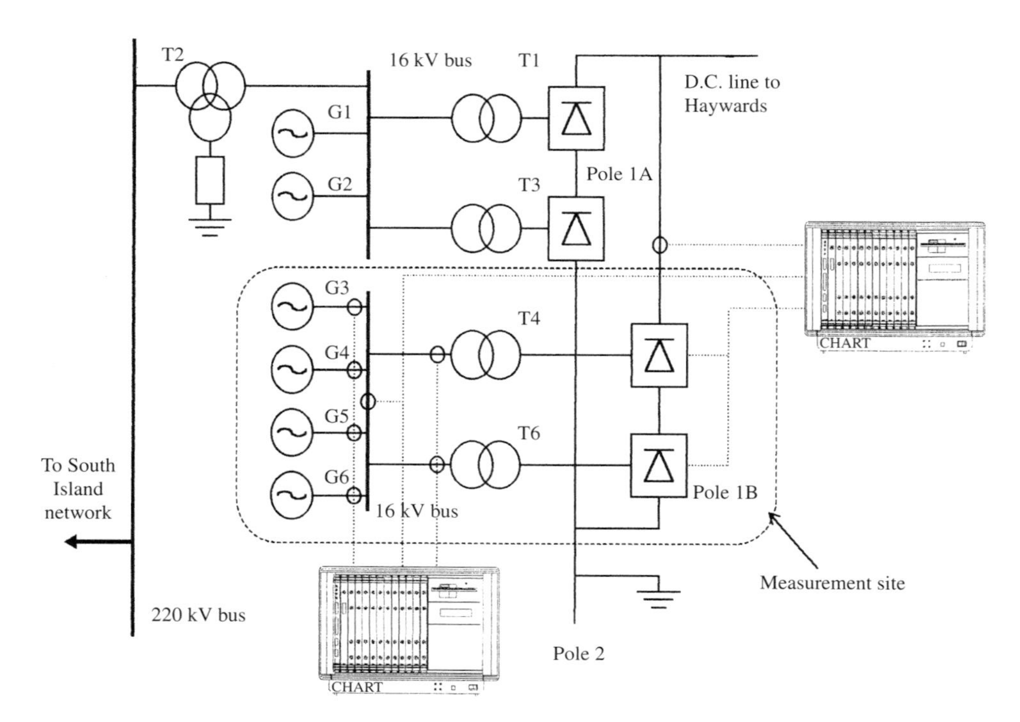

Figure 5.43 CHART measurement points in the Benmore test

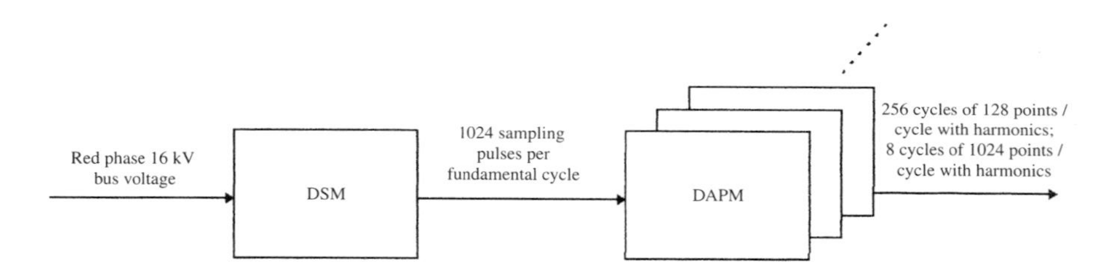

Figure 5.44 Setup of DSM and DAPM application software for the Benmore test

The two main parts of CHART that need to be addressed in this test are the DSM application and the DAPM application. Figure 5.44 shows the set-up of the DSM and DAPM in this test. The DSM application, designed for the synchronised test described earlier, is also used in this test.

The DSM is programmed to generate 1024 sampling pulses per fundamental cycle with the fundamental frequency referenced from the red phase voltage of the 16 kV busbar. Adjustments are made if the fundamental frequency drifts away from its previously measured values so as to maintain exactly 1024 pulses within a single fundamental cycle throughout the measurement.

The DAPM application software is designed to operate on a cycle-by-cycle basis with 1024 samples designated as one cycle. Figure 5.45 shows how the DAPM application processes and generates the wanted data as outlined above. The numbers contained

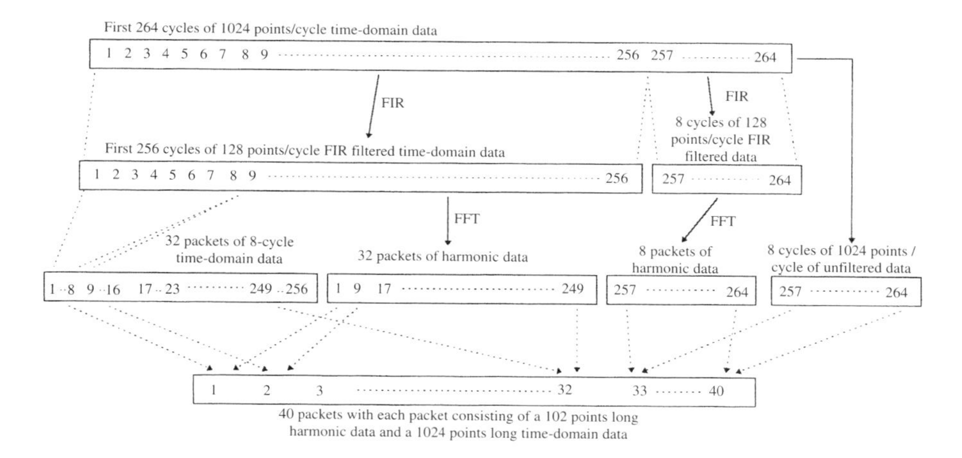

Figure 5.45 Data processing carried out by the DAPM in the Benmore test

in the rectangular boxes refer to a fundamental cycle (1024 raw samples) and the computation part of the application is activated at the end of each cycle. Only the first 264 cycles of data are used to produce the wanted data and the cycle number is reset back to 1 when it reaches 3000, which corresponds to nearly one minute if the fundamental frequency remains at about 50 Hz.

The first 256 cycles of time-domain data are passed through a FIR filter with 8:1 decimation reducing the number of samples per cycle from 1024 to 128 points. The resultant 256 cycles of 128-point time-domain data are packed in groups of 8 cycles producing a total of 32 packets. Furthermore, every 8th cycle (1st, 9th, 17th, etc.) the samples are Fourier transformed to extract their harmonic contents. The resultant harmonic packet is 102 points in length, the first 51 points taken by the magnitude of the harmonics from d.c. to the 50th, and the rest by the phase angles. Each of these harmonic packets, as well as the corresponding 8-cycle time-domain data, were used for the verification of the computer simulation models.

The second set of data is simply the raw 1024 points per cycle time-domain samples. However, the harmonic content of this data may be useful for future analysis and thus cycles 257 to 264 are also FIR filtered and Fourier transformed to obtain the harmonics. The computed harmonic information is combined with the corresponding single-cycle time-domain data to form the second set.

The acquired data has to be sampled simultaneously in all channels to be useful for verifying the simulation models. This is achieved by utilising the DSM as the sole source of sampling pulses in each CHART unit and a synchronisation process incorporating a GPS system is employed in each unit to link the operation of the two DSMs together. The DSM application allows the dispatching of the sampling pulses to be deferred until a pre-defined time is reached. An identical starting time is set up in both DSMs and with similar time information received by the DSMs from the accurate GPS satellite system, the first sampling pulse is issued at the same instant in both units. The same red phase 16 kV bus voltage is fed into both units to provide a common reference for the fundamental frequency, thereby ensuring that the sampling process in both units takes place concurrently.

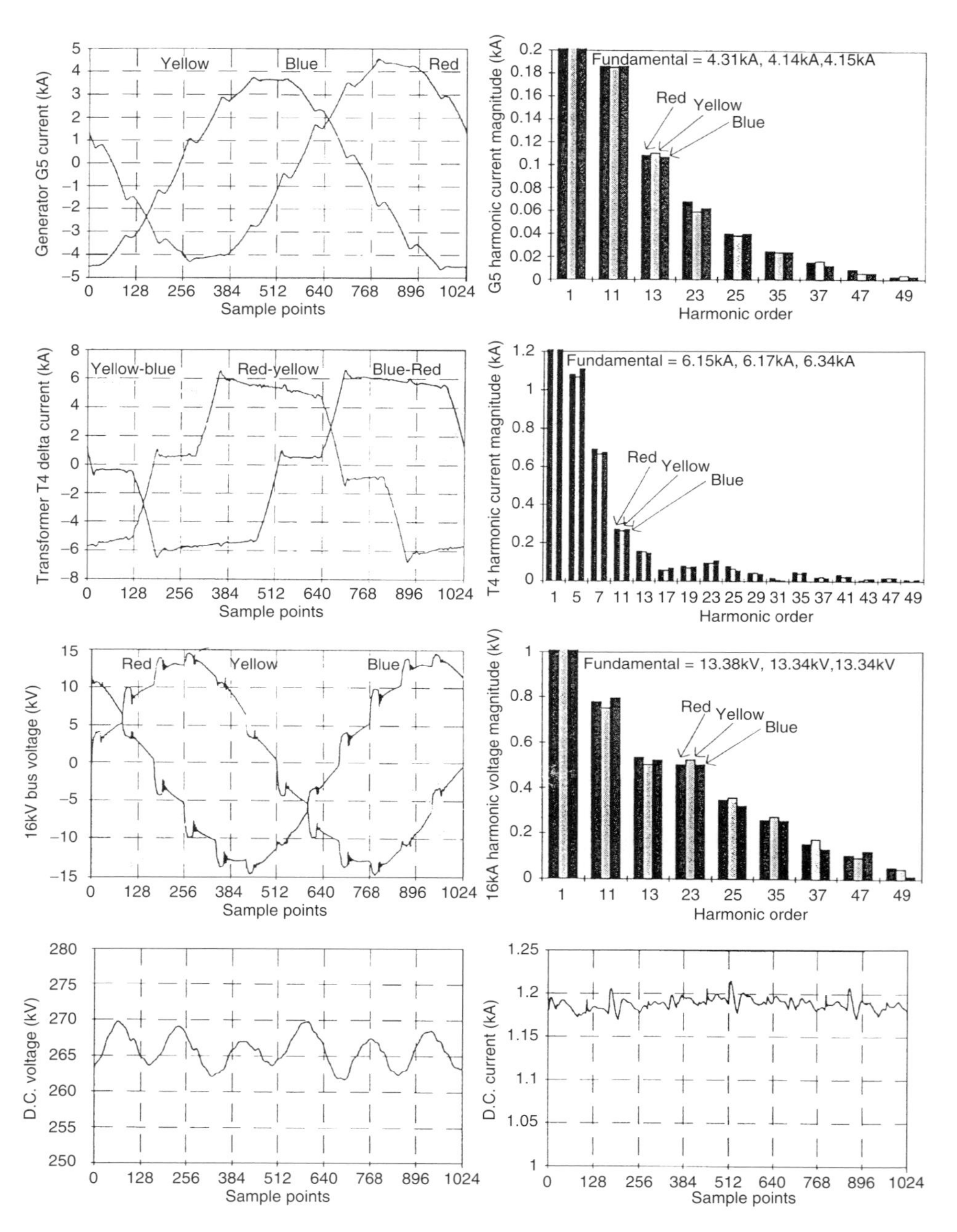

Figure 5.46 Selection of data acquired in the Benmore islanded converter test

Summary of Test Results The test was carried out over a five-hour period with the d.c. current and a.c. generation settings varied approximately every 30 minutes. This allowed several sets of data to be collected for verifying the computer simulation models under different operating conditions.

Although the DAPM application is capable of generating the intended data every 3000 cycles, approximately every minute, it was decided that there would be too much redundant information. Instead, the DAPM application was set up to generate the data every 9000 cycles, corresponding to approximately every three minutes. This would produce at least three sets of data for each d.c. current and a.c. generation setting, while allowing 10–15 min for the system to settle to a steady state after each configuration change.

Changes in the d.c. current and a.c. generation were achieved smoothly, with the Pole 1A converter covering any shortfall as a result of the variations in the Pole 1B converter. A total of over 200 Mbytes of data was recorded providing more than 24 sets of usable data with eight different d.c. current and a.c. generation settings.

Figure 5.46 shows a selection of recorded data over a fundamental cycle at several measurement points, under a particular configuration. The generator G5 current and the voltage waveforms contain a substantial amount of characteristic harmonics as expected in a scheme without harmonic filters. The T4 transformer delta current shows the presence of the expected harmonics. Lastly, the presence of a 6th harmonic distortion is clearly evident in the d.c. voltage waveform, as well as a considerable distortion in the d.c. current.

5.10 Summary

The main components of power system waveform monitoring, i.e. transducers and instrumentation, have been critically reviewed for their ability to assess power quality. In particular, the inadequacy of conventional capacitor voltage transformers has been considered and possible alternatives to improve the transducers' response at harmonic frequencies have been discussed.

As the only practical alternative technology, most of the chapter has been devoted to the implementation of digital processing in power quality instruments and, in particular, to the FFT. Both hard and software system requirements have been described for use in transient events characterisation as well as in real-time waveform processing. Single busbar and system-wide assessment have been discussed, the latter requiring perfect sampling synchronisation at geographically separated buses. The adaptation of digital instruments to voltage flicker assessment has also been described.

Finally, examples of local and system-wide field test monitoring, using advanced digital instrumentation, have been included to illustrate their capability in the real environment.

5.11 References

1. Douglass, D A, (1981). Current transformer accuracy with asymmetric and high frequency fault currents, *IEEE Transactions*, **PAS-100**, pp. 1006–1012.
2. Malewski, R and Douville, J, (1976). *Measuring Properties of Voltage and Current Transformers for the Higher Harmonics Frequencies*, Paper presented at the Canadian Communication and Power Conference, Montreal.
3. Mouton, L, Stalewski, A and Bullo, P, (1978). Non-conventional current and voltage transformers, *Electra*, **59**, pp. 91–122.

4. Rogers, A J, (1979). Optical measurement of current and voltage on power system, *IEE Conference Publication 174*, pp. 22–26.
5. Douglass, D A, (1981). Potential transformer accuracy at 60 Hz voltages above and below rating and at frequencies about 60 Hz, *IEEE Transactions* **PAS-100**, pp. 1370–1375.
6. CIGRE Working Group 36–05 (Disturbing loads), (1981). Harmonics characteristics parameters, methods of study, estimating of existing values in the network, *Electra*, **77**, pp. 35–54.
7. Lisser, J, (1976). High accuracy with amplifier type voltage transformer, *Electrical Review*, **198**(21), pp. 31–33.
8. Gray, F M, Hughes, M A and Stalewsky, A, (1975). Monitoring of transmission line voltage for protective relaying purposes using capacitive dividers, *IEE Conference Publication 125*, pp 214–221.
9. Dewe, M B, Arrillaga, J and Arnold, C P, (1995). Accurate harmonic assessment of randomly varying non-linear loads, *CIGRE Study Committee 36 (Power system electromagnetic compatibility) Colloquium and Meeting*, Foz Do Iguacu, Brazil.
10. Miller, A J V and Dewe, M B, (1993). The application of multi-rate digital signal processing techniques to the measurement of power system harmonic levels, *IEEE Transactions on Power Delivery*, **8**(2), pp. 531–539.
11. Chen, S and Dewe, M B, (1996). System and hardware considerations of a flexible multi-channel real time data acquisition system for power quality, *IEEE Conference on Harmonics and Quality of Power* (*ICHQP*). Las Vegas, pp. 189–195.
12. Chen, S and Dewe, M B, (1996). Virtual operating system design *via* flexible multi-channel real time data acquisition system for power quality assessment, *IEEE Conference on Harmonics and Quality of Power* (*ICHQP*). Las Vegas, pp. 182–188.
13. Miller, A J V, Lake, C B and Dewe, M B, (1990). Multichannel real-time harmonic analysis using the Intel Multibus II bus architecture, *Proceedings of the 7th International Intel Real-Time Users Group Conference*, St. Louis, Missouri, USA, pp. 11–24.
14. Turpin, Brad, (1998). Advanced synchronisation techniques for data acquisition, *National Instruments Application Note 128.*
15. Working group H-7 of the Relaying Channels Subcommittee of the IEEE Power System Relaying Committee (Phadke A.G.–Chairman), (1994). Synchronised sampling and phasor measurements for relaying and control, *IEEE Transactions on Power Delivery*, **9**(1), pp. 442–452.
16. Burnett, R O, Jr., Butts, M M, Cease, T W, Centeno, V, Michel, G, Murphy, R J and Phadke, A G, (1994). Synchronised phasor measurements of a power system event, *IEEE Transactions on Power Systems*, **9**(3), pp. 1643–1650.
17. The Electricity Division of the Ministry of Energy, (1983). Limitation of harmonic levels, Issue 2, New Zealand.
18. Lake, C B, Dewe, M B, Miller, A J V, Shurety, M R, Cusdin, M J and Arrillaga, J, (1990). Multi-channel continuous real time harmonic monitoring, *Proceedings of the 4th International Conference on Harmonics in Power Systems* (*ICHPS*), Budapest, Hungary, pp. 424–430.
19. Brigham, E O, (1974). *The Fast Fourier Transform*, Prentice-Hall Inc.
20. Crochiere, R E and Rabiner, L R, (1983). *Multirate Digital Signal Processing*, Prentice-Hall Inc.
21. IEC 61000-4-7, (1991). Testing and measurement techniques — general guide on harmonics and interharmonics measurements and instrumentation for power supply systems and equipment connected thereto.
22. Barker, P P, Mancao, R T, Kvaltine, D J and Parrish, D E, (1993). Characteristics of lightning surges measured at metal oxide distribution arresters, *IEEE Transactions on Power Delivery*, **8**(1).
23. Hydro-Quebec, (1996). Power quality measurement protocol; CEA guide to performing power quality surveys, Document 220 D 711.

24. Poisson, O, Rioual, P and Meunier, M, (1998). Detection and measurement of power quality disturbances using wavelet transform, *IEEE Conference on Harmonics and Quality of Power (ICHQP)* Athens, pp. 1125–1130.
25. Savary, P and Bechler, A, (1994). L'indicateur de Qualité de Fourniture, a simple and cheap device for wide quality measurement, Paper E-1.07, *PQA 94*, Amsterdam.
26. UIE, (1991). Flicker measurement and evaluation, *Union Internationale d'Electrothermie*.
27. Sakulin, M, Renner, H, Bergeron, R, Key, T and Nastasi, D, (1996). International recommendation for universal use of the UIE/IEC flickermeter, *UIE Congress*, Birmingham.
28. Keppler, T, (1998). Flicker measurement and propagation in power systems, Ph.D. thesis, University of Canterbury, New Zealand.
29. Mombauer, W, (1990). Flicker caused by interharmonics, *etz Archiv.*, **12**(12), pp. 391–396.
30. IEC, (1986). Flickermeter —functional and design specification.
31. Perl J and Scharf, L L, (1977). Covariance–invariant digital filtering, *IEEE Transactions on Acoustics, Speech and Signal Processing*, **25**(2), pp. 143–151.
32. Sorensen, H V and Burrus, C S, (1993). Efficient computation of the DFT with only a subset of input or output points, *IEEE Transactions on Signal Processing*, **41**(3), pp. 1184–1200.
33. Manitoba HVDC Research Centre, (1994). PSCAD/EMTDC power system simulation software tutorial manual.

6

EVALUATION OF POWER SYSTEM HARMONIC DISTORTION

6.1 Introduction

When the assessment of power quality is done purely by measurements the information obtained is limited to the monitoring points and therefore deriving system-wide information becomes an expensive exercise. For more effective assessment the measurements should be complemented with as much information as possible from the system configuration and operating conditions. In this respect, the network configuration, including generation, transmission and a number of load equivalent impedances, are readily available. Also, the system state estimation provides fundamental frequency information at specified instants of time.

It will be shown in Chapter 7 that by combining a number of strategically placed monitoring channels with network simulation, it is possible to derive system-wide harmonic information. Thus, as a preliminary step to harmonic state estimation, this chapter describes the network simulation techniques applicable to harmonic frequencies. This is often named harmonic flow or harmonic penetration.

The term *harmonic* used here is not restricted to integer multiples of the fundamental frequencies. The analysis and algorithms described are equally applicable to other frequencies in the region of interest such as interharmonics and subharmonics.

The simplest harmonic flow involves a single harmonic source and single-phase network analysis. This model is commonly used to derive the system harmonic impedances at the point of common coupling in filter design. In general, however, the network will be unbalanced and may contain several harmonic sources. Therefore, the derivation of the harmonic voltages and currents requires multi-source three-phase harmonic analysis. If the harmonic content injected by a non-linear component is independent of the distortion level in the a.c. system, calculation of the harmonic sources can be decoupled from the analysis of harmonic penetration and a direct (nodal) solution is possible. Since most non-linearities manifest themselves as harmonic current sources, this is normally called the current injection method. In such cases, the expected voltage levels or the results of a fundamental frequency load flow are used to derive the current waveforms, and thus the harmonic content, of the non-linear components.

Following a description of the direct harmonic analysis algorithm, the following sections discuss the modelling of network components and the formulation of the nodal admittance matrix, as well as the computer implementation of the harmonic flow algorithm.

6.2 Direct Harmonic Analysis

The distribution of voltage and current harmonics throughout a linear power network containing one or more harmonic current sources is normally carried out using nodal analysis [1]. The asymmetry inherent in transmission systems cannot be studied with any simplification by using the symmetrical component frame of reference; therefore, phase components are used.

The nodal admittance matrix of the network at a frequency f is of the form:

$$[Y_f] = \begin{bmatrix} Y_{11} & Y_{12} & \cdots & Y_{1i} & \cdots & Y_{1k} & \cdots & Y_{1n} \\ Y_{21} & Y_{22} & \cdots & Y_{2i} & \cdots & Y_{2k} & \cdots & Y_{2n} \\ \vdots & \vdots & \ddots & \vdots & \ddots & \vdots & \ddots & \\ Y_{i1} & Y_{i2} & \cdots & Y_{ii} & \cdots & Y_{ik} & \cdots & Y_{in} \\ \vdots & \vdots & \ddots & \vdots & \ddots & \vdots & \ddots & \\ Y_{k1} & Y_{k2} & \cdots & Y_{ki} & \cdots & Y_{kk} & \cdots & Y_{kn} \\ \vdots & \vdots & \ddots & \vdots & \ddots & \vdots & \ddots & \\ Y_{n1} & Y_{n2} & \cdots & Y_{ni} & \cdots & Y_{nk} & \cdots & Y_{nn} \end{bmatrix}. \quad (6.1)$$

where Y_{ki} = mutual admittance between busbars k and i at frequency f, and Y_{ii} = self-admittance busbar i at frequency f.

A separate system admittance matrix is generated for each frequency of interest. The main difficulty is to determine which model best represents the various system components at the required frequency and obtain appropriate parameters for them. With this information, it is straightforward to build up the system fundamental and harmonic frequency admittance matrices.

The three-phase nature of the power system always results in some load or transmission line asymmetry, as well as circuit coupling. These effects give rise to unbalanced self and mutual admittances of the network elements.

For the three-phase system, the elements of the admittance matrix are themselves 3×3 matrices consisting of self and transfer admittances between phases, i.e.

$$Y_{ii} = \begin{bmatrix} Y_{\text{aa}} & Y_{\text{ab}} & Y_{\text{ac}} \\ Y_{\text{ba}} & Y_{\text{bb}} & Y_{\text{bc}} \\ Y_{\text{ca}} & Y_{\text{cb}} & Y_{\text{cc}} \end{bmatrix}. \quad (6.2)$$

Figure 6.1 shows a case of two three-phase harmonic sources and an unbalanced a.c. system. The current injections, i.e. I_{1h}–I_{3h} and I_{4h}–I_{6h}, can be unbalanced in magnitude and phase angle.

The system harmonic voltages are calculated by direct solution of the linear equation

$$[I_h] = [Y_h][V_h], \qquad \text{for } h \neq 1, \quad (6.3)$$

where $[Y_h]$ is the system admittance matrix.

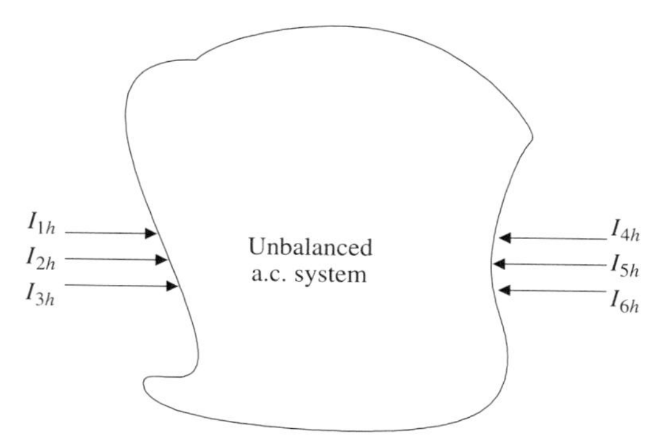

Figure 6.1 Unbalanced current injection into an unbalanced a.c. system

Therefore, the direct solution involves $h - 1$ independent sets of linear simultaneous equations, i.e.

$$\begin{aligned} [I_h] &= [Y_h]\ [V_h] \\ \vdots\ \ &\qquad \vdots \qquad \vdots \\ [I_3] &= [Y_3]\ [V_3] \\ [I_2] &= [Y_2]\ [V_2]. \end{aligned} \tag{6.4}$$

The injected currents at most a.c. busbars will be zero, since the sources of the harmonic considered are the non-linear devices. To calculate an admittance matrix for the reduced portion of a system comprising of just the injection busbars, the admittance matrix is formed with those buses at which harmonic injection occurs ordered last. Advantage is taken of the symmetry and sparsity of the admittance matrix [2], by using a row-ordering technique to reduce the amount of off-diagonal element build-up. The matrix is triangulated using Gaussian elimination, down to, but excluding, the rows of the specified buses.

The resulting matrix equation for an n-node system with $n - j + 1$ injection points is

$$\begin{bmatrix} 0 \\ \bullet \\ \bullet \\ \bullet \\ 0 \\ \hline I_j \\ \bullet \\ \bullet \\ \bullet \\ I_n \end{bmatrix} = \begin{bmatrix} \ddots & & 0 & \\ \hline 0 & Y_{jj} & \cdots & Y_{jn} \\ & \vdots & & \vdots \\ & Y_{nj} & \cdots & Y_{nm} \end{bmatrix} \bullet \begin{bmatrix} V_1 \\ \bullet \\ \bullet \\ \bullet \\ V_{j-1} \\ \hline V_j \\ \bullet \\ \bullet \\ \bullet \\ V_n \end{bmatrix} \tag{6.5}$$

As a consequence, $I_j \cdots I_n$ remain unchanged since the currents above these in the current vector are zero. The reduced matrix equation is

$$\begin{bmatrix} I_j \\ \vdots \\ I_n \end{bmatrix} = \begin{bmatrix} Y_{jj} & \cdots & Y_{jn} \\ \vdots & \cdots & \vdots \\ Y_{nj} & \cdots & Y_{nn} \end{bmatrix} \cdot \begin{bmatrix} V_j \\ \vdots \\ V_n \end{bmatrix} \tag{6.6}$$

and the order of the admittance matrix is three times the number of injection busbars for a three-phase system. The elements are the self and transfer admittances of the reduced system as viewed from the injection busbars. Whenever required, the impedance matrix may be obtained for the reduced system by matrix inversion.

Reducing a system to provide an equivalent admittance matrix, as viewed from a specific bus, is an essential part of filter design. The reduction of the admittance matrix to a set of busbars where non-linearities exist is an essential step to allow accurate a.c. system representation in many other types of analysis, such as iterative harmonic analysis or as frequency-dependent equivalents in time-domain analysis.

6.2.1 Incorporation of Harmonic Voltage Sources

Most power system non-linearities manifest themselves as harmonic current sources, but sometimes harmonic voltage sources are used to represent the distortion background present in the network prior to the installation of the new non-linear load. Moreover, the latest generation of power electronic devices being applied to transmission and distribution systems use Gate Turn-off (GTO) and Insulated Gate Bipolar Transistor (IGBT) devices, and act as a voltage source behind an impedance.

A system containing harmonic voltages at some busbars and harmonic current injections at other busbars is solved by partitioning the admittance matrix and performing a partial inversion. This then allows the unknown busbar voltages and unknown harmonic currents to be found. If V_2 represents the known voltage sources, then I_2 are the unknown variables. The remaining busbars are represented as a harmonic current injection I_1 (which can be either zero or specified by an harmonic current source) and the corresponding harmonic voltage vector V_1 represents the unknown variables.

Partitioning the matrix equation to separate the two types of nodes gives

$$\begin{bmatrix} Y_{11} & Y_{12} \\ Y_{21} & Y_{22} \end{bmatrix} \begin{bmatrix} V_1 \\ V_2 \end{bmatrix} = \begin{bmatrix} I_1 \\ I_2 \end{bmatrix}. \tag{6.7}$$

The unknown voltage vector V_1 is found by solving

$$[Y_{11}][V_1] = [I_1] - [Y_{12}][V_2]. \tag{6.8}$$

The harmonic currents injected by the harmonic voltage sources are then found by solving

$$[Y_{21}][V_1] + [Y_{22}][V_2] = [I_2]. \tag{6.9}$$

With this formulation some extra processing is required to obtain the reduced admittance matrix, which is not generated as part of the solution.

6.2.2 Cascading Sections

Mutually coupled transmission lines with different tower geometries over the line length need special consideration.If only terminal voltage information is required, the line sections may be combined into one equivalent section using *ABCD* parameters see Figure 6.2.

All the individual sections must contain the same number of mutually coupled three-phase elements to ensure that the parameter matrices are of the same order and that matrix multiplications are executable. In this respect, uncoupled sections will use the coupled format with zero coupling elements to maintain the correct dimensions.

For the case of a non-homogeneous line with n different sections:

$$\begin{bmatrix} V_S \\ I_S \end{bmatrix} = \begin{bmatrix} [A_1] & [B_1] \\ [C_1] & [D_1] \end{bmatrix} \times \begin{bmatrix} [A_2] & [B_2] \\ [C_2] & [D_2] \end{bmatrix} \times \cdots \times \begin{bmatrix} [A_n] & [B_n] \\ [C_n] & [D_n] \end{bmatrix} \begin{bmatrix} V_R \\ -I_R \end{bmatrix},$$

$$\begin{bmatrix} V_S \\ I_S \end{bmatrix} = \begin{bmatrix} [A] & [B] \\ [C] & [D] \end{bmatrix} \begin{bmatrix} V_R \\ -I_R \end{bmatrix}. \tag{6.10}$$

It must be noted that in general $[A] \neq [D]$ for a non-homogeneous line.

Once the resultant *ABCD* parameters have been found, the equivalent nodal admittance matrix for the subsystem can be calculated from

$$[Y] = \begin{bmatrix} [D]\,[B]^{-1} & [C] - [D]\,[B]^{-1}\,[A] \\ [B]^{-1} & -[B]^{-1}\,[A] \end{bmatrix}. \tag{6.11}$$

If, however, extra information along the line is required, rather than creating fictitious nodes and increasing the harmonic analysis computation, post-processing can be performed (see Section 6.6).

$$\begin{bmatrix} I_S \\ I_1 \\ I_2 \\ \vdots \\ I_n \\ I_R \end{bmatrix}_h = \begin{bmatrix} [Y_{SS}] & -[Y_{S1}] & & & & \\ -[Y_{1S}] & [Y_{SS}]+[Y_{11}] & -[Y_{12}] & & & \\ & -[Y_{21}] & [Y_{11}]+[Y_{22}] & \ddots & & \\ & & \ddots & \ddots & \ddots & \\ & & & \ddots & [Y_{nn}]+[Y_{RR}] & -[Y_{nR}] \\ & & & & -[Y_{Rn}] & [Y_{RR}] \end{bmatrix}_h \begin{bmatrix} V_S \\ V_1 \\ V_2 \\ \vdots \\ V_n \\ V_R \end{bmatrix}_h \tag{6.12}$$

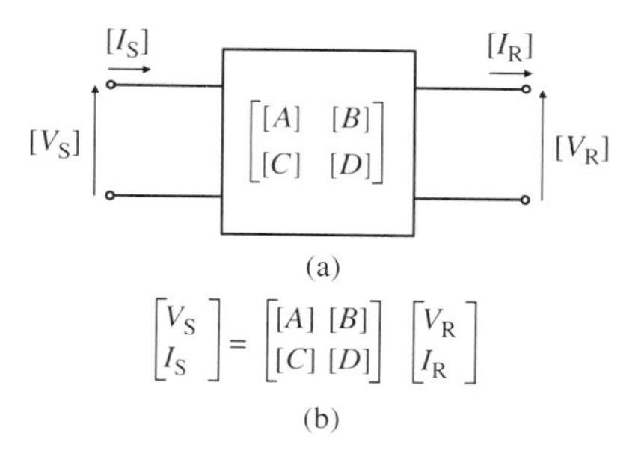

$$\begin{bmatrix} V_S \\ I_S \end{bmatrix} = \begin{bmatrix} [A] & [B] \\ [C] & [D] \end{bmatrix} \begin{bmatrix} V_R \\ I_R \end{bmatrix}$$

(b)

Figure 6.2 Two-port network transmission parameters: (a) multi-two-port network; (b) matrix transmission parameters

6.3 Experimental Derivation of the Network Harmonic Impedances

In the absence of more accurate information, existing harmonic standards and recommendations often refer to harmonic impedance sources derived from the balanced short-circuit impedance of the system at fundamental frequency. The inadequacy of such an approach is becoming apparent with the help of on-line tests and computer studies.

The availability of voltage and current transducers throughout the power system provides the basis for indirect derivation of harmonic impedances. Their assessment is therefore dependent on the performance of such transducers, which may not have been designed to respond accurately to harmonic frequencies.

The present techniques for the derivation of harmonic impedances can be divided into three groups, depending on the origin of harmonics or interharmonics.

1 Use of existing harmonics sources (non-invasive measurements),
2 Direct injection (invasive measurements),
3 Analysis of transient waveforms (non-invasive measurements).

6.3.1 Use of Existing Sources (On-line Non-invasive Tests)

In the non-invasive test, the information required is obtained purely from measurements of existing waveforms, i.e. using the harmonic content already present in the system. This is the simplest and most commonly used technique.

Electricité de France [3] has proposed two alternative ways of deriving information on harmonic impedances, in the form of sequence components, using the principle of digital filtering rather than Fourier analysis. One method uses numerical techniques to perform the digital filtering of the physical input values, and retains only the frequencies contained in a selected bandwidth. The information is then used to identify the network with a simple impedance model which is only valid over a relatively narrow frequency range. The identification is carried out separately for the zero sequence and positive (negative) sequence values. The second method uses electronic filters to transform the six voltage and current values into four (two zero sequence and two positive sequence), after eliminating the 50 Hz component.

The harmonic content produced by an existing high-voltage d.c. converter station has been used [4] to obtain the harmonic impedances directly from the ratios of voltage

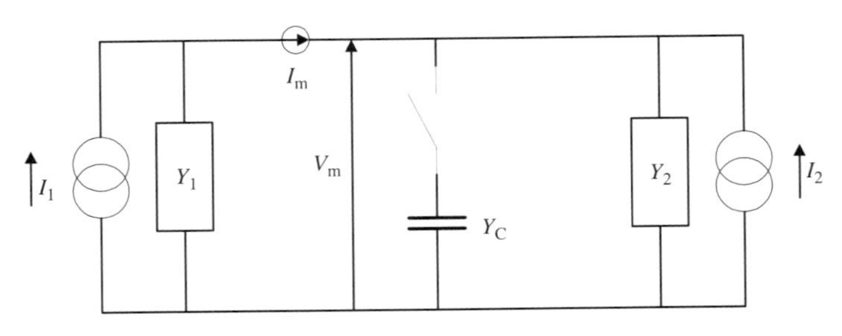

Figure 6.3 Identification of the main harmonic source by capacitor switching

and current readings. This assumes that all other important harmonic sources in the power system are disconnected; however, measurements taken without the existing high-voltage d.c. converter station operating, or operating at a different operating point, could have been used to account for the other harmonic sources in the system.

When the harmonic voltage levels are high and affected by the connection of a shunt capacitor bank, the method illustrated in Figure 6.3 has been used to identify the main source of the harmonics.

The harmonic current injection and harmonic impedances (for the Norton equivalents) are obtained by solving the linear set of equations

$$\begin{bmatrix} 1 & 0 & -V'_{\mathrm{m}} & 0 \\ 0 & -1 & 0 & V'_{\mathrm{m}} \\ 1 & 0 & -V''_{\mathrm{m}} & 0 \\ 0 & -1 & 0 & V''_{\mathrm{m}} \end{bmatrix} \begin{bmatrix} I_1 \\ I_2 \\ Y_1 \\ Y_2 \end{bmatrix} = \begin{bmatrix} I'_{\mathrm{m}} \\ I'_{\mathrm{m}} \\ I''_{\mathrm{m}} \\ I''_{\mathrm{m}} - Y_c V''_{\mathrm{m}} \end{bmatrix}, \tag{6.13}$$

where V'_{m} and I'_{m} are the measured harmonic voltage and current prior to connecting the capacitor and V''_{m} and I''_{m} after.

It must be pointed out that this method assumes that the Norton equivalents are not affected by the switching operation.

6.3.2 Direct Injection (On-line Invasive Tests)

In principle, it is possible to design a source of harmonic power which absorbs fundamental frequency power and converts it into harmonic power of the appropriate frequencies, magnitudes and phases. However, the measurement of harmonic impedance requires a distorting source of considerable capacity.

In practice, measurements are usually taken at points where there is considerable distortion due to existing harmonic sources in the network. The effect of injecting a further source needed for harmonic measurement is superimposed with these. The result is a low signal-to-noise ratio, which makes the measurement unreliable.

To overcome this problem, the Electricity Council Centre (Capenhurst, UK) [5] have designed a system which generates power at frequencies midway between the characteristic harmonic frequencies of interest, i.e. at the odd multiples of 25 Hz, on the assumption that interpolation between these frequencies is justifiable. The power ratings of the distorting source for measurements at 11 kV, 33 kV and 132 kV are 9 kW, 36 kW and 180 kW, respectively, and the units consist of a switching modulator in series with a resistive load in the form of a fan heater. The 11 kV and 33 kV systems are portable and the 132 kV system is located in a purpose-built van.

Figure 6.4 illustrates the results of typical measurements carried out with this equipment in combination with a harmonic impedance instrument [6], specially designed to provide simultaneous information of voltage and current with their phase relationship.

6.3.3 Analysis of Transient Waveforms (On-line Non-invasive Tests)

The methods discussed above involve measurement of the steady-state levels of particular frequencies. Alternatively, the network frequency characteristics can be derived

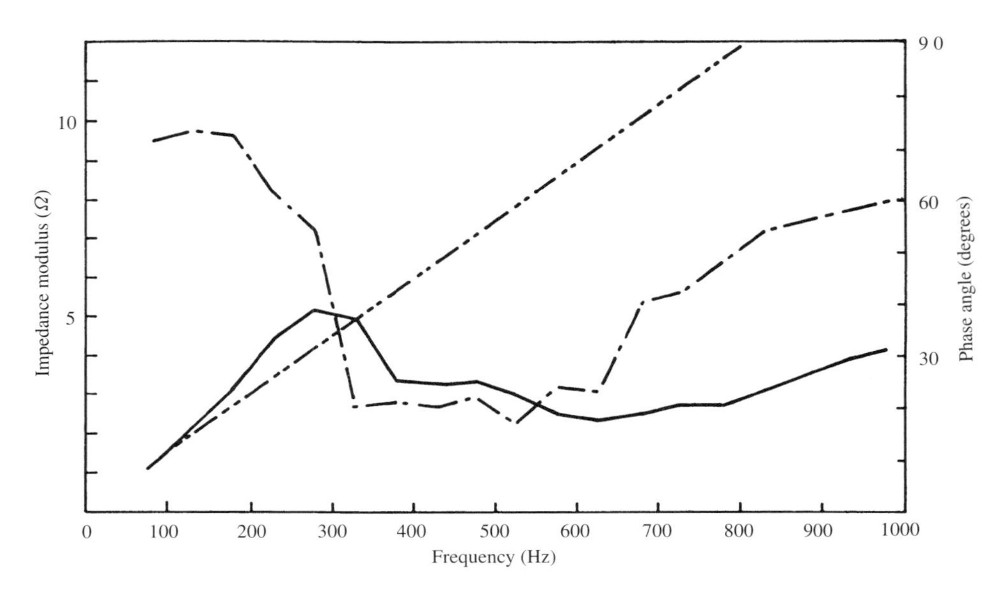

Figure 6.4 Source impedance seen from an 11 kV busbar supplying a commercial load: —‐‐‐—, obtained from the short-circuit impedance at 50 Hz; —, measured impedance magnitude; —‐—, measured impedance phase angle

from the transient voltage and current waveforms produced by normal switching operations. These include the switching of capacitor banks and transformers or even natural system variations [7].

The advantage of capacitor banks is their widespread use and their frequent switching which produce a rich spectrum of interharmonics. However, the resulting harmonic currents are symmetrical, of short duration and depend on the point of wave of switching.

The transient inrush current produced by transformer switching produces high harmonic current levels compared with existing harmonics with a spectrum, ranging from 100 Hz to about 7000 Hz. Moreover, the signals are present for several seconds. Again, the currents are highly symmetric and depend on the switching moment and core remanence.

The use of natural system variations for spectral analysis can also provide time-dependent system impedances. Although this method is generally applicable, it can only achieve good precision in the presence of some predominant disturbing load.

The accuracy of applying the FFT to the voltage and current time-domain recordings in the presence of noise can be improved by correlation analysis in conjunction with spectral analysis. Spectral analysis of the auto- and cross-power spectra has been used to determine the frequency response of two 26.6 kV feeders [8] and also to determine the 3×3 impedance matrices [9]. With these methods, correlation indices are used to reject measurements where the signal-to-noise causes the calculated values to be suspect. With least squares estimates, the matrix condition number also indicates the accuracy of the answer.

Identification techniques, which do not require the use of FFT, have also been applied to time-domain responses. Prony analysis and Direct ARMA Method are two identification techniques suitable for determining frequency characteristics [10] from time domain waveforms.

The main advantage of these non-invasive techniques is that they can be readily applied without the need for special injection equipment; however, they are equally applicable to invasive testing, where an impulse or other frequency-rich signal is injected.

6.4 Representation of Individual Power System Components

6.4.1 The Overhead Transmission Line

A transmission line consists of distributed inductance and capacitance which represent the magnetic and electrostatic conditions of the line, and resistance and conductance which represent the losses. These electrical parameters are calculated from the line geometry and conductor data. The effects of ground currents and earth wires are included in the calculation of these parameters. The calculated parameters are expressed as a series impedance and shunt admittance per unit length. A simple representation of the line includes the line total inductance, capacitance, resistance and conductance as lumped parameters, as shown in Figure 6.5 (nominal PI model). However, when the line length becomes an appreciable part of the wavelength of the frequency of interest, errors become apparent. Subdividing the line and using cascaded nominal PI sections to represent the line can alleviate this problem. The more nominal PI sections used, the closer the model represents the distributed nature of the line, and hence the more accurate the model. However, computational burden also greatly increases due to the increase in the number of busbars and lumped elements.

Before considering long line effects in detail, the lumped component representation of a three-phase transmission line with ground wire and earth return, suitable for inclusion in the system admittance matrix, will be considered next.

The impedance of a three-phase transmission line with an overhead earth wire is illustrated in Figure 6.6. Each conductor has resistance, inductance, and capacitance, and is mutually coupled to the others.

With respect to Figure 6.6, the following equation can be written for the series impedance equivalent of phase a:

$$\begin{aligned} V_{\mathrm{a}} - V'_{\mathrm{a}} &= I_{\mathrm{a}}(R_{\mathrm{a}} + j\omega L_{\mathrm{a}}) + I_{\mathrm{b}}(j\omega L_{\mathrm{ab}}) + I_{\mathrm{c}}(j\omega L_{\mathrm{ac}}) \\ &\quad + j\omega L_{\mathrm{ag}} I_{\mathrm{g}} - j\omega L_{\mathrm{an}} I_{\mathrm{n}} + V_{\mathrm{n}}, \end{aligned} \tag{6.14}$$

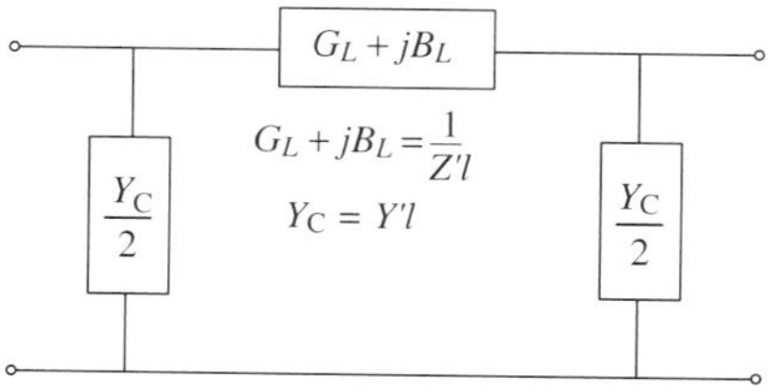

$G_L + jB_L$

$G_L + jB_L = \frac{1}{Z'l}$

$Y_C = Y'l$

$\frac{Y_C}{2}$ $\frac{Y_C}{2}$

Figure 6.5 Nominal PI representation of a transmission line

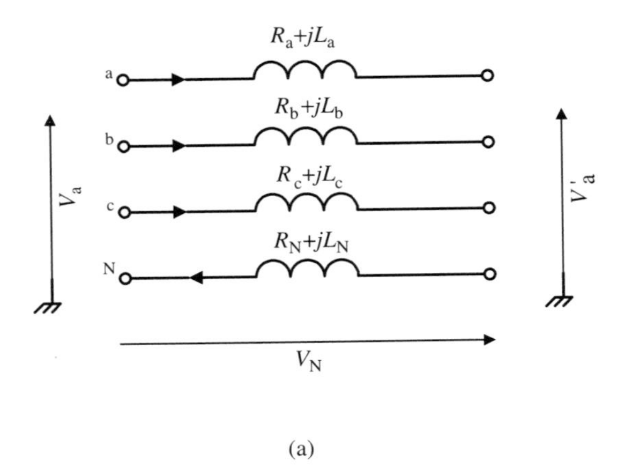

(a)

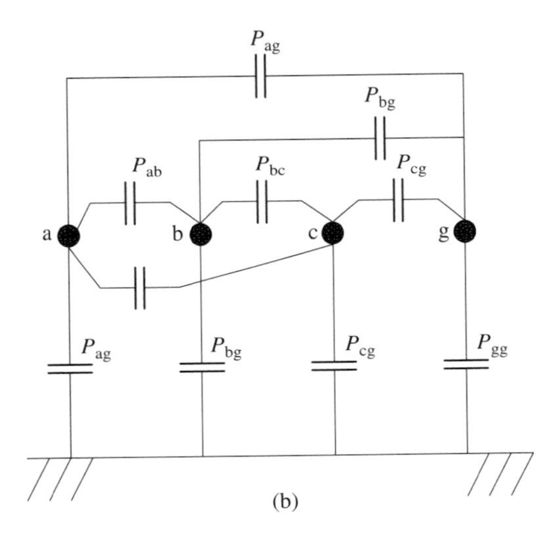

(b)

Figure 6.6 (a) Three-phase transmission series impedance equivalent; (b) three-phase transmission shunt impedance equivalent

where

$$V_n = I_n(R_n + j\omega L_n) - I_a j\omega L_{na} - I_b j\omega L_{nb} - I_c j\omega L_{nc} - I_g j\omega L_{ng} \tag{6.15}$$

and substituting

$$I_n = I_a + I_b + I_c + I_g \tag{6.16}$$

gives

$$\begin{aligned} V_a - V'_a = {} & I_a(R_a + j\omega L_a) + I_b j\omega L_{ab} + I_c j\omega L_{ac} \\ & + j\omega L_{ag} I_g - j\omega L_{an}(I_a + I_b + I_c + I_g) + V_n. \end{aligned} \tag{6.17}$$

Regrouping and substituting for V_n, i.e.

$$\begin{aligned}\Delta V_a &= V_a - V'_a \\ &= I_a(R_a + j\omega L_a - j\omega L_{an} + R_n + j\omega L_n - j\omega L_{na}) \\ &\quad + I_b(j\omega L_{ab} - j\omega L_{an} + R_n + j\omega L_n - j\omega L_{nb}) \\ &\quad + I_c(j\omega L_{ac} - j\omega L_{an} + R_n + j\omega L_n - j\omega L_{nc}) \\ &\quad + I_g(j\omega L_{ag} - j\omega L_{an} + R_n + j\omega L_n - j\omega L_{ng}), \end{aligned} \tag{6.18}$$

$$\begin{aligned}\Delta V_a &= I_a(R_a + j\omega L_a - 2j\omega L_{an} + R_n + j\omega L_n) \\ &\quad + I_b(j\omega L_{ab} - j\omega L_{bn} - j\omega L_{an} + R_n + j\omega L_n) \\ &\quad + I_c(j\omega L_{ac} - j\omega L_{cn} - j\omega L_{an} + R_n + j\omega L_n) \\ &\quad + I_g(j\omega L_{ag} - j\omega L_{gn} - j\omega L_{an} + R_n + j\omega L_n), \end{aligned} \tag{6.19}$$

or

$$\Delta V_a = Z_{aa-n} I_a + Z_{ab-n} I_b + Z_{ac-n} I_c + Z_{ag-n} I_g, \tag{6.20}$$

and writing similar equations for the other phases and earth wire, the following matrix equation results:

$$\begin{bmatrix} \Delta V_a \\ \Delta V_b \\ \Delta V_c \\ \Delta V_g \end{bmatrix} = \begin{bmatrix} Z_{aa-n} Z_{ab-n} Z_{ac-n} & Z_{ag-n} \\ Z_{ba-n} Z_{bb-n} Z_{bc-n} & Z_{bg-n} \\ Z_{ca-n} Z_{cb-n} Z_{cc-n} & Z_{cg-n} \\ Z_{ga-n} Z_{gb-n} Z_{gc-n} & Z_{gg-n} \end{bmatrix} \begin{bmatrix} I_a \\ I_b \\ I_c \\ I_g \end{bmatrix}. \tag{6.21}$$

Usually we are interested only in the performance of the phase conductors, and it is more convenient to use a three-conductor equivalent for the transmission line. This is achieved by writing matrix equation (6.21) in partitioned form as follows.

$$\begin{bmatrix} \Delta V_{abc} \\ \Delta V_g \end{bmatrix} = \begin{bmatrix} Z_A & Z_B \\ Z_C & Z_D \end{bmatrix} \begin{bmatrix} I_{abc} \\ I_g \end{bmatrix}. \tag{6.22}$$

From (6.22)

$$[\Delta V_{abc}] = [Z_A][I_{abc}] + [Z_B][I_g], \tag{6.23}$$

$$[\Delta V_g] = [Z_C][I_{abc}] + [Z_D][I_g]. \tag{6.24}$$

From Equations (6.23) and (6.24), and assuming that the earth wire is at zero potential,

$$[\Delta V_{abc}] = [Z_{abc}][I_{abc}], \tag{6.25}$$

where

$$[Z_{abc}] = [Z_A] - [Z_B][Z_D]^{-1}[Z_C] = \begin{bmatrix} Z'_{aa-n} & Z'_{ab-n} & Z'_{ac-n} \\ Z'_{ba-n} & Z'_{bb-n} & Z'_{bc-n} \\ Z'_{ca-n} & Z'_{cb-n} & Z'_{cc-n} \end{bmatrix} \tag{6.26}$$

With reference to Figure 6.6(b), the potentials of the line conductors are related to the conductor charges by the matrix equation [11]

$$\begin{bmatrix} V_a \\ V_b \\ V_c \\ V_g \end{bmatrix} = \begin{bmatrix} P_{aa} & P_{ab} & P_{ac} & P_{ag} \\ P_{ba} & P_{bb} & P_{bc} & P_{bg} \\ P_{ca} & P_{cb} & P_{cc} & P_{cg} \\ P_{ga} & P_{gb} & P_{gc} & P_{gg} \end{bmatrix} \begin{bmatrix} Q_a \\ Q_b \\ Q_c \\ Q_g \end{bmatrix}. \tag{6.27}$$

Considerations, as for the series impedance matrix, lead to

$$[V_{abc}] = [P'_{abc}][Q_{abc}], \tag{6.28}$$

where P'_{abc} is a 3×3 matrix which includes the effects of the earth wire. The capacitance matrix of the transmission line of Figure 6.6 is given by

$$[C'_{abc}] = [P'_{abc}]^{-1} = \begin{bmatrix} C_{aa} & -C_{ab} & -C_{ac} \\ -C_{ba} & C_{bb} & -C_{bc} \\ -C_{ca} & -C_{cb} & C_{cc} \end{bmatrix}. \tag{6.29}$$

The series impedance and shunt admittance lumped-PI model representation of the three-phase line is shown in Figure 6.7(a) and its matrix equivalent is illustrated in Figure 6.7(b). These two matrices can also be represented by compound admittances [12] (Figure 6.7(c)).

Using the compound component concept, the nodal injected currents of Figure 6.7(c) are related to the nodal voltages by the equation

$$\underset{6\times 1}{\begin{bmatrix} [I_i] \\ [I_k] \end{bmatrix}} = \underset{6\times 6}{\begin{bmatrix} [Z]^{-1} + [Y]/2 & -[Z]^{-1} \\ -[Z]^{-1} & [Z]^{-1} + [Y]/2 \end{bmatrix}} \underset{6\times 1}{\begin{bmatrix} [V_i] \\ [V_k] \end{bmatrix}}. \tag{6.30}$$

This forms the element admittance matrix representation for the short line between busbars i and k in terms of 3×3 matrix quantities.

Mutually Coupled Three-Phase Lines When two or more transmission lines occupy the same right of way for a considerable length, the electrostatic and electromagnetic coupling between those lines must be taken into account.

Consider the simplest case of two mutually coupled single-circuit three-phase lines. The two coupled lines are considered to form one subsystem composed of four system busbars. The coupled lines are illustrated in Figure 6.8, where each element is a 3×3 compound admittance and all voltages and currents are 3×1 vectors.

The coupled series elements represent the electromagnetic coupling, while the coupled shunt elements represent the capacitive or electrostatic coupling. These coupling parameters are lumped in a similar way to the standard line parameters.

With the admittances labelled as in Figure 6.8, and applying the rules of linear transformation for compound networks, the admittance matrix for the subsystem is

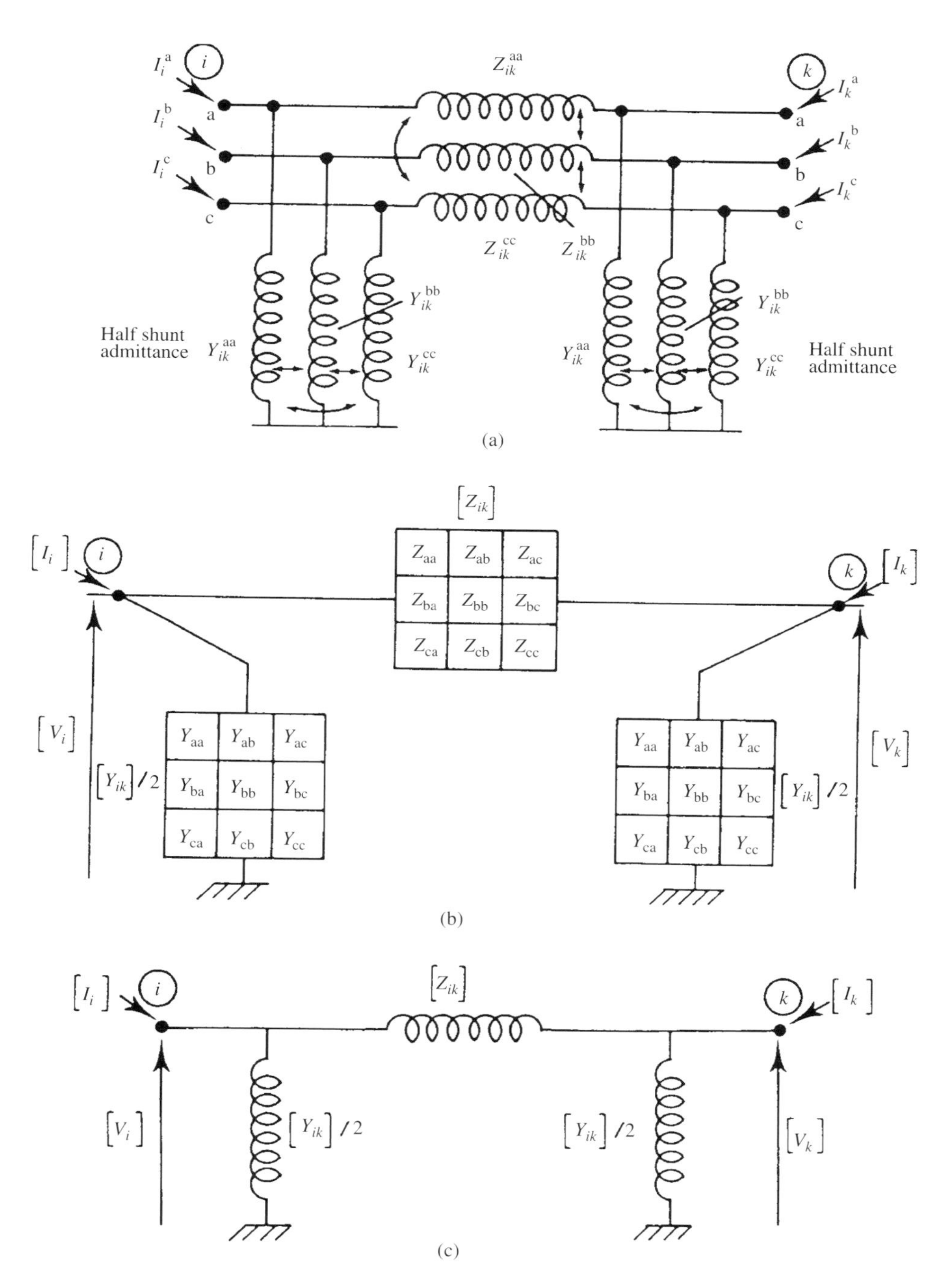

Figure 6.7 Lumped PI model of a short three-phase line series impedance: (a) full circuit representation; (b) matrix equivalent; (c) using three-phase compound admittances

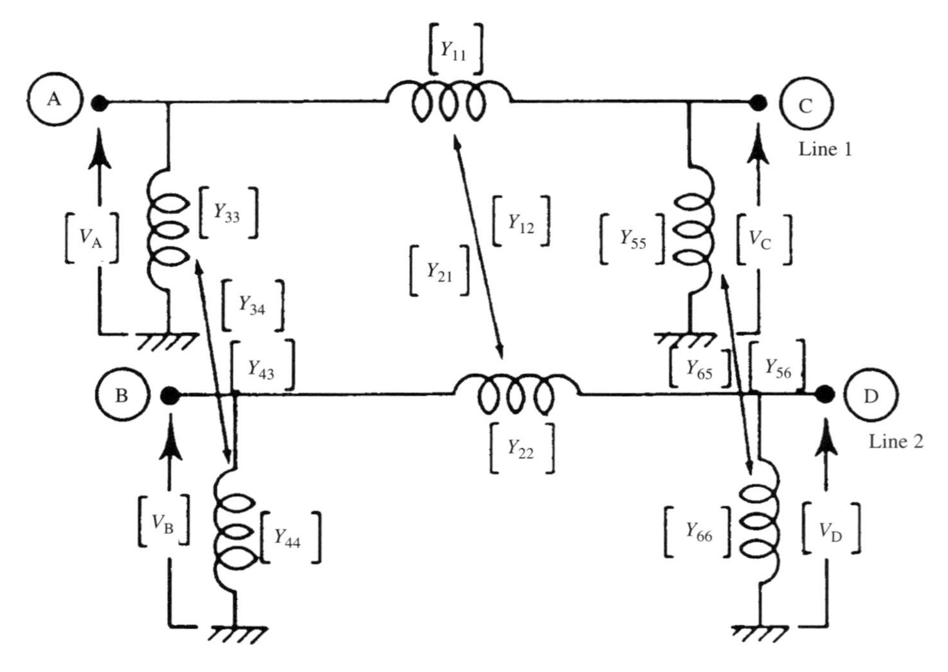

Figure 6.8 Twocoupled three-phase lines

defined as follows.

$$\begin{bmatrix} I_A \\ I_B \\ I_C \\ I_D \\ 12 \times 1 \end{bmatrix} = \begin{bmatrix} Y_{11}+Y_{33} & Y_{12}+Y_{34} & -Y_{11} & -Y_{12} \\ Y_{12}^T+Y_{34}^T & Y_{22}+Y_{44} & -Y_{12}^T & -Y_{22} \\ -Y_{11} & -Y_{12} & Y_{11}+Y_{55} & Y_{12}+Y_{56} \\ -Y_{12}^T & -Y_{22} & Y_{12}^T+Y_{56}^T & Y_{22}+Y_{66} \\ & 12 \times 12 & & \end{bmatrix} \begin{bmatrix} V_A \\ V_B \\ V_C \\ V_D \\ 12 \times 1 \end{bmatrix}. \tag{6.31}$$

It is assumed here that the mutual coupling is bilateral. Therefore $Y_{21} = Y_{12}^T$, etc.

The subsystem may be redrawn as in Figure 6.9. The pairs of coupled 3×3 compound admittances are now represented as a 6×6 compound admittance. The matrix representation is also shown. Following this representation and the labelling of the admittance block in the figure, the admittance matrix may be written in terms of the 6×6 compound coils as

$$\begin{bmatrix} \begin{bmatrix} I_A \\ I_B \end{bmatrix} \\ \begin{bmatrix} I_C \\ I_D \end{bmatrix} \\ 12 \times 1 \end{bmatrix} = \begin{bmatrix} [Z_s]^{-1}+[Y_{s1}] & -[Z_s]^{-1} \\ -[Z_s]^{-1} & [Z_s]^{-1}+[Y_{s2}] \\ 12 \times 12 & \end{bmatrix} \begin{bmatrix} \begin{bmatrix} V_A \\ V_B \end{bmatrix} \\ \begin{bmatrix} V_C \\ V_D \end{bmatrix} \\ 12 \times 1 \end{bmatrix}. \tag{6.32}$$

This is clearly identical to Equation (6.31) with the appropriate matrix partitioning.

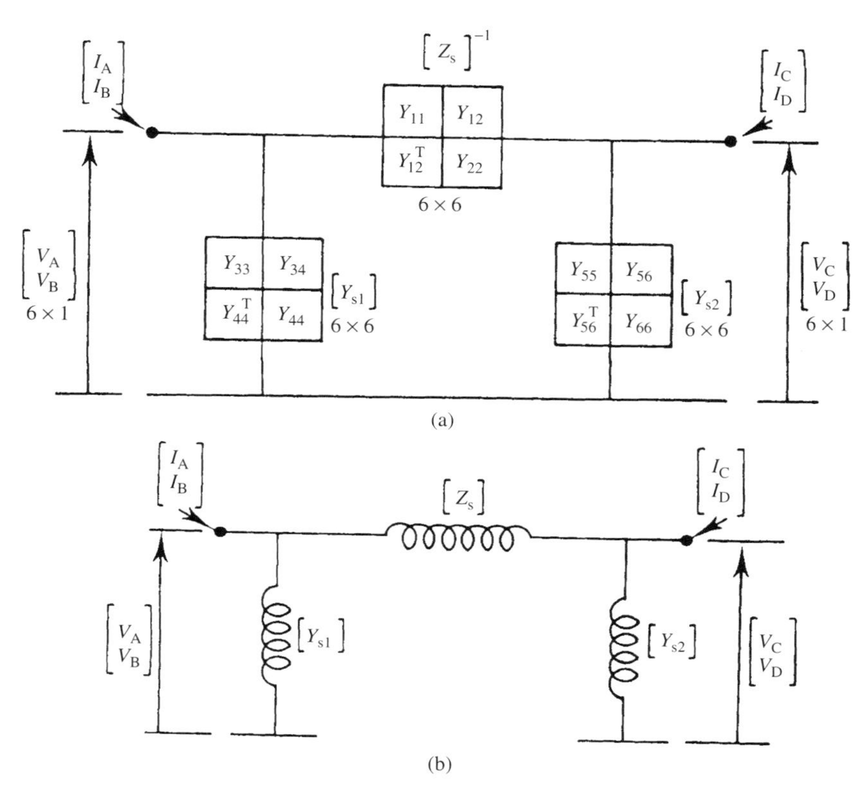

Figure 6.9 A 6 × 6 compound admittance representation of two coupled three-phase lines: (a) 6 × 6 matrix representation; (b) 6 × 6 compound admittance representation

The representation of Figure 6.9 is more concise and the formation of Equation (6.32) from this representation is straightforward, being exactly similar to that which results from the use of 3 × 3 compound admittances for the normal single three-phase line.

The data which must be available to enable coupled lines to be treated in a similar manner to single lines is the series impedance and shunt admittance matrices. These matrices are of order 3 × 3 for a single line, 6 × 6 for two coupled lines, 9 × 9 for three and 12 × 12 for four coupled lines.

Once the matrices $[Z_s]$ and $[Y_s]$ are available, the admittance matrix for the subsystem is formed by application of Equation (6.32)

When all the busbars of the coupled lines are distinct, the subsystem may be combined directly into the system admittance matrix. However, if the busbars are not distinct, then the admittance matrix as derived from Equation (6.32) must be modified. This is considered in the following section.

Consideration of Terminal Connections The admittance matrix as derived above must be reduced if there are different elements in the subsystem connected to the same busbar. As an example, consider two parallel transmission lines as illustrated in Figure 6.10.

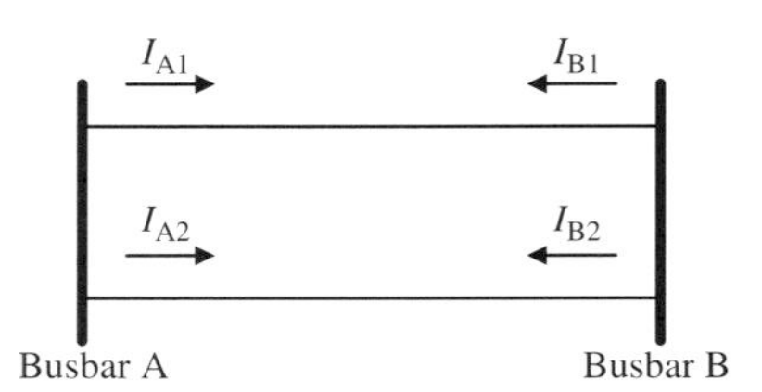

Figure 6.10 Mutually coupled parallel transmission lines

The admittance matrix derived previously related the currents and voltages at the four busbars A1, A2, B1 and B2. This relationship is given by

$$\begin{bmatrix} I_{A1} \\ I_{A2} \\ I_{B1} \\ I_{B2} \end{bmatrix} = [Y_{A1A2B1B2}] \begin{bmatrix} V_{A1} \\ V_{A2} \\ V_{B1} \\ V_{B2} \end{bmatrix}, \tag{6.33}$$

The nodal injected current at busbar A, I_A, is given by

$$I_A = I_{A1} + I_{A2}; \tag{6.34}$$

similarly

$$I_B = I_{B1} + I_{B2}. \tag{6.35}$$

Also, from inspection of Figure 6.10

$$V_A = V_{A1} = V_{A2}, \quad V_B = V_{B1} = V_{B2}. \tag{6.36}$$

The required matrix equation relates the nodal injected currents, I_A and I_B, to the voltages at these busbars. This is readily derived from Equation (6.33) and the conditions specified above. This is simply a matter of adding appropriate rows and columns, and yields

$$\begin{bmatrix} I_A \\ I_B \end{bmatrix} = [Y_{AB}] \begin{bmatrix} V_A \\ V_B \end{bmatrix} \tag{6.37}$$

where $[Y_{AB}]$ is the required nodal admittance matrix for the subsystem.

It should be noted that the matrix in Equation (6.33) must be retained, since it is required in the calculation of the individual line currents.

Equivalent PI Model For long lines, a number of nominal PI models are connected in series to improve the accuracy of voltages and currents, which are affected by standing wave effects. For example, a three-section PI model provides an accuracy to 1.2% for a quarter wavelength line (a quarter wavelength corresponds with 1500 km and 1250 km at 50 Hz and 60 Hz, respectively).

As the frequency increases, the number of nominal PI sections to maintain a particular accuracy increases proportionally, e.g. a 300 km line requires 30 nominal PI sections to maintain the 1.2% accuracy for the 50th harmonic. However, near resonance, the accuracy departs significantly from an acceptable value.

The computational effort can be greatly reduced and the accuracy improved with the use of an equivalent PI model derived from the solution of the second-order linear differential equations describing wave propagation along transmission lines [13].

The solution of the wave equations at a distance x from the sending end of the line is

$$V(x) = \exp(-\gamma \cdot x)V_{\mathrm{i}} + \exp(\gamma \cdot x)V_{\mathrm{r}}, \tag{6.38}$$

$$I(x) = (Z')^{-1}\gamma[\exp(-\gamma \cdot x)V_{\mathrm{i}} - \exp(\gamma \cdot x)V_{\mathrm{r}}], \tag{6.39}$$

where $\gamma = \sqrt{Z'Y'} = \alpha + j\beta$ is the propagation constant, $Z' = r + j2\pi fL$ is the series impedance per unit length, $Y' = g + j2\pi fC$ is the shunt admittance per unit length, and V_i and V_r the forward and reverse travelling voltages, respectively.

Depending on the problem in hand, e.g. if the evaluation of terminal quantities only is required, then it is more convenient to formulate a solution using two-port matrix equations. This leads to the equivalent PI model, shown in Figure 6.11, where

$$Z = Z_{\mathrm{o}} \sinh(\gamma \cdot l) \tag{6.40}$$

$$Y_1 = Y_2 = \frac{1}{Z_{\mathrm{o}}} \frac{\cosh(\gamma \cdot l) - 1}{\sinh(\gamma \cdot l)} = \frac{1}{Z_{\mathrm{o}}} \tanh \frac{(\gamma \cdot l)}{2}. \tag{6.41}$$

and

$$Z_{\mathrm{o}} = \sqrt{Z'/Y'} \text{ is the characteristic impedance of the line.} \tag{6.42}$$

Owing to the standing wave effect of voltages and currents on transmission lines, the maximum values of these are likely to occur at points other than at the receiving end or sending end busbars. These local maxima could result in insulation damage, overheating or electromagnetic interference. It is thus important to calculate the maximum values of currents and voltages along a line and the points at which these occur.

In the case of multiconductor transmission lines, the nominal PI series impedance and shunt admittance matrices per unit distance, $[Z']$ and $[Y']$, respectively, are square and their size is fixed by the number of mutually coupled conductors.

The derivation of the equivalent PI model for harmonic penetration studies from the nominal PI matrices is similar to that of the single phase lines, except that it involves the evaluation of hyperbolic functions of the propagation constant which is now a matrix:

$$[\gamma] = ([Z'][Y'])^{\frac{1}{2}}. \tag{6.43}$$

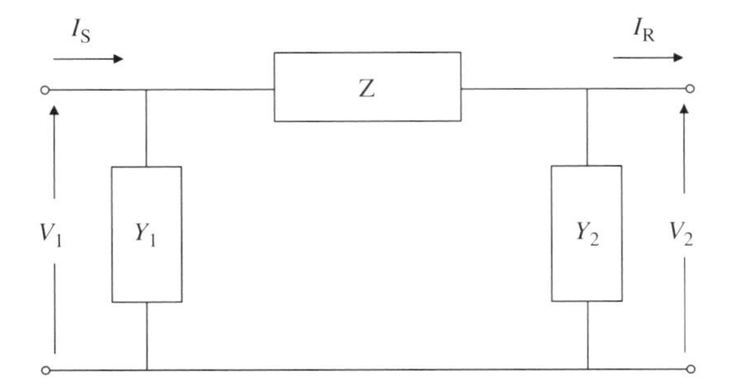

Figure 6.11 The equivalent PI model of a long transmission line

There is no direct way of calculating sinh or tanh of a matrix, thus a method using eigenvalues and eigenvectors, called *modal analysis* is employed [14] that leads to the following expressions for the series and shunt components of the equivalent PI circuit [15]:

$$[Z]_{\mathrm{EPM}} = l[Z'][M]\left[\frac{\sinh \gamma l}{\gamma l}\right][M]^{-1}, \tag{6.44}$$

where l is the transmission line length, $[Z]_{\mathrm{EPM}}$ is the equivalent PI series impedance matrix, [M] is the matrix of normalised eigenvectors,

$$\left[\frac{\sinh \gamma l}{\gamma l}\right] = \begin{bmatrix} \dfrac{\sinh \gamma_1 l}{\gamma_1 l} & 0 & \cdots & 0 \\ 0 & \dfrac{\sinh \gamma_2 l}{\gamma_2 l} & \cdots & 0 \\ \vdots & \vdots & & \vdots \\ 0 & 0 & \cdots & \dfrac{\sinh \gamma_j l}{\gamma_j l} \end{bmatrix} \tag{6.45}$$

and γ_j is the jth eigenvalue for $j/3$ mutually coupled circuits. Similarly

$$\frac{1}{2}[Y]_{\mathrm{EPM}} = \frac{1}{2}l[M]\left[\frac{\tanh(\gamma l/2)}{\gamma l/2}\right][M]^{-1}[Y'], \tag{6.46}$$

where $[Y]_{\mathrm{EPM}}$ is the equivalent PI shunt admittance matrix.

Computer derivation of the correction factors for conversion from the nominal PI to the equivalent PI model, and their incorporation into the series impedance and shunt admittance matrices, is carried out as indicated in the structure diagram of Figure 6.12. The LR2 algorithm of Wilkinson and Reinsch [16] is used for accurate calculations in the derivation of the eigenvalues and eigenvectors.

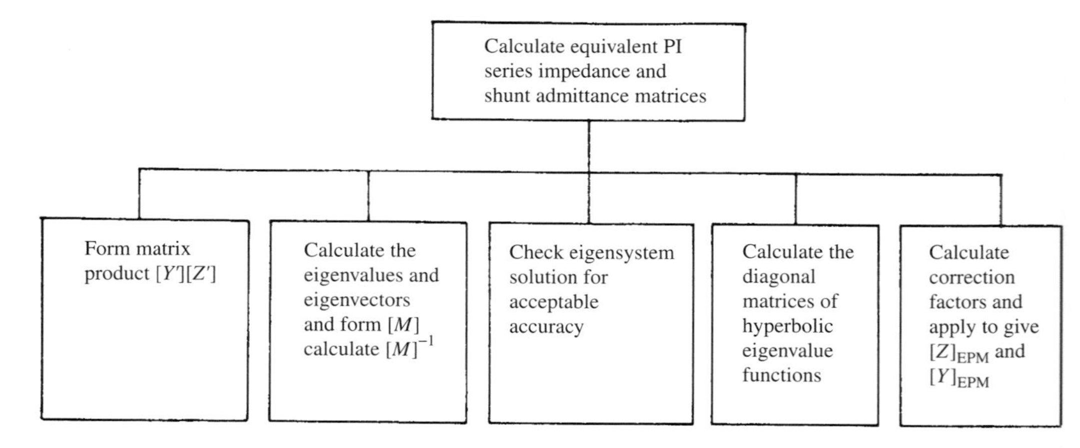

Figure 6.12 Structure diagram for calculation of the equivalent PI model

6.4.2 Evaluation of Transmission Line Parameters

The lumped series impedance matrix $[Z]$ of a transmission line consists of three components, while the shunt admittance matrix $[Y]$ contains one.

$$[Z] = [Z_e] + [Z_g] + [Z_c], \tag{6.47}$$

$$[Y] = [Y_g], \tag{6.48}$$

where

$[Z_c]$ = the internal impedance of the conductors $(\Omega \cdot \text{km}^{-1})$,
$[Z_g]$ = the impedance due to the physical geometry of the conductor's arrangement $(\Omega \cdot \text{km}^{-1})$,
$[Z_e]$ = the earth return path impedance $(\Omega \cdot \text{km}^{-1})$, and
$[Y_g]$ = the admittance due to the physical geometry of the conductor $(\Omega^{-1} \cdot \text{km}^{-1})$.

In multiconductor transmission all *primitive matrices* (the admittance matrices of the unconnected branches of the original network components) are symmetric and, therefore, the functions that define the elements need only be evaluated for elements on or above the leading diagonal.

Earth Impedance Matrix $[Z_e]$ The impedance due to the earth path varies with frequency in a non-linear fashion. The solution of this problem, under idealised conditions, has been given in the form of either an infinite integral or an infinite series [17].

As the need arises to calculate ground impedances for a wide spectrum of frequencies, the tendency is to select simple formulations aiming at a reduction in computing time, while maintaining a reasonable level of accuracy.

Consequently, what was originally a heuristic approach [18], is becoming the more favoured alternative, particularly at high frequencies.

Based on Carson's work, the ground impedance can be concisely expressed as

$$z_e = 1000J(r, \theta)(\Omega \cdot \text{km}^{-1}), \tag{6.49}$$

where

$$\begin{aligned}
z_e &\in [Z_e], \\
J(r, \theta) &= \frac{\omega \mu_a}{\pi}\{P(r, \theta) + jQ(r, \theta)\}, \\
r_{ij} &= \sqrt{\frac{\omega \mu_a}{\rho}}\ D_{ij}, \\
D_{ij} &= \sqrt{(h_i + h_j)^2 + d_{ij}^2}, \quad \text{for } i \neq j, \\
D_{ij} &= 2h_i, \quad \text{for } i = j, \\
\theta_{ij} &= \arctan \frac{d_{ij}}{h_i + h_j}, \quad \text{for } i \neq j, \\
\theta_{ij} &= 0, \quad \text{for } i = j, \\
\omega &= 2\pi f (\text{rad} \cdot \text{s}^{-1}),
\end{aligned}$$

f = frequency (Hz),
h_i = height of conductor i(m),
d_{ij} = horizontal distance between conductors i and j(m),
μ_a = permeability of free space=$7\pi \times 10^{-7}(\mathrm{H \cdot m^{-1}})$,
ρ = earth resistivity $(\Omega \cdot \mathrm{m})$.

Carson's solution to Equation (6.49) is defined by eight different infinite series which converge quickly for problems related to transmission line parameter calculation, but the number of required computations increases with frequency and separation of the conductors.

More recent literature has described closed form formulations for the numerical evaluation of line-ground loops, based on the concept of a mirroring surface beneath the earth at a certain depth. The most popular complex penetration model which has had more appeal is that of C. Dubanton [19], due to its simplicity and high degree of accuracy for the whole frequency span for which Carson's equations are valid.

Dubanton's formulae for the evaluation of the self and mutual impedances of conductors i and j are

$$Z_{ii} = \frac{j\omega\mu_o}{2\pi} \times \ln \frac{2(h_i + p)}{r_i}, \tag{6.50}$$

$$Z_{ij} = \frac{j\omega\mu_o}{2\pi} \times \ln \frac{\sqrt{2(h_i + p)}}{\sqrt{(h_i - h_j)^2 + d_{ij}^2}}, \tag{6.51}$$

where $p = 1/\sqrt{j\omega\mu_o\sigma}$ is the complex depth below the earth at which the mirroring surface is located.

An alternative and very simple formulation has been proposed recently by Acha [20], which for the purpose of harmonic penetration yields accurate solutions when compared with those obtained using Carson's equations.

Geometrical Impedance Matrix $[Z_g]$ and Admittance Matrix $[Y_g]$ If the conductors and the earth are assumed to be equipotential surfaces, the geometrical impedance can be formulated in terms of potential coefficients theory.

The self-potential coefficient ψ_{ii} for the ith conductor and the mutual potential coefficient ψ_{ij} between the ith and jth conductors are defined as follows.

$$\psi_{ii} = \ln(2h_i/r_i), \tag{6.52}$$

$$\psi_{ij} = \ln(D_{ij}/d_{ij}), \tag{6.53}$$

where r_i is the radius of the ith conductor (m), while the other variables are as defined earlier.

Potential coefficients depend entirely on the physical arrangement of the conductors and need only be evaluated once.

For practical purposes the air is assumed to have zero conductance and

$$[Z_g] = j\omega K'[\psi]\Omega/\mathrm{km}, \tag{6.54}$$

where $[\psi]$ is a matrix of potential coefficients and $K' = 2 \times 10^{-4}$.

The lumped shunt admittance parameters $[Y]$ are completely defined by the inverse relation of the potential coefficients matrix, i.e.

$$[Y_g] = 1000 j\omega 2\pi \varepsilon_a [\psi]^{-1}, \tag{6.55}$$

where ε_a = permittivity of free space $= 8.857 \times 10^{-12} (\mathrm{F} \cdot \mathrm{m}^{-1})$.

Since $[Z_g]$ and $[Y_g]$ are linear functions of frequency, they need only be evaluated once and scaled for other frequencies.

Conductor Impedance Matrix $[Z_c]$ This term accounts for the internal impedance of the conductors. Both resistance and inductance have a non-linear frequency dependence. Current tends to flow on the surface of the conductor, this *skin effect* increases with frequency and needs to be computed at each frequency. An accurate result for a homogeneous non-ferrous conductor of annular cross-section involves the evaluation of long equations based on the solution of Bessel functions,

$$Z_c = \frac{j\omega\mu_o}{2\pi} \frac{1}{x_e} \frac{J_o(x_e) N'_o(x_i) - N_o(x_e) J'_o(x_i)}{J'_o(x_e) N'_o(x_i) - N'_o(x_e) J'_o(x_i)}, \tag{6.56}$$

where

$x_e = j\sqrt{j\omega\mu_o\sigma_c}\, r_e$,
$x_i = j\sqrt{j\omega\mu_o\sigma_c}\, r_i$,
r_e = external radius of the conductor (m),
r_i = internal radius of the conductor (m),
J_o = Bessel function of the first kind and zero order,
J'_o = derivative of the Bessel function of the first kind and zero order,
N_o = Bessel function of the second kind and zero order,
N'_o = derivative of the Bessel function of the second kind and zero order,
σ_c = conductivity of the conductor material at the average conductor temperature.

The Bessel functions and their derivatives are solved, within a specified accuracy, by means of their associated infinite series. Convergence problems are frequently encountered at high frequencies and low ratios of conductor thickness to external radius i.e. $(r_e - r_i)/r_e$, necessitating the use of asymptotic expansions.

A new closed form solution has been proposed based on the concept of complex penetration, Semlyen [18]; unfortunately, errors of up to 6.6% occur in the region of interest.

To overcome the difficulties of slow convergence of the Bessel function approach and the inaccuracy of the complex penetration method at relatively low frequency, an alternative approach based upon curve fitting to the Bessel function formula has been proposed by Acha [20]

Lewis and Tuttle [21] presented a practical method for calculating the skin effect resistance ratio by approximating aluminium conductor steel reinforced (ACSR) conductors to uniform tubes having the same inside and outside diameters as the aluminium conductors, see Figure 6.13. Figure 6.14 illustrates the skin effect ratio for different models and various tube ratios for ACSR conductors. Skin effect modelling is important for long lines. Although the series resistance of a transmission line is typically a small component of the series impedance, it dominates its value at resonances.

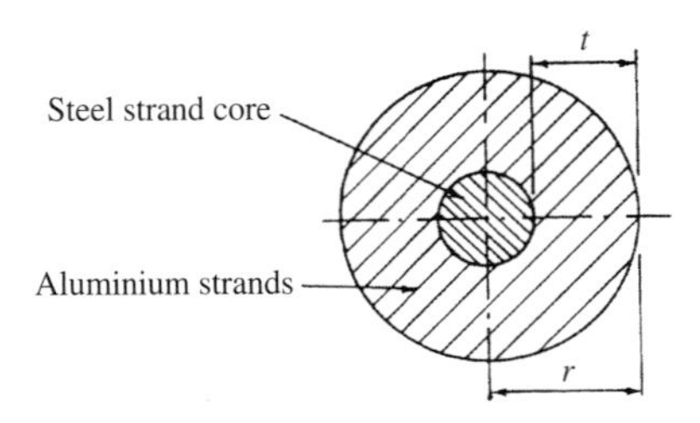

Figure 6.13 ACSR hollow tube conductor representation

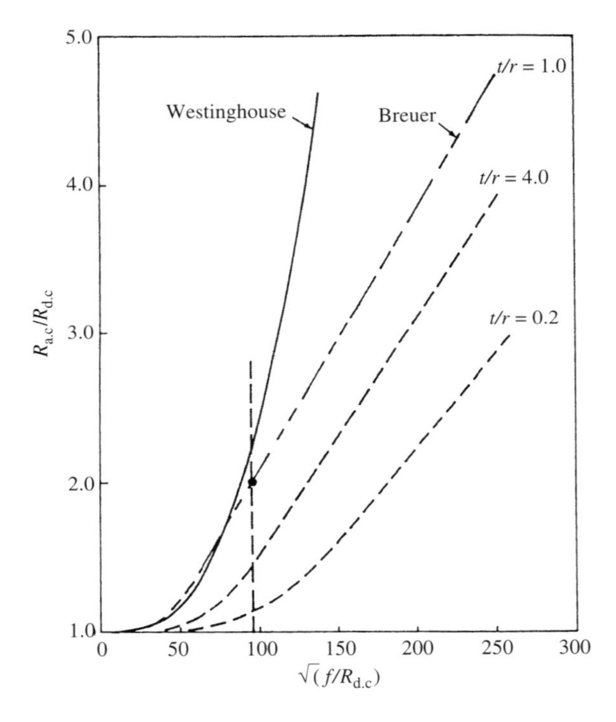

Figure 6.14 Skin effect resistance for different models

6.4.3 Underground and Submarine Cables [22]

A unified solution similar to that of overhead transmission is difficult for underground cables because of the great variety in their construction and layouts.

The cross-section of a cable, although extremely complex, can be simplified to that of Figure 6.15 and its series per unit length harmonic impedance is calculated by the following set of loop equations.

$$-\begin{bmatrix} dV_1/dx \\ dV_2/dx \\ dV_3/dx \end{bmatrix} = \begin{bmatrix} Z'_{11} & Z'_{12} & 0 \\ Z'_{21} & Z'_{22} & Z'_{23} \\ 0 & Z'_{32} & Z'_{33} \end{bmatrix} \begin{bmatrix} I_1 \\ I_2 \\ I_3 \end{bmatrix}, \tag{6.57}$$

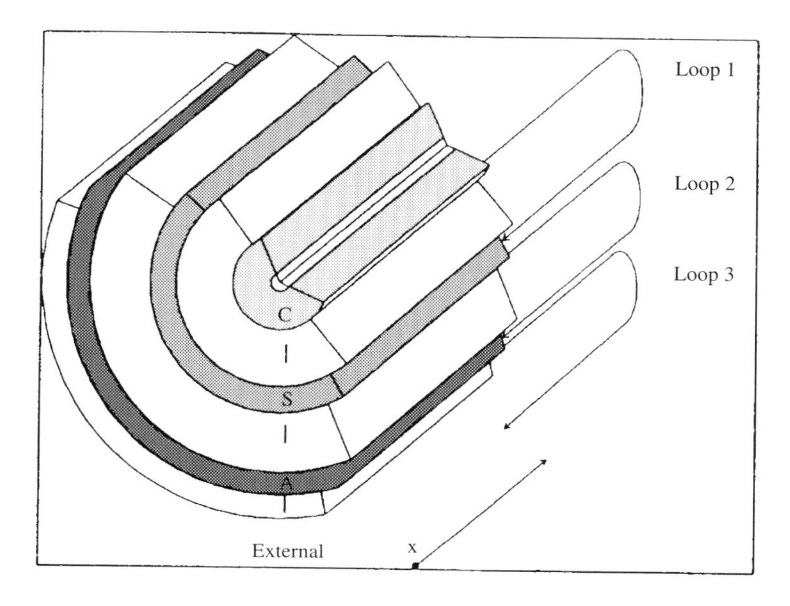

Figure 6.15 Cable cross-section

where

Z'_{11} = the sum of the following three component impedances

$Z'_{\text{core — outside}}$ = the internal impedance of the core with the return path outside the core

$Z'_{\text{core — insulation}}$ = the impedance of the insulation surrounding the core

$Z'_{\text{sheath — inside}}$ = the internal impedance of the sheath with the return path inside the sheath.

Similarly

$$Z'_{22} = Z'_{\text{sheath — outside}} + Z'_{\text{sheath/armour — insulation}} + Z'_{\text{armour — inside}} \tag{6.58}$$

and

$$Z'_{33} = Z'_{\text{armour — outside}} + Z'_{\text{armour/earth — insulation}} + Z'_{\text{earth — inside}}. \tag{6.59}$$

The coupling impedances $Z'_{12} = Z'_{21}$ and $Z'_{23} = Z'_{32}$ are negative because of opposing current directions (I_2 in negative direction in loop 1, and I_3 in negative direction in loop 2), i.e.

$$Z'_{12} = Z'_{21} = -Z'_{\text{sheath — mutual}} \tag{6.60}$$

$$Z'_{23} = Z'_{32} = -Z'_{\text{armour — mutual}} \tag{6.61}$$

where

$Z'_{\text{sheath−mutual}}$ = mutual impedance (per unit length) of the tubular sheath between inside loop 1 and the outside loop 2

$Z'_{\text{armour−mutual}}$ = mutual impedance (per unit length) of the tubular armour between the inside loop 2 and the outside loop 3.

Finally, $Z'_{13} = Z'_{31} = 0$ because loop 1 and loop 3 have no common branch.

The impedances of the insulation are given by

$$Z'_{\text{insulation}} = j\omega \frac{\mu}{2\pi} \ln \frac{r_{\text{outside}}}{r_{\text{inside}}} \quad \text{in } \Omega/\text{m}, \tag{6.62}$$

where

μ = permeability of insulation in H/m,
r_{outside} = outside radius of insulation,
r_{inside} = inside radius of insulation.

If there is no insulation between the armour and earth, then Z' insulation $= 0$.

The internal impedances and the mutual impedance of a tubular conductor are a function of frequency, and can be derived from Bessel and Kelvin functions:

$$Z'_{\text{tube — inside}} = \frac{\sqrt{j}.\omega.\mu}{2\pi.mq.D}\left[I_o\left(\sqrt{j}mq\right)K_1\left(\sqrt{j}mr\right) + K_o\left(\sqrt{j}mq\right)I_1\left(\sqrt{j}mr\right)\right], \tag{6.63a}$$

$$Z'_{\text{tube — outside}} = \frac{\sqrt{j}.\omega.\mu}{2\pi.mr.D}\left[I_o\left(\sqrt{j}mr\right)K_1\left(\sqrt{j}mq\right) + K_o\left(\sqrt{j}mr\right)I_1\left(\sqrt{j}mq\right)\right], \tag{6.63b}$$

$$Z'_{\text{tube−mutual}} = \frac{\omega.\mu}{2\pi.mq.mr.D}, \tag{6.63c}$$

with

$$D = I_1\left(\sqrt{j}mr\right)K_1\left(\sqrt{j}mq\right) - I_1\left(\sqrt{j}mq\right)K_1\left(\sqrt{j}mr\right), \tag{6.63d}$$

where

$$mr = \sqrt{K\frac{1}{1-s^2}}, \tag{6.64}$$

$$mq = \sqrt{K\frac{s^2}{1-s^2}}, \tag{6.65}$$

with

$$K = \frac{8\pi.10^{-4.}f.\mu_r}{R'_{\text{dc}}}, \tag{6.66}$$

$$s = \frac{q}{r}, \tag{6.67}$$

where

q = inside radius,
r = outside radius
R'_{dc} = d.c. resistance in Ω/km.

The only remaining term is $Z'_{\text{earth — inside}}$ in Equation (6.59), which is the earth return impedance for underground cables, or the sea return impedance for submarine cables.

The earth return impedance can be calculated approximately with Equation (6.63a) by letting the outside radius go to infinity. This approach, also used by Bianchi and Luoni [23] to find the sea return impedance, is quite acceptable considering the fact that sea resistivity and other input parameters are not known accurately.

Equation (6.57) is not in a form compatible with the solution used for overhead conductors, where the voltages with respect to local ground and the actual currents in the conductors are used as variables. Equation (6.57) can easily be brought into such a form by introducing the appropriate terminal conditions; namely with

$$V_1 = V_{\text{core}} - V_{\text{sheath}}, \qquad I_1 = I_{\text{core}},$$
$$V_2 = V_{\text{sheath}} - V_{\text{armour}}, \qquad I_2 = I_{\text{core}} + I_{\text{sheath}},$$

and

$$V_3 = V_{\text{armour}}, \qquad I_3 = I_{\text{core}} + I_{\text{sheath}} + I_{\text{armour}}.$$

Equation (6.57) can be rewritten as

$$-\begin{bmatrix} dV_{\text{core}}/dx \\ dV_{\text{sheath}}/dx \\ dV_{\text{armour}}/dx \end{bmatrix} = \begin{bmatrix} Z'_{\text{cc}} & Z'_{\text{cs}} & Z'_{\text{ca}} \\ Z'_{\text{sc}} & Z'_{\text{ss}} & Z_{\text{sa}} \\ Z'_{\text{ac}} & Z'_{\text{as}} & Z'_{\text{aa}} \end{bmatrix} \begin{bmatrix} I_{\text{core}} \\ I_{\text{sheath}} \\ I_{\text{armour}} \end{bmatrix}, \tag{6.68}$$

where

$$\begin{aligned} Z'_{\text{cc}} &= Z'_{11} + 2Z'_{12} + Z'_{22} + 2Z'_{23} + Z'_{33}, \\ Z'_{\text{cs}} &= Z'_{\text{sc}} = Z'_{12} + Z'_{22} + 2Z'_{23} + Z'_{33}, \\ Z'_{\text{ca}} &= Z'_{\text{ac}} = Z'_{\text{sa}} = Z'_{\text{as}} = Z'_{23} + Z'_{33}, \\ Z'_{\text{ss}} &= Z'_{22} + 2Z'_{23} + Z'_{33}, \\ Z'_{\text{aa}} &= Z'_{33}. \end{aligned}$$

Because a good approximation for many cables having bonding between the sheath and the armour, and the armour earthed to the sea, is $V_{\text{sheath}} = V_{\text{armour}} = 0$, the system can be reduced to

$$-dV_{\text{core}}/dx = ZI_{\text{core}}, \tag{6.69}$$

where Z is a reduction of the impedance matrix of Equation (6.68).

Similarly, for each cable the per unit length harmonic admittance is

$$-\begin{bmatrix} dI_1/dx \\ dI_2/dx \\ dI_3/dx \end{bmatrix} = \begin{bmatrix} j\omega C'_1 & 0 & 0 \\ 0 & j\omega C'_2 & 0 \\ 0 & 0 & j\omega C'_3 \end{bmatrix} \begin{bmatrix} V_1 \\ V_2 \\ V_3 \end{bmatrix}, \tag{6.70}$$

where $C'_i = 2\pi\,\varepsilon_o\,\varepsilon_r/l_n(r/q)$. Therefore, when converted to core, sheath and armour quantities,

$$-\begin{bmatrix} dI_{\text{core}}/dx \\ dI_{\text{sheath}}/dx \\ dI_{\text{armour}}/dx \end{bmatrix} = \begin{bmatrix} Y'_1 & -Y'_1 & 0 \\ -Y'_1 & Y' + Y'_2 & -Y'_2 \\ 0 & -Y'_2 & Y'_2 + Y'_3 \end{bmatrix} \begin{bmatrix} V_{\text{core}} \\ V_{\text{sheath}} \\ V_{\text{armour}} \end{bmatrix}, \tag{6.71}$$

Table 6.1 Corrections for skin effect in Cables

Company	Voltage (kV)	Harmonic order	Resistance
NGC	400, 275 (Based on 2.5 sq in conductor at 5 in spacing between centres)	$h \geq 1.5$	$0.74R_1(0.267 + 1.073\sqrt{h})$
	132	$h \geq 2.35$	$R_1(0.187 + 0.532\sqrt{h})$
EDF	400, 225	$h \geq 2$	$0.74R_1(0.267 + 1.073\sqrt{h})$
	150, 90	$h \geq 2$	$R_1(0.187 + 0.532\sqrt{h})$

where $Y'_i = j\omega l_i$. If, as before, $V_{\text{sheath}} = V_{\text{armour}} = 0$, Equation (6.71) reduces to

$$-\mathrm{d}I_{\text{core}}/\mathrm{d}x = Y'_1 V_{\text{core}}. \tag{6.72}$$

Therefore, for frequencies of interest, the cable per unit length harmonic impedance, Z', and admittance, Y', are calculated with both the zero and positive sequence values being equal to the Z in Equation (6.69), and the Y' in Equation (6.72), respectively.

In the absence of rigorous computer models, such as described above, power companies often use approximations to the skin effect by means of correction factors. Typical corrections used by the NGC(UK) and EDF(France) are given in Table 6.1.

6.4.4 Three-phase Transformer Models

The basic three-phase two-winding transformer is shown in Figure 6.16. Its primitive network, on the assumption that the flux paths are symmetrically distributed between all windings, is represented by the equation

$$\begin{bmatrix} I_1 \\ I_2 \\ I_3 \\ I_4 \\ I_5 \\ I_6 \end{bmatrix} = \begin{bmatrix} Y_p & Y'_m & Y'_m & -Y_m & Y''_m & Y''_m \\ Y'_m & Y_p & Y'_m & Y''_m & -Y_m & Y''_m \\ Y'_m & Y'_m & Y_p & Y''_m & Y''_m & -Y_m \\ -Y_m & Y''_m & Y''_m & Y_s & Y'''_m & Y'''_m \\ Y''_m & -Y_m & Y''_m & Y'''_m & Y_s & Y'''_m \\ Y''_m & Y''_m & -Y_m & Y'''_m & Y'''_m & Y_s \end{bmatrix} \begin{bmatrix} V_1 \\ V_2 \\ V_3 \\ V_4 \\ V_5 \\ V_6 \end{bmatrix}, \tag{6.73}$$

where Y'_m is the mutual admittance between primary coils, Y''_m is the mutual admittance between primary and secondary coils on different cores, and Y'''_m is the mutual admittance between secondary coils.

If a tertiary winding is also present, the primitive network consists of nine (instead of six) coupled coils and its mathematical model will be a 9×9 admittance matrix.

The interphase coupling can usually be ignored (e.g. the case of three single-phase separate units) and all the primed terms are effectively zero.

The connection matrix $[C]$ between the primitive network and the actual transformer buses is derived from the transformer connection.

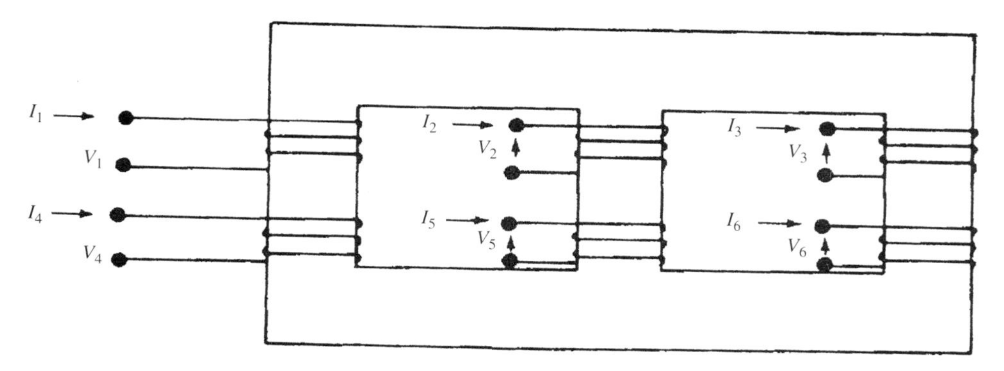

Figure 6.16 Diagrammatic representation of a two-winding transformer

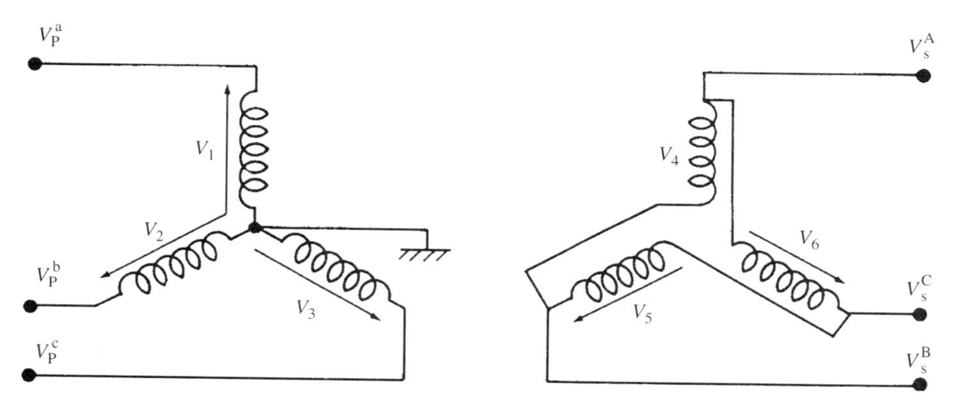

Figure 6.17 Network connection diagram for a Wye G-delta transformer

By way of example, consider the Wye G-Delta connection of Figure 6.17. The following connection matrix applies.

$$\begin{bmatrix} V_1 \\ V_2 \\ V_3 \\ V_4 \\ V_5 \\ V_6 \end{bmatrix} = \begin{bmatrix} 1 & 0 & 0 & 0 & 0 & 0 \\ 0 & 1 & 0 & 0 & 0 & 0 \\ 0 & 0 & 1 & 0 & 0 & 0 \\ 0 & 0 & 0 & 1 & -1 & 0 \\ 0 & 0 & 0 & 0 & 1 & -1 \\ 0 & 0 & 0 & -1 & 0 & 1 \end{bmatrix} \begin{bmatrix} V_p^a \\ V_p^b \\ V_p^c \\ V_s^A \\ V_s^B \\ V_s^C \end{bmatrix} \tag{6.74}$$

or

$$[V]_{\text{branch}} = [C][V]_{\text{node}}. \tag{6.75}$$

We can also write

$$[Y]_{\text{NODE}} = [C]^{\text{T}}[Y]_{\text{PRIM}}[C] \tag{6.76}$$

and using $[Y]_{\text{PRIM}}$ from Equation (6.73)

$$[Y]_{\text{NODE}} = \begin{bmatrix} Y_p & Y'_m & Y'_m & -(Y_m + Y''_m) & (Y_m + Y''_m) & 0 \\ Y'_m & Y_p & Y'_m & 0 & -(Y_m + Y''_m) & (Y_m + Y''_m) \\ Y'_m & Y'_m & Y_p & (Y_m + Y''_m) & 0 & -(Y_m + Y''_m) \\ -(Y_m + Y''_m) & 0 & (Y_m + Y''_m) & 2(Y_s - Y'''_m) & -(Y_s - Y'''_m) & -(Y_s - Y'''_m) \\ (Y_m + Y''_m) & -(Y_m + Y''_m) & 0 & -(Y_s - Y'''_m) & 2(Y_s - Y'''_m) & -(Y_s - Y'''_m) \\ 0 & (Y_m + Y''_m) & -(Y_m + Y''_m) & -(Y_s - Y'''_m) & -(Y_s - Y'''_m) & 2(Y_s - Y'''_m) \end{bmatrix} \begin{matrix} A \\ B \\ C \\ A \\ B \\ C \end{matrix} . \quad (6.77)$$

If the primitive admittances are expressed in per unit the upper right and lower left quadrants of matrix (6.77) must be divided by $\sqrt{3}$ and the lower right quadrant by 3. Then, in the absence of inter-phase coupling the nodal admittance matrix equation of the Wye G-delta connection becomes

$$\begin{bmatrix} I_p^a \\ I_p^b \\ I_p^c \\ I_s^A \\ I_s^B \\ I_s^C \end{bmatrix} = \begin{bmatrix} y & & & -y/\sqrt{3} & y/\sqrt{3} & \\ & y & & & -y/\sqrt{3} & y/\sqrt{3} \\ & & y & y/\sqrt{3} & & -y/\sqrt{3} \\ -y/\sqrt{3} & & y/\sqrt{3} & 2/3y & -1/3y & -1/3y \\ y/\sqrt{3} & -y/\sqrt{3} & & -1/3y & 2/3y & -1/3y \\ & y/\sqrt{3} & -y/\sqrt{3} & -1/3y & -1/3y & 2/3y \end{bmatrix} \begin{bmatrix} V_p^a \\ V_p^b \\ V_p^c \\ V_s^A \\ V_s^B \\ V_s^C \end{bmatrix}, \quad (6.78)$$

where y is the transformer leakage admittance in per unit, which is approximated by

$$Y_{\text{th}} = \frac{1}{R\sqrt{h} + jX_1 h}, \quad (6.79)$$

where R = the resistance at fundamental frequency and X_1 = the transformer's leakage reactance.

An example of a typical variation of the inductive coefficient of a transformer with frequency is shown in Figure 6.18.

The magnetising admittance is usually ignored since under normal operating conditions its contribution is not significant. If, however, the transformer is under severe saturation, appropriate current harmonic sources must be added at the transformer terminals.

In general, any two-winding three-phase transformer may be represented by two coupled compound coils as shown in Figure 6.19 where $[Y_{sp}] = [Y_{ps}]^T$.

If the parameters of the three phases are assumed balanced, all the common three-phase connections can be modelled by three basic submatrices. The submatrices $[Y_{pp}]$, $[Y_{ps}]$, etc. are given in Table 6.2 for the common connections in terms of the following matrices.

$$Y_1 = \begin{bmatrix} y_t & & \\ & y_t & \\ & & y_t \end{bmatrix}, \quad Y_{11} = \begin{bmatrix} 2y_t & -y_t & -y_t \\ -y_t & 2y_t & -y_t \\ -y_t & -y_t & 2y_t \end{bmatrix}, \quad Y_{111} = \begin{bmatrix} -y_t & y_t & \\ & -y_t & y_t \\ y_t & & -y_t \end{bmatrix}.$$

For transformers with neutrals connected through an impedance, an extra coil is added to the primitive network for each unearthed neutral and the primitive admittance

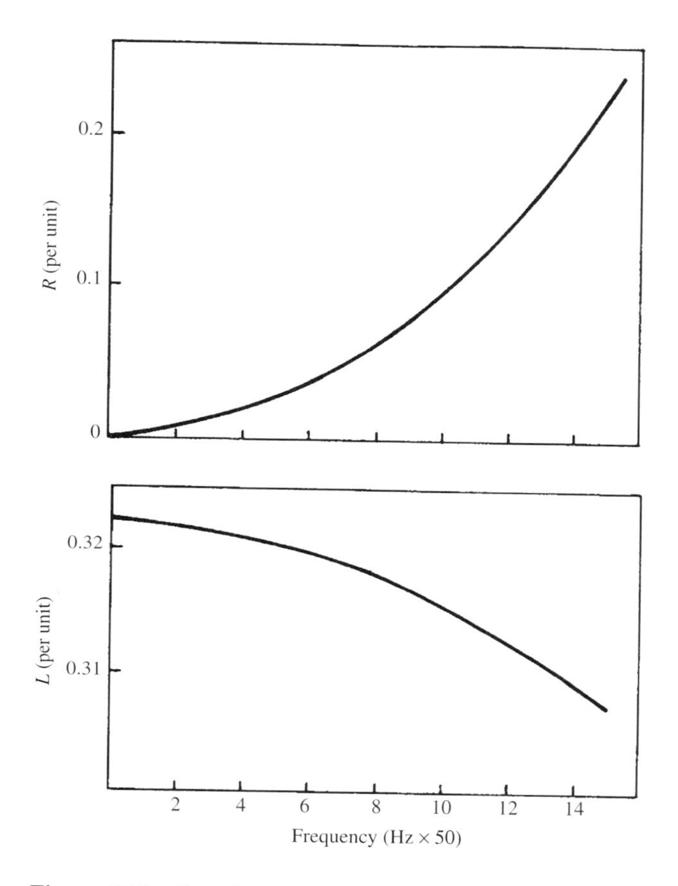

Figure 6.18 Transformer parameter frequency dependency

increases in dimension. However, by noting that the injected current in the neutral is zero (no direct connection), these extra terms can be eliminated from the connected network admittance matrix. This results in the matrix.

$$Y = \begin{bmatrix} y_t - c & -c & -c \\ -c & y_t - c & -c \\ -c & -c & y_t - c \end{bmatrix}$$

where $c = y_t.y_t/(3.y_t + y_n)$.

Once the admittance matrix has been formed for a particular connection it represents a simple subsystem composed of the two busbars interconnected by the transformer.

6.4.5 Generator Modelling

For the purpose of determining the network harmonic admittances, the generators can be modelled as a series combination of resistance and inductive reactance, i.e.

$$Y_{gh} = \frac{1}{R\sqrt{h} + jX''_d h}, \tag{6.80}$$

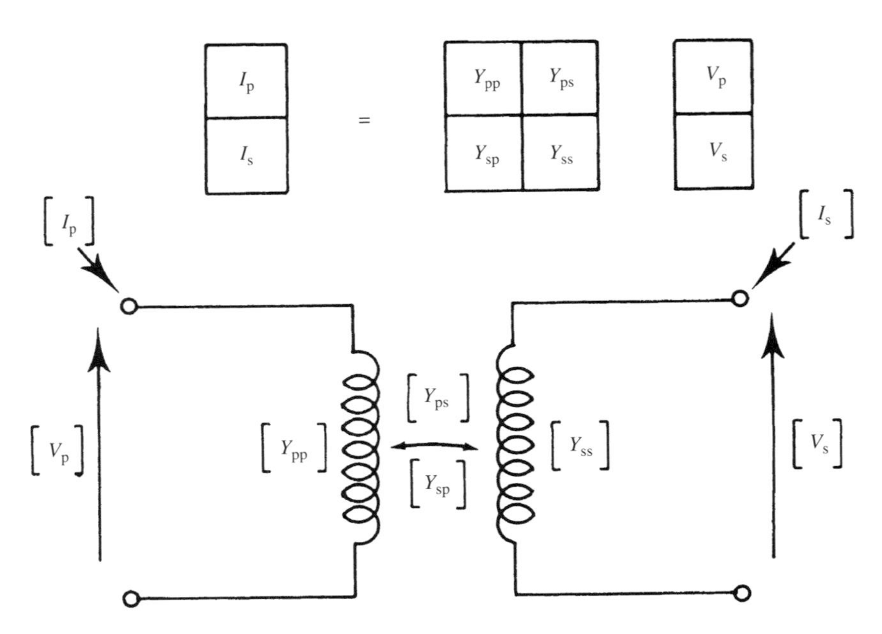

Figure 6.19 Two-winding three-phase transformer as two coupled compound coils

Table 6.2 Characteristic submatrices used in forming the transformer admittance matrices

Transformer connection		Self-admittance		Mutual admittance
Bus P	Bus S	Y_{pp}	Y_{ss}	Y_{ps}, Y_{sp}
Wye G	Wye G	Y_1	Y_1	$-Y_1$
Wye G	Wye	Y_1	$Y_{11/3}$	$-Y_{11/3}$
Wye G	Delta	Y_1	Y_{11}	Y_{111}
Wye	Wye	$Y_{11/3}$	$Y_{11/3}$	$-Y_{11/3}$
Wye	Delta	$Y_{11/3}$	Y_{11}	Y_{111}
Delta	Delta	Y_{11}	Y_{11}	$-Y_{11}$

where $R =$ derived from the machine power losses and $X''_{\mathrm{d}} =$ the generator's subtransient reactance.

A frequency-dependent multiplying factor can be added to the reactance terms to account for the skin effect. It should be noted that this is not valid at fundamental frequency since the positive sequence component still sees the synchronous impedance due to the flux not rotating with respect to the rotor.

6.4.6 Shunt Elements

Shunt reactors and capacitors are used in a transmission system for reactive power control. The data for these elements is usually given in terms of their rated megavoltamps and rated kilovolts and the equivalent phase admittance in per unit is calculated from this data.

The coupled admittances to ground at bus k are formed into a 3×3 admittance matrix as shown in Figure 6.20, and this reduces to the compound admittance representation indicated. The admittance matrix is incorporated directly into the system admittance matrix, contributing only to the self-admittance of the particular bus.

While provision for off-diagonal terms exists, the admittance matrix for shunt elements is usually diagonal, since there is normally no coupling between the components of each phase.

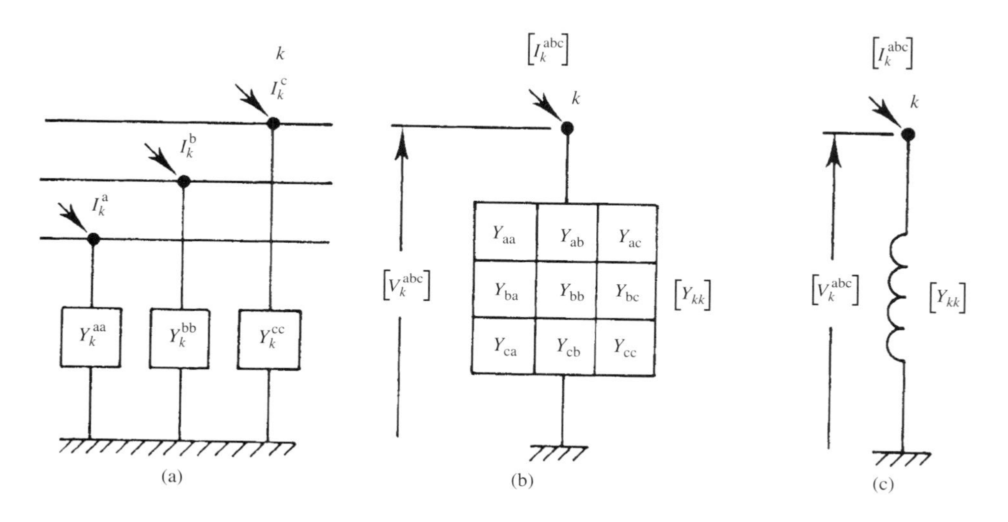

Figure 6.20 Representation of a shunt element: (a) coupled admittance; (b) admittance matrix; (c) compound admittance

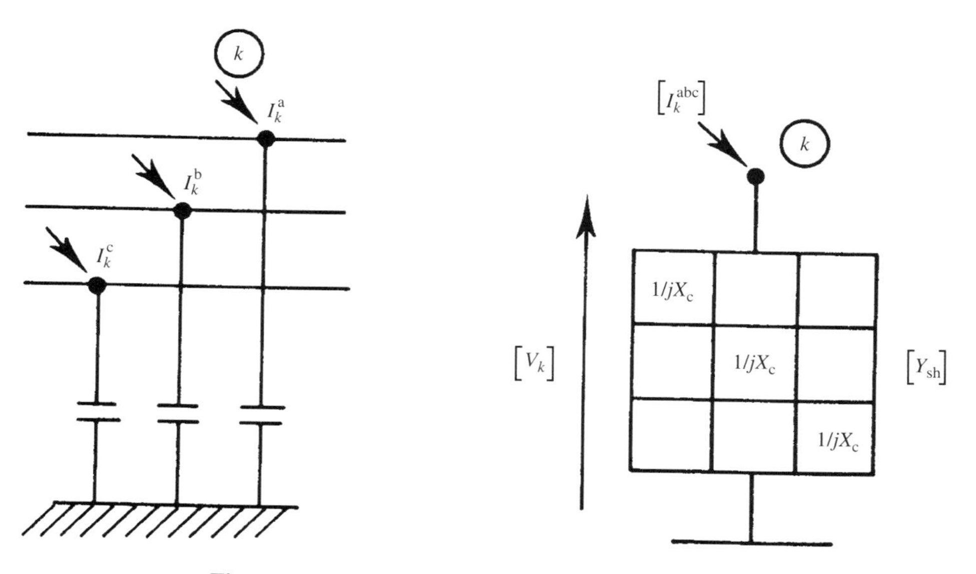

Figure 6.21 Representation of a shunt capacitor bank

Consider, as an example, the three-phase capacitor bank shown in Figure 6.21. A 3×3 matrix representation similar to that for a line section is illustrated.

The megavolt-amp rating at fundamental frequency (Q) and the nominal voltage (V) are normally used to calculate the capacitive reactance at the n harmonic, i.e. $X_c = V^2/nQ$.

In terms of *ABCD* parameters, the matrix equation of a shunt element is

$$\begin{bmatrix} V_s \\ I_s \end{bmatrix} = \begin{bmatrix} [U] & \\ [Y_{sh}] & [U] \end{bmatrix} \times \begin{bmatrix} V_r \\ -I_r \end{bmatrix}, \tag{6.81}$$

where $[Y_{sh}] =$ diag (shunt admittance of each phase) and $[U] =$ identity matrix.

However, in harmonic analysis, any added inductance, often placed in series with shunt capacitors, must be explicitly represented. For floating star or delta-connected configurations, the procedure used in section 6.4.4. for the transformer representation should be followed.

6.4.7 Series Elements

Series elements are connected directly between two buses and for modelling purposes they constitute a subsystem in the network subdivision.

A three-phase coupled series admittance between two busbars i and k is shown in Figure 6.22(a), as well as its reduced nodal admittance matrix (Figure 6.22(b)) and compound admittance (Figure 6.22(c)).

The series capacitor, used for transmission line reactance compensation, is an example of an uncoupled series element; in this case, the admittance matrix is diagonal. For a lumped series element, the *ABCD* parameter matrix equation is

$$\begin{bmatrix} V_s \\ I_s \end{bmatrix} = \begin{bmatrix} [U] & [Z_{se}] \\ & [U] \end{bmatrix} \times \begin{bmatrix} V_r \\ -I_r \end{bmatrix}, \tag{6.82}$$

where $[Z_{se}] =$ diag (series impedance of each phase) and $[U] =$ identity matrix.

6.4.8 Distribution and Load System Modelling

The harmonic impedances seen from primary transmission system buses are greatly affected by the degree of representation of the distribution system and the consumer loads fed radially from each busbar.

A typical simplified *dominant* configuration of a distribution feeder is shown in Figure 6.23. Generally, the bulk of the load fed from distribution feeders is located behind two transformers downstream. Thus, to calculate the harmonic impedances seen from the high-voltage primary transmission side it may be sufficient to use a discrete model of the composite effect of many loads and distribution system lines and transformers at the high-voltage side of the main distribution transformers; typically, the 110 kV in a system using 400 kV and 220 kV transmission.

The aggregate nature of the load makes it difficult to establish models based purely on theoretical analysis. Attempts to deduce models from measurements have been

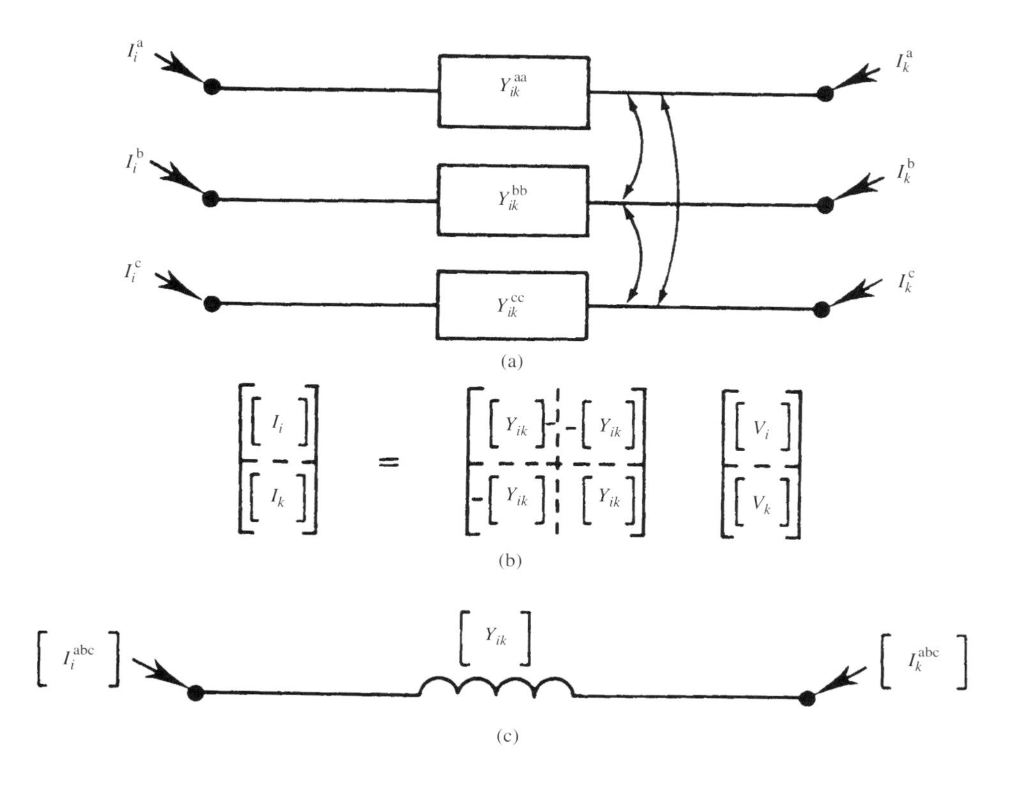

Figure 6.22 Representation of a series element: (a) coupled admittances; (b) admittance matrix; (c) compound admittance

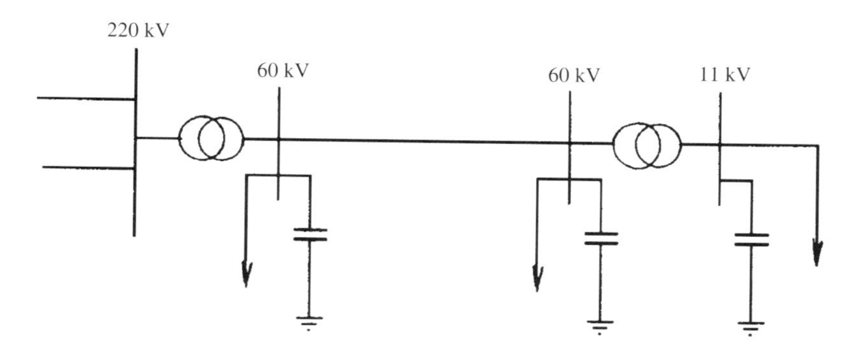

Figure 6.23 Typical distribution feeder

made [24] but lack general applicability. Utilities should be encouraged to develop a database of their electrical *regions*, with as much information as possible to provide accurate equivalent harmonic impedances for future studies.

The following guidelines are recommended for the derivation of distribution feeder equivalents.

- Distribution lines and cables (e.g. 69–33 KV) should be represented by an equivalent-π model. For short lines, the total capacitance at each voltage level should be estimated and connected at that busbar.
- Transformers between distribution voltage levels should be represented by an equivalent element.
- As the active power absorbed by rotating machines does not correspond to a damping value, the active and reactive power demand at the fundamental frequency may not be used in a straightforward manner. Alternative models for load representation should be used according to their composition and characteristics.
- Power factor correction (PFC) capacitance should be estimated as accurately as possible and allocated at the corresponding voltage level.
- Other elements, such as transmission line inductors, filters and generators, should be represented according to their actual configuration and composition.
- The representation should be more detailed nearer the points of interest. Simpler equivalents for the transmission and distribution systems should be used only for remote points.
- All elements should be uncoupled three-phase branches, including unbalanced phase parameters.

There are no *generally acceptable* load equivalents for harmonic analysis [25]. In each case, the derivation of equivalent conductance and susceptance harmonic bandwidths from specified P (active) and Q (reactive) power flows will need extra information on the actual composition of the load. Power distribution companies will have a reasonable idea of the proportion of each type in their system depending on the time of day and should provide such information.

Consumers' loads constitute not only the main element of the damping component but may affect the resonance conditions, particularly at higher frequencies. Some early measurements [26] showed that maximum plant conditions can result in reduced impedances at lower frequencies and increased impedances at higher frequencies. Simulation studies [27] have also demonstrated that the addition of detailed load representation can result in either an increase or decrease in harmonic flow.

Consequently, an adequate representation of the system loads is needed. There are basically three types of loads — passive, motive and power electronic.

1 Predominantly passive loads (typically domestic) can be represented approximately by a series R, X impedance, i.e.

$$Z_{\mathrm{r}}(\omega) = R_{\mathrm{r}}\sqrt{h} + jX_{\mathrm{r}}h, \tag{6.83}$$

where

R_{r} = load resistance at the fundamental frequency,
X_{r} = load reactance at the fundamental frequency,
h = harmonic orders (ω/ω_1).

The weighting coefficient $\sqrt{h}$, used above for frequency dependence of the resistive component, is different in different models; for instance, [6] uses a factor of 0.6 $\sqrt{h}$

instead. The equivalent inductance represents the relatively small motor content when known.

In studies concerning mainly the transmission network the loads are usually equivalent parts of the distribution network, specified by the consumption of active and reactive power. Normally a parallel model is used, i.e.

$$Y_{\mathrm{L}}(\omega) = 1/R_{\mathrm{p}} + j1/(X_{\mathrm{p}}h), \tag{6.84}$$

where

R_{p} = load resistance at the fundamental frequency
X_{p} = load reactance at the fundamental frequency
h = harmonic orders (ω/ω_1).

$$X_P = \frac{V^2}{Q}, \qquad R = \frac{V^2}{P}. \tag{6.85}$$

There are many variations of this parallel form of load representation. For example, the parallel load model suggested by [25] is a parallel connection of inductive reactance and resistance whose values are

$$X = j\frac{V^2}{(0.1h + 0.9)Q}, \qquad R = \frac{V^2}{(0.1h + 0.9)P}, \tag{6.86}$$

where P and Q are fundamental frequency active and reactive powers.

2 Various models of predominantly motive loads have been suggested using resistive–inductive equivalents, their differences being often due to the boundary of system representation. A detailed analysis of the induction motor response to harmonic frequencies, leading to a relatively simple model, is described later in this section.

3 Modelling the power electronic loads is a more difficult problem because, besides being harmonic sources, these loads do not present a constant *R, L, C* configuration and their non-linear characteristics cannot fit within the linear harmonic equivalent model. In the absence of detailed information, the power electronic loads are often left open-circuited when calculating harmonic impedances. However, their effective harmonic impedances need to be considered when the power ratings are relatively high, such as arc furnaces, aluminium smelters, etc.

An alternative approach to explicit load representation based on detailed information is the use of empirical models derived from measurements.

In particular, information obtained from harmonic current and voltage measurements with different operating conditions; for example, by switching a shunt capacitor (as described in section 6.3.1 and Figure 6.3), can be used to derive approximate Norton harmonic equivalents of the load or group of loads connected at a distribution bus. By way of example, Figure 6.24 shows measured and calculated harmonic voltages at four time intervals for an 11 kV distribution bus fed from the 220 kV transmission system *via* a 40 MVA transformer [28]. Clearly, the Norton approach gives a better estimation of the harmonic voltages than the constant current source.

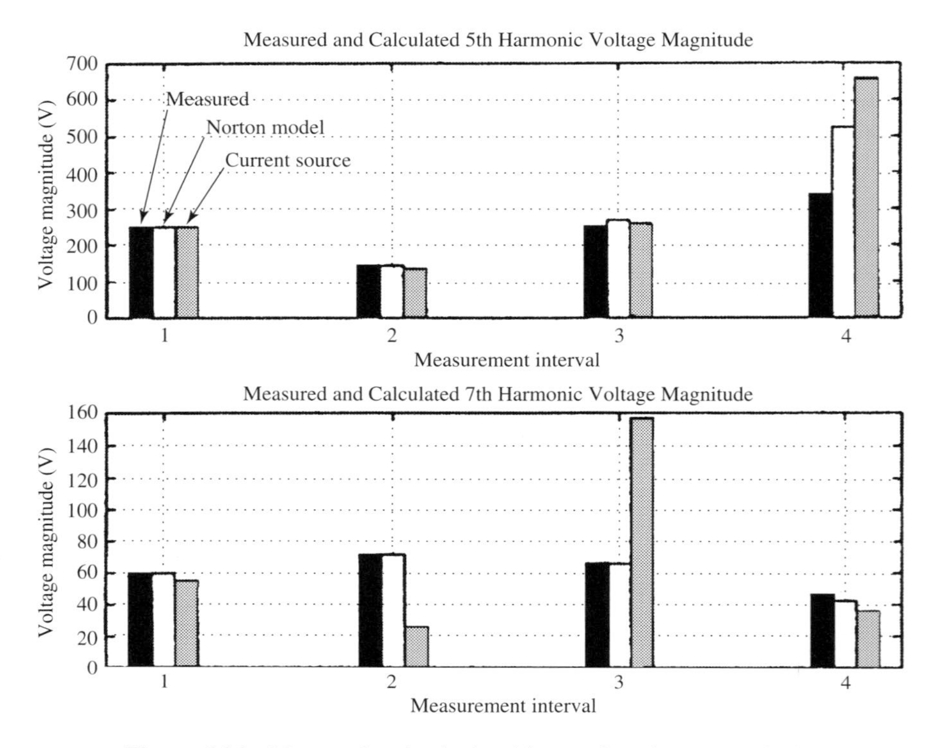

Figure 6.24 Measured and calculated harmonic voltage magnitude

Induction Motor Model The circuit shown in Figure 6.25 is an approximate representation of the induction motor, with the magnetising impedance ignored.

The motor impedance at any frequency can be expressed as:

$$Z_{\mathrm{m}}(\omega) = R_{\mathrm{mh}} + jX_{\mathrm{mh}}. \tag{6.87}$$

At the fundamental frequency ($h = 1$)

$$X_{\mathrm{m}1} = X_1 + X_2 = X_{\mathrm{B}}, \tag{6.88}$$

$$R_{\mathrm{m}1} = R_1 + \frac{R_2}{S} = R_{\mathrm{B}}\left(a + \frac{b}{S}\right), \tag{6.89}$$

where

R_{B} = total motor resistance with the rotor locked,
R_1 = stator resistance related to R_{B} by coefficient a (which is typically 0.45),
R_2 = rotor resistance related to R_{B} by coefficient b (which is typically 0.55),
X_{B} = total motor reactance with the rotor locked,
S = slip = $\dfrac{\omega_{\mathrm{s}} - \omega_{\mathrm{r}}}{\omega_{\mathrm{s}}}$.

At harmonic frequencies:

$$X_{\mathrm{mh}} = h \cdot X_{\mathrm{B}}, \tag{6.90}$$

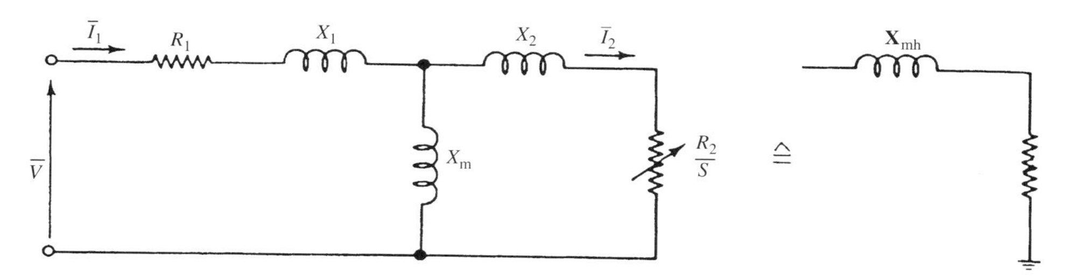

Figure 6.25 Approximate representation of the induction motor

$$R_{mh} = R_B \left(a \cdot k_a + \frac{b}{S_h} \cdot k_b \right), \tag{6.91}$$

where

k_a, k_b = correction factors to take into account skin effect in the stator and rotor, respectively

S_h = apparent slip at the superimposed frequency, i.e.

$$S_h = \frac{\pm h\omega_s - \omega_r}{\pm h\omega_s} \tag{6.92}$$

i.e.

$$S_h \approx 1 - \frac{\omega_r}{h\omega_s} \quad \text{for the positive sequence harmonics,}$$

$$S_h \approx 1 + \frac{\omega_r}{h\omega_s} \quad \text{for the negative sequence harmonics,}$$

Assuming an exponential variation of the resistances with frequency, i.e.

$$k_b = h^{\propto},$$

$$k_b = (\pm h - 1)^{\propto},$$

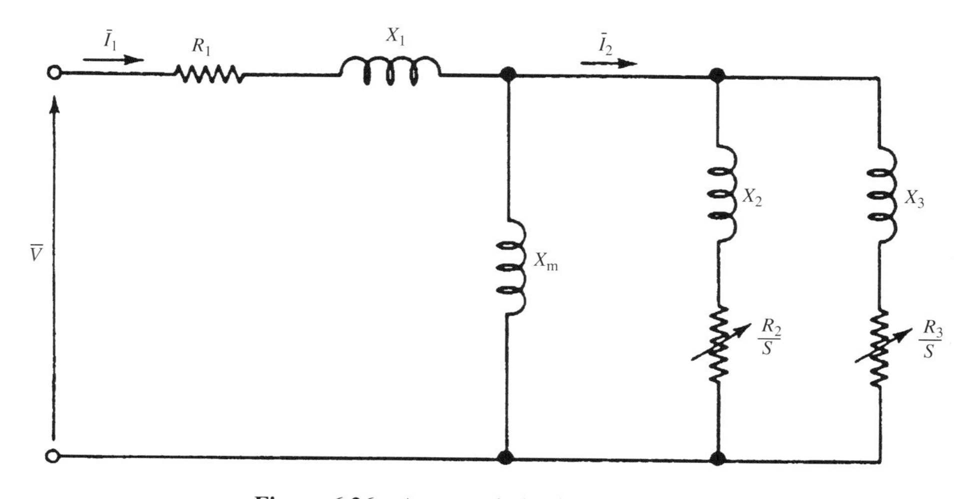

Figure 6.26 Accurate induction motor model

the motor equivalent resistance for $\alpha = 0.5$ becomes

$$R_{\mathrm{mh}} = R_{\mathrm{B}}[a\sqrt{h} + (\pm h.b\sqrt{\pm h - 1})/(\pm h - 1)]. \tag{6.93}$$

An accurate model of a double cage induction motor is shown in Figure 6.26.

6.5 Implementation of the Harmonic Analysis

The evolution of computer technology has removed many of the limitations that affected implementation decisions in the past.

Earlier implementations were restricted by the use of main frame computers, and limitations in graphical support, memory and storage space.

The main factor affecting recent implementations has been the acceptability of the personal computer (PC) as the main computing platform in terms of capability and price. Other important developments have been the availability of ample computer bandwidths, reduced cost of memory modules making memory limitation secondary, cheaper storage modules with much larger capacity, great improvement in software support and development support tools.

To cope with the larger size and complexity of computer programs, new facilities have become available to make the software more modular and easier to maintain. Three important examples are multi-tasking operating systems, graphical user interfaces, and object-oriented design methodology.

6.5.1 Harmonic Penetration Overview

A modern harmonic penetration program includes a graphical user interface, the simulation algorithm engine and a database handling-data structure.

The graphical user interface (GUI) is used for data entry (component parameters and network topology), and for the presentation of the simulation results.

The simulation algorithm engine performs the calculations required and has traditionally been written in the FORTRAN language, although recently object-oriented languages have also been used. Mixed language programming makes it possible to use FORTRAN's advantage of built-in complex number manipulation, with other languages to benefit from the strengths of each. However, in the race to add new desirable features to the computer languages, bugs are introduced that reduce the reliability of the compiler implementations.

Database handling routines are used to store and retrieve the power system data from disk. However, the traditional fixed format, still widely used today, is very rigid and inflexible. The commercial databases used in the information technology field can also be employed for storing power system data. Although these are largely platform dependent, there are several proprietary versions that support several platforms. In this text, the use of a platform-independent ASCII text format, taken from the Windows™ INI file ideas, has been found to be extremely flexible.

6.5.2 Computer Implementation

The first consideration must be the computing environment on which the software will be executed. The general acceptance of the PC is now affecting the power system area.

This application has been enhanced by the use of several graphical operating systems, and particularly Microsoft Windows™, in the PC platform.

Other factors to be taken into consideration at the outset are:

- the availability of software tools, advanced programming languages and object-oriented methodologies specifically targeting the application software.
- the rapid improvements made in computer hardware that have practically eliminated the traditional limitations of computer capability, memory and storage.
- the presence of graphical user-interface facilities under all computer environments.
- the multi-tasking environment, realisable with practically no effort even on the PC platform.

With more specific reference to the algorithm implementation, the software and user-interface development should aim at simplifying the setting up and maintenance of the source code and databases. Of course, the main consideration should be to achieve reliable evaluation and accuracy in the simulation results.

In line with the steps described in the previous sections of the chapter, the implementation will involve the following stages:

- computation of the admittance matrices of individual components at the specified harmonic frequencies,
- formation of the system admittance matrices at individual frequencies according to the network topology,
- calculation of the system harmonic voltages at all the system nodes given the harmonic current injections at the nodes containing non-linear plant components.

6.5.3 Program Structure

Lack of memory space has forced simulation packages to be fragmented into several programs so that the executable binary and the corresponding data can be accommodated by the computer systems. Moreover, lack of computer power has often made it necessary to store intermediate results rather than recalculate them as and when needed. Hence, proper procedures must be followed when performing the simulation and extra vigilance is required to ensure that each of the fragments performs correctly, since any mistake in one of the processes will render the entire effort futile.

The structure of a typical software package for harmonic analysis is illustrated in Figure 6.27. The computational effort required to generate the harmonic admittance matrices of transmission lines and cables usually exceeds that of forming the system admittance matrix and solving for the harmonic voltages and currents. Therefore, these matrices are usually computed separately and stored in disk files, which are then read by the main simulation program when forming the system admittance matrices. The modelling of non-linear loads has also been separated from the main program; the calculated current injections are then stored in disk files and read by the main program when performing the harmonic penetration analysis. The results of the harmonic penetration study are also outputted in disk files, typically in row and column text format which can be imported into plotting tools for presentation. These can also be used by other tools such as MATLAB™ for further analysis or spread-sheets for reporting purposes.

Although the above subdivisions have enabled complex analysis to be performed successfully despite the limitations presented by earlier computers, their main drawback is the use of many intermediary files. Maintenance of these intermediary files becomes laborious when the simulated system is large, since all these files must be updated before the succeeding program is activated. Moreover, the conversion of complex or floating numbers into the text format typically used in these files introduces truncation or round-off errors. These errors may distort the final results, especially at high harmonic frequencies when the distortion levels are generally low, particularly when the analysis process involves many such conversions. The use of unformatted data has the advantages of reduced file size, and faster reading and writing to disk operations without introducing truncation errors; however, the data in the file cannot be readily inspected.

Typically, a graphical data entry or editor is added to the package to help the user to construct the simulation cases. These editors are usually separate applications that are able to generate the necessary data files required in the simulation. However, these applications employ their own *graphic* files for maintaining the details of the constructed cases. The simulation results are imported into plotting applications specifically designed for presenting harmonic results, or other applications such as MATLAB™ or Microsoft Excel™ for further analysis or reporting purposes.

Several advancements in the computer technology have made it possible to integrate all the above processes into a single framework without the need to combine them all into a single huge executable binary. The most important developments are:

- abundance of cheap memory and new operating systems that permit the computers to accommodate large executable with extensive data storage. In addition, the utilisation of virtual memory technology (paging to hard disks) has eliminated the traditional limitation of insufficient memory space for executable binary and simulation data. Furthermore, the dynamic memory allocation facility has enabled optimised usage of the memory space, since memory space is only reserved when needed, being released immediately when it is no longer required.

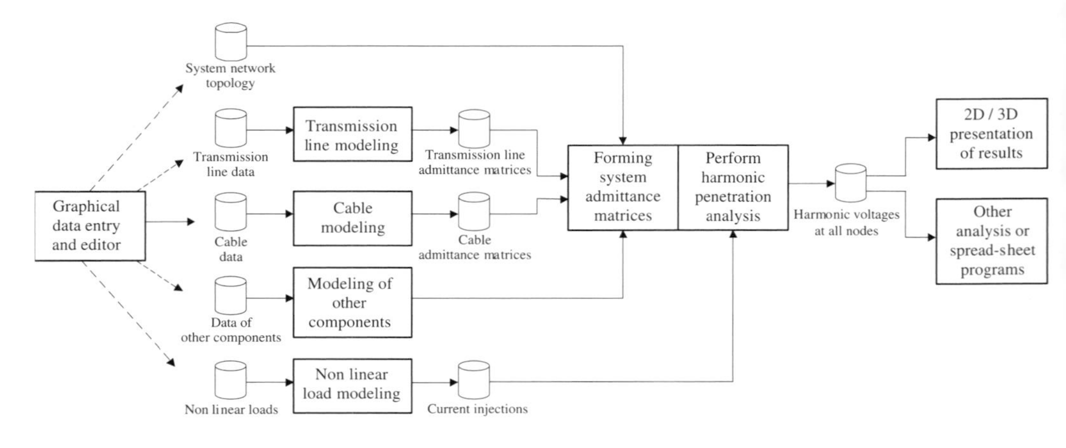

Figure 6.27 Structure of harmonic analysis software

- application code that can be made relocatable and dynamically loaded into the memory and called from the main program only when it is needed. These binaries are generally referred to as dynamic linked library (DLL) and they usually contain procedures and data types that are common to a certain specific function.
- wide availability of object-oriented development tools that permits building modular structures for the complicated simulation programs. This object-oriented technique is usually combined with the dynamic memory allocation facility to enable the creation and deletion of objects on the fly. This combination forms a potent tool for tackling the memory problem in sophisticated or comprehensive software applications.
- a new technology largely known in the software community as object linking and embedding (OLE) that allows useful and powerful software objects to be included into the user's application program. A typical example of such binary is the 2D/3D plotting facility.
- mixed language programming that enables different parts of the package to remain with their most suitable programming languages, but allows them to interface with each other without using intermediary disk files and error-prone data conversions.

Making use of these new technologies, Figure 6.28 depicts how the harmonic penetration simulation program is put together. The mathematical modelling of the power system components including non-linear loads, and the harmonic penetration are programmed as relocatable dynamic linked libraries and these are made accessible to the graphical user interface. Through this graphical user interface, users construct the power system simulation network by joining different power system components

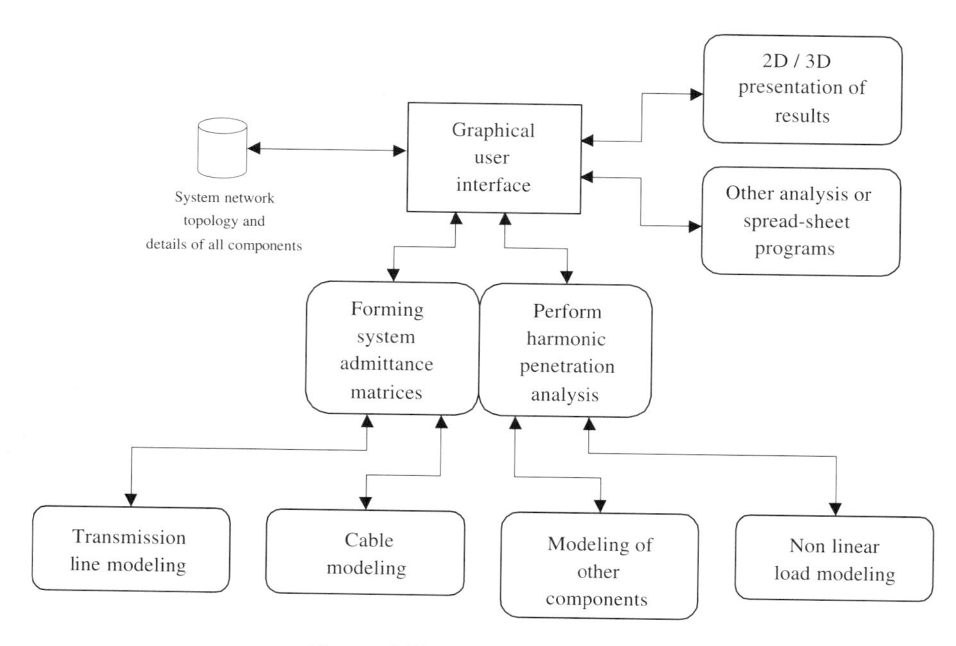

Figure 6.28 Functional overview

together and specifying the component parameter settings. The constructed simulation network consists of lists of power system components and their settings; these are passed to the harmonic penetration library to carry out the tasks required by the simulation study, as described in section 6.5.2.

The next section on data structure elaborates further on the handling of the power system components for the simulation study. The graphical user interface makes use of commercial 2D and 3D plotting OLE facilities for presenting the simulation results. The results can also be manipulated using other commercial applications that support OLE, from within the graphical user interface. This implementation, depicted in Figure 6.28, enables a single interface to be used for constructing the simulation cases, for executing the simulation and for the presentation of the simulation results. It avoids the use of intermediary disk files and hence eliminates the possible errors caused by data conversions and truncations.

6.5.4 Data Structure

The essence of an object-oriented application is a family of classes that make up the application. Each of the classes has its own unique properties or settings and includes procedures to alter the settings. At execution, objects of the classes are created when they are needed to perform certain tasks and they are usually deleted upon completion of these tasks. Some of the objects may be based on the same class and hence have similar properties. However, the properties may be set with different values and hence differentiate one object from another object of a similar class. Consequently, an object-oriented application basically consists of numerous objects of the classes defined in the application.

The data used in the harmonic penetration study includes the power system components and the topology by which they are connected to form a system. Therefore, based on the object-oriented methodology, a simulation case can be regarded as a drawing page or canvas object, which acts as a container for all the power system components objects making up the system. The various classes of objects making up the object-oriented version of harmonic analysis are summarised in Figure 6.29.

The figure shows hierarchically the classes that are defined in the harmonic analysis software, with the child classes resting on top of the parent classes. The two

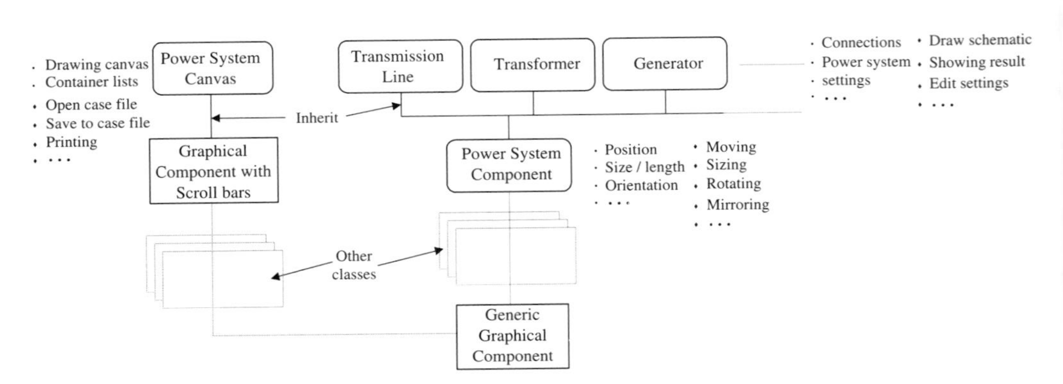

Figure 6.29 Power system component models

main classes are the Power System Canvas and the Power System Component classes. Both classes are inherited from the Generic Graphical Component, which provides the interfaces to the low-level graphical functions of the GUI operating system (which is Microsoft Windows^{TM} in this implementation). Hence, these components are able to determine how they are to be displayed without having to manipulate the low-level graphical functions.

The power system canvas class, inherited directly from the scroll box class, already has the scrolling bars programmed into the class. Hence, the power system canvas class is able to make use of these scrolling bars without having to be concerned about their creation, deletion and display. It merely sets the position of the scrolling thumb tab to indicate the position of its current view within the canvas. Among the properties of this canvas class are the container lists for storing the power system component objects that make up the simulation network, and a canvas object for displaying these components. However, it does not possess the function needed to draw each of the power system components on the canvas. These drawing functions reside with the individual power system component class.

The power system component class acts as the parent for all power system components. It consists of several general properties pertaining to the display of any component. Such properties include the component size or length (for transmission line and cable), component position within the canvas, its orientation. It also contains several general functions for manipulating the display of the component, such as mirroring, rotating, sizing. Each specific power system component such as transmission line, transformer and generator are defined as separate classes inheriting from the same power system component class. These classes have their individual settings such as the winding configuration in the transformer class, line height and ground resistivity in the transmission line class. They also contain functions specific to the component itself, such as the drawing function for displaying the component, and a function to compute the component admittance matrices. In this way, a simulation case will contain objects of these classes (which are maintained as several container lists as mentioned above) and the properties of each of these objects are set according to the makeup of the network.

Figure 6.30 illustrates the use of the object-oriented method in building a network for simulation studies. It shows a three busbar power system consisting of a generator and a load connected through a two winding transformer, and a double circuit transmission line. All of these components are inherited from the common power system component class. They differ from each other by the functions used to draw them resulting in different images and requiring different editing forms, as shown for the generator and transformer in the figure. Consequently, the simulation test case is made up of three objects of busbar class and one object each of the generator, transformer, transmission lines and load classes. The busbar objects are differentiated from each other by their busbar name property.

In this object-oriented manner, a power system network is described by lists of these power system component objects. There are altogether five container lists in the example above, corresponding to the five different classes. These lists can be created automatically as the network is built up using a graphical editor, or they are constructed according to the information stored in the data file. These lists are then passed to the simulation engine, which constructs the system admittance matrices at the respective

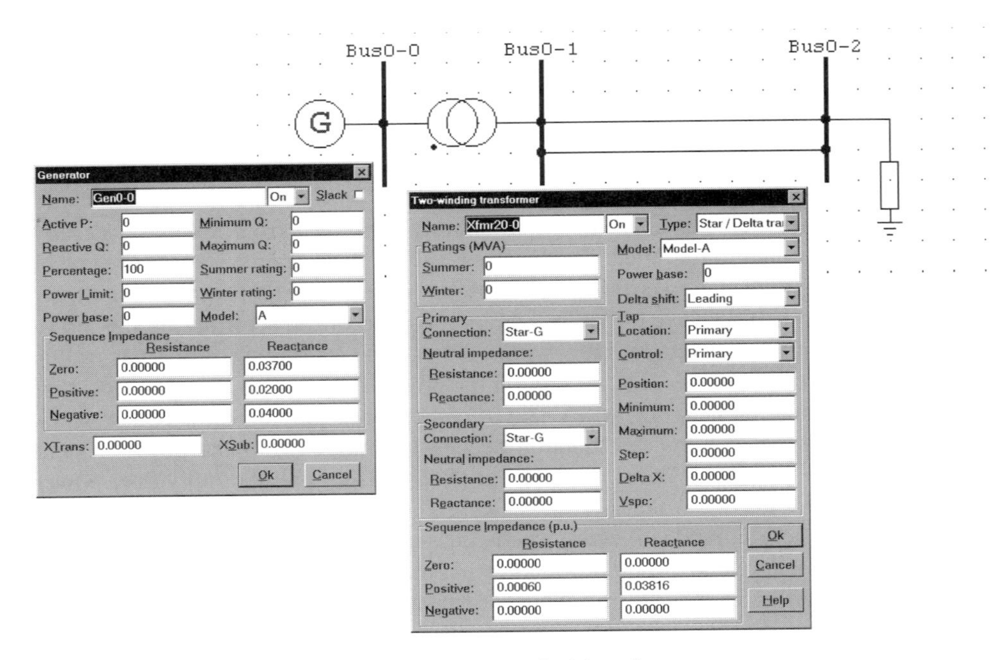

Figure 6.30 Typical interface

harmonic frequencies, and carries out the harmonic penetration studies as outlined in earlier sections. Each of the power system component objects also contains properties for recording simulation results. Hence, the graphical user interface can then extract the result values from the lists of objects and display them on the canvas as shown in Figure 6.31. The presentation of the results is customised, allowing them to be presented differently according to the component type. Harmonic voltages are only shown on the busbar objects while multi-terminal objects such as the transformer and transmission line will show the harmonic currents at the terminals, with the direction of the flow indicated using arrows. Finally, the colour of the arrows and the result texts can be used to indicate whether the values exceed certain limits.

6.5.5 Database Format

The database used in harmonic penetration analysis can be divided into graphical and power system data. The graphical database contains information required for displaying the power system network in a graphical user interface. The power system database, on the other hand, contains network topology information that describes the make-up of the simulated power system network. Although it is possible to combine both databases, hence dealing with only one file, there are compelling advantages in separating them. First and foremost is the close association of the graphical user interface with the underlying operating system. Hence, there is no advantage in making the database platform independent. On the other hand, the simulation engine is not operating system dependent and hence making this database platform independent would ease the future possible migration of the simulation engine to other operating systems.

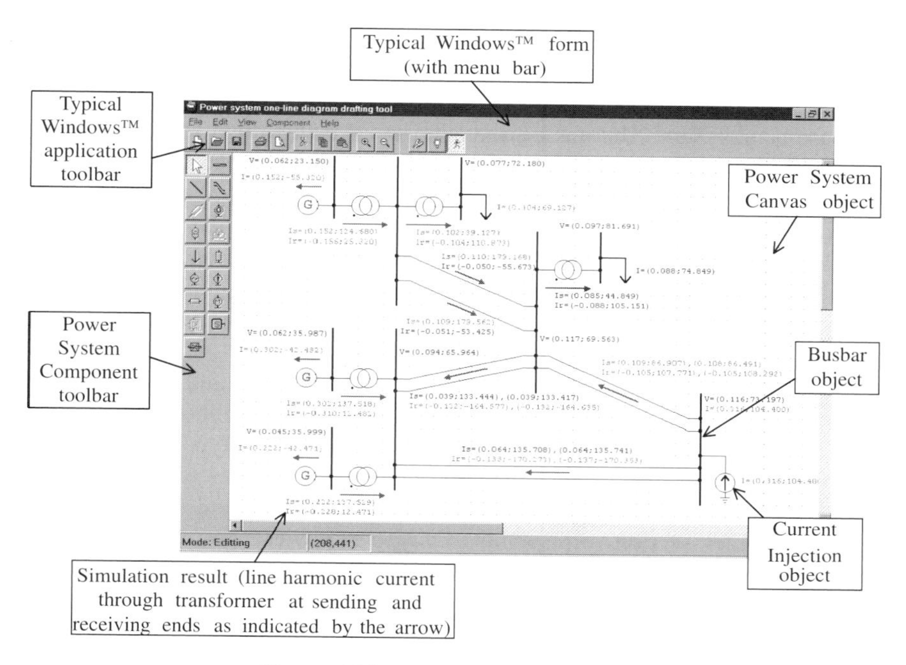

Figure 6.31 View of solution results

Thus, the implementation of harmonic penetration analysis discussed here separates the storing of these two databases.

In this alternative, each simulation case contains a graphical database file, which in turn contains a reference to a power system database file. Each time a simulation case is loaded from the graphical database file, the power system component settings are then retrieved from the corresponding power system database file. Hence, several simulation cases, each with their own graphical database file, may refer to a common power system database. This arrangement facilitates the centralised control and maintenance of the power system database. Moreover, new simulation cases, including simplified cases, can be constructed by referring to the centralised power system database and hence eliminating the tedious task of entering the component settings.

Generally, the power system database has a longer life span than its graphical counterpart, which depends on the survival of the graphical user interface (which in turn depends on the usability of the operating system). Moreover, a common power system database can be shared among different simulation programs with different graphical user interfaces. For instance, the three-phase graphical display used in the harmonic state estimation program, described in Chapter 7, can share the same power system database as the single-line diagram used by the harmonic penetration software.

The graphical database consists of a fixed binary format file and file-stream objects; these are available through the graphical design package and are used for writing to and reading from the database file. The file stream objects are incorporated into the Power System Canvas object and hence, user interfaces for any power system analysis software designed with this Canvas object have easy and transparent access to the graphical database.

On the other hand, the power system database is designed to be a text file so that it is easier to transfer the database from one operating system to another. Moreover, the text format will make the database file human readable, enabling manual checking of the database if necessary. The main disadvantages of using a text format are increased file size and increased accessing (both reading from and writing to file) time. Although these disadvantages have been reduced with the availability of abundant disk storage space and fast disk access speed, data reading is still a considerable proportion of the solution time; for a 110 bus 50 harmonic solution, data reading is about 20% of the total. However, a new GUI could reference the old graphical database and thus avoid re-entering the graphical system. As computer systems are upgraded, a system-independent graphical database will facilitate the task.

Generally, power systems have a fixed and rigid structure and their design takes advantage of data formats such as the well-known IEEE common data format and the PTI (Power Technology Incorporated) data format. These types of database typically start with several lines of comments or general descriptions of the power system network. These are followed by sections of lines delimited by some code word such as '999' or 'End'. Each section corresponds to a type of power system components. Each line (or several subsequent lines) within a section has a fixed format with multiple fixed-width columns for specifying the setting values of a particular component. An example of this is the IEEE common data format illustrated in Figure 6.32.

Some database formats, such as PTIs, also require the section to be arranged in a particular order. In PTI format, the first section contains busbar data and is followed by that of generators. Each section is then terminated (and hence denotes the start of another section) by a component with an index number of zero.

This type of format is easy to implement and its popularity is largely due to the rather rigid file access routines in the FORTRAN programming language, the dominant language used in power system analysis. However, it is inflexible, and it is difficult to add new settings or change the layout of existing settings. This is because all software modules that make use of the database would have to be modified when the format was changed. Although, modular programming techniques can alleviate this deficiency, the fixed nature of the format makes it highly vulnerable to human errors. Missing spaces or extra spaces between columns may cause incomplete setting values to be read, resulting in erroneous simulation results. In fact, an incorrect database due to an incorrect number of spaces between columns is the most common obstacle faced by users of simulation software, including the experienced ones.

The design of database format must be considered in line with the routines used to access it. The compromise between the complexity of the access routines must be taken into account when considering the level of flexibility to be included in the database. A fixed format as described above would have minimal flexibility but the accessing routines would be trivial. On the other hand, an extendable format may require parser or decoder routines when reading the database. Similarly, an encoder will be needed to create the database in the first place. Figure 6.33 shows a simple but flexible format used in the harmonic analysis software. The main features of this format are:

- comment line starting with a ';' symbol,
- each record delimited by the '[and']' symbols,
- each setting having a keyword associated with it, i.e. Keyword = value.

```
09/25/93 UW ARCHIVE          100.0  1962 W IEEE 14 Bus Test Case
BUS DATA FOLLOWS                          14 ITEMS
   1 Bus 1     HV  1  1  3 1.060    0.0     0.0     0.0   232.4   -16.9    0.0  1.060   0.0    0.0
0.0    0.0         0
   2 Bus 2     HV  1  1  2 1.045  -4.98    21.7    12.7    40.0    42.4    0.0  1.045  50.0  -40.0
0.0    0.0         0
   3 Bus 3     HV  1  1  2 1.010 -12.72    94.2    19.0     0.0    23.4    0.0  1.010  40.0    0.0
0.0    0.0         0
   4 Bus 4     HV  1  1  0 1.019 -10.33    47.8    -3.9     0.0     0.0    0.0  0.0     0.0    0.0
0.0    0.0         0
   5 Bus 5     HV  1  1  0 1.020  -8.78     7.6     1.6     0.0     0.0    0.0  0.0     0.0    0.0
0.0    0.0         0
   6 Bus 6     LV  1  1  2 1.070 -14.22    11.2     7.5     0.0    12.2    0.0  1.070  24.0   -6.0
0.0    0.0         0
   7 Bus 7     ZV  1  1  0 1.062 -13.37     0.0     0.0     0.0     0.0    0.0  0.0     0.0    0.0
0.0    0.0         0
   8 Bus 8     TV  1  1  2 1.090 -13.36     0.0     0.0     0.0    17.4    0.0  1.090  24.0   -6.0
0.0    0.0         0
   9 Bus 9     LV  1  1  0 1.056 -14.94    29.5    16.6     0.0     0.0    0.0  0.0     0.0    0.0
0.0    0.19        0
  10 Bus 10    LV  1  1  0 1.051 -15.10     9.0     5.8     0.0     0.0    0.0  0.0     0.0    0.0
0.0    0.0         0
  11 Bus 11    LV  1  1  0 1.057 -14.79     3.5     1.8     0.0     0.0    0.0  0.0     0.0    0.0
0.0    0.0         0
  12 Bus 12    LV  1  1  0 1.055 -15.07     6.1     1.6     0.0     0.0    0.0  0.0     0.0    0.0
0.0    0.0         0
  13 Bus 13    LV  1  1  0 1.050 -15.16    13.5     5.8     0.0     0.0    0.0  0.0     0.0    0.0
0.0    0.0         0
  14 Bus 14    LV  1  1  0 1.036 -16.04    14.9     5.0     0.0     0.0    0.0  0.0     0.0    0.0
0.0    0.0         0
-999
BRANCH DATA FOLLOWS                        20 ITEMS
   1    2  1  1 1 0  0.01938   0.05917   0.0528   0   0   0   0 0  0.0     0.0 0.0   0.0
0.0    0.0   0.0
   1    5  1  1 1 0  0.05403   0.22304   0.0492   0   0   0   0 0  0.0     0.0 0.0   0.0
0.0    0.0   0.0
   2    3  1  1 1 0  0.04699   0.19797   0.0438   0   0   0   0 0  0.0     0.0 0.0   0.0
0.0    0.0   0.0
   2    4  1  1 1 0  0.05811   0.17632   0.0374   0   0   0   0 0  0.0     0.0 0.0   0.0
0.0    0.0   0.0
   2    5  1  1 1 0  0.05695   0.17388   0.0340   0   0   0   0 0  0.0     0.0 0.0   0.0
0.0    0.0   0.0
   3    4  1  1 1 0  0.06701   0.17103   0.0346   0   0   0   0 0  0.0     0.0 0.0   0.0
0.0    0.0   0.0
   4    5  1  1 1 0  0.01335   0.04211   0.0128   0   0   0   0 0  0.0     0.0 0.0   0.0
0.0    0.0   0.0
   4    7  1  1 1 1  0.0       0.20912   0.0      0   0   0   0 0  0.978   0.0 0.0   0.0
0.0    0.0   0.0
   4    9  1  1 1 1  0.0       0.55618   0.0      0   0   0   0 0  0.969   0.0 0.0   0.0
0.0    0.0   0.0
   5    6  1  1 1 1  0.0       0.25202   0.0      0   0   0   0 0  0.932   0.0 0.0   0.0
0.0    0.0   0.0
   6   11  1  1 1 0  0.09498   0.19890   0.0      0   0   0   0 0  0.0     0.0 0.0   0.0
0.0    0.0   0.0
   6   12  1  1 1 0  0.12291   0.25581   0.0      0   0   0   0 0  0.0     0.0 0.0   0.0
0.0    0.0   0.0
   6   13  1  1 1 0  0.06615   0.13027   0.0      0   0   0   0 0  0.0     0.0 0.0   0.0
0.0    0.0   0.0
   7    8  1  1 1 0  0.0       0.17615   0.0      0   0   0   0 0  0.0     0.0 0.0   0.0
0.0    0.0   0.0
   7    9  1  1 1 0  0.0       0.11001   0.0      0   0   0   0 0  0.0     0.0 0.0   0.0
0.0    0.0   0.0
   9   10  1  1 1 0  0.03181   0.08450   0.0      0   0   0   0 0  0.0     0.0 0.0   0.0
0.0    0.0   0.0
   9   14  1  1 1 0  0.12711   0.27038   0.0      0   0   0   0 0  0.0     0.0 0.0   0.0
0.0    0.0   0.0
  10   11  1  1 1 0  0.08205   0.19207   0.0      0   0   0   0 0  0.0     0.0 0.0   0.0
0.0    0.0   0.0
  12   13  1  1 1 0  0.22092   0.19988   0.0      0   0   0   0 0  0.0     0.0 0.0   0.0
0.0    0.0   0.0
  13   14  1  1 1 0  0.17093   0.34802   0.0      0   0   0   0 0  0.0     0.0 0.0   0.0
0.0    0.0   0.0
-999
LOSS ZONES FOLLOWS                     1 ITEMS
  1 IEEE 14 BUS
-99
INTERCHANGE DATA FOLLOWS                 1 ITEMS
 1    2 Bus 2     HV    0.0  999.99  IEEE14  IEEE 14 Bus Test Case
TIE LINES FOLLOES                       0 ITEMS
-999
END OF DATA
```

Figure 6.32 IEEE common data format example

```
; Power System Database: C:\POWERSYS\Software\Delphi\PSDraft\NZSouthS.dat
; Copyright 1998 Dept. of Electrical & Electronic Engineering
; University of Canterbury, Christchurch, New Zealand

; System general information
;   Desc=system description
;   PBase=power base
;   VBase=voltage base
;   Fund=fundamental frequency
[ System=NZSouth Desc=New Zealand South Island System PBase=100.00
  VBase=220.00 Fund=50.00 ]

; Busbar
[ Bus=Roxburgh 1011 Vsys=11.00 ]
[ Bus=Manapouri 1014 Vsys=14.00 ]
[ Bus=Invercargill 220 Vsys=220.00 ]

; Generators
[ Gen=Roxburgh-G1 Bus=Roxburgh 1011 Pgen=200.000 Per=50.00
  Zp=(0.00000,0.02000) Zn=(0.00000,0.04000) Zz=(0.00000,0.06200) ]
[ Gen=Manapouri-G1 Bus=Manapouri 1014 Zp=(0.00000,0.02000)
  Zn=(0.00000,0.04000) Zz=(0.00000,0.03700) ]

; Two-winding transformer
[ Xfmr2=Roxburgh-T1 Bus=Roxburgh 1011,Roxburgh 220 Model=C Ph=Lagging
  Tap=2.50 Conn=Star-G,Delta Zp=(0.00060,0.03816) ]
```

Figure 6.33 Example of flexible data format

The two delimiters make each individual record independent from others and therefore it is possible to add a record for a new component without affecting the existing records. Secondly, since each setting is indicated by a keyword, new keywords can be added for new settings without changing the locations of existing settings. The setting values do not have to be placed in any specific column as in the fixed format database. Although the records do not have to be grouped into sections as shown, they are grouped in that way here for clarity purposes. The parser would have to be programmed so that it does not rely on the records being grouped in a certain order. With this database format, a parser or decoder routine has been programmed to read in the records and arrange each record as a list of keywords with corresponding text values. An encoder routine has also been created to accept lists of keywords and corresponding text values, and to output these records to file according to the above format.

The main disadvantage of this format is the increased use of disk storage space. It is possible to reduce the disk space required by having default values for the settings. Hence, those settings with default values do not have to be included in the file. The default values can be programmed into the analysis software itself and be included as one record in the file. A record with a predefined keyword value can be regarded as the default record, whereby all components of this type will take the default values from the settings values in this particular record. However, with the availability of abundant disk space, it is questionable whether such a disk space saving exercise is required. Another important feature is the ability to load multiple database formats.

6.5.6 Development Environment

Traditionally, power system programs have been designed using a single programming language, FORTRAN being the dominant language. However, there is little graphic support available through FORTRAN and hence many power system programs lack a GUI. Therefore, a text-based database file is commonly used, allowing users to construct the database by typing in the power system data using an ordinary text editor. Manual database editing has the advantage that data errors are often immediately obvious.

Recent advancements in software development tools have made it possible for the program binary generated by different compilers on different programming languages to be able to interact with each other. This enables programmers to exploit the strengths of different languages in the development of single computer software. As mentioned in preceding sections, the harmonic analysis software can be separated into a GUI and a simulation engine. The GUI and the simulation engine have different development tools requirements and hence different languages are chosen to construct them. FORTRAN is retained as the language of the simulation engine, due mainly to its implicit handling of complex-number arithmetic. The simulation engine has also been programmed using basic FORTRAN commands (i.e. avoiding extended features of some FORTRAN compilers) so that it is transportable across different operating systems/platforms.

Microsoft Windows™ is an obvious choice of GUI as the operating environment. A graphical *pick and drop* development tool called Delphi™ is also a good choice to build the user interface. The graphical user interface for harmonic analysis, shown in Figure 6.30, consists of the Power System Canvas object which has been dropped onto a Windows™ form. This Canvas enables power system network diagrams to be constructed graphically. By making use of the dynamic linked library feature provided in Windows™ environment, the Delphi™ programmed user interface is able to access the simulation routines programmed using FORTRAN. Data retrieval routines have also been programmed alongside the simulation routines so as to enable the GUI to obtain the simulation results from the engine.

6.6 Post-Processing and Display of Results

Preliminary factors influencing the post-processing software and display of harmonic information are:

- simplicity of usage, maintenance and updating,
- flexibility to select equally spaced and non-equally spaced data, making use of curve fitting and interpolation techniques,
- flexibility to drive the various printers and plotters available,
- possibility of data comparisons from different files (simulations),

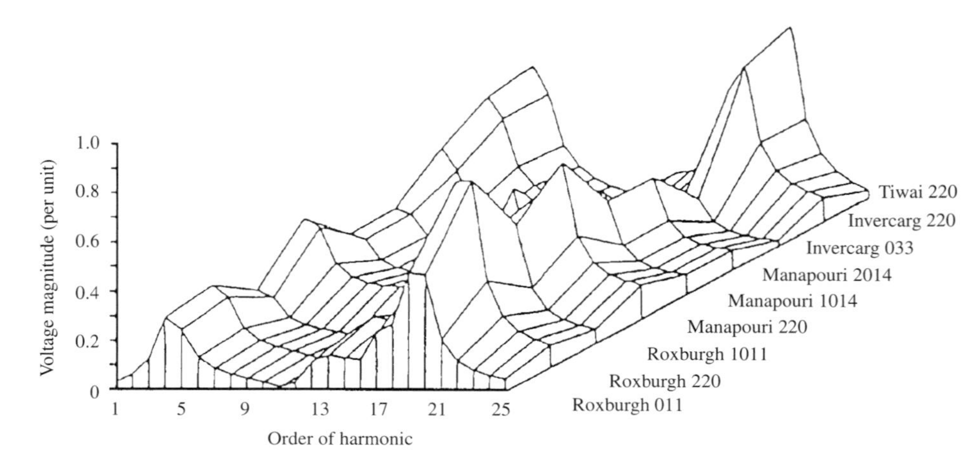

Figure 6.34 Three-dimensional plot of harmonic voltage profile

- flexibility of view harmonic information in phase and sequence components at any part of the system and select appropriate ranges.

Harmonic Voltage and Current Displays The display format can be two-or Three-dimensional. Three-dimensional plots are useful to give an overview of the harmonic levels throughout the system; an example of this application with reference to the lower part of the New Zealand system is shown in Figure 6.34. These plots, however, do not provide detailed quantitative information. The 2-D plots, such as shown in Figure 6.35, are more informative in this respect.

A typical histogram of the harmonic current at a system branch is illustrated in Figure 6.36.

Harmonic Impedance Displays Harmonic impedance information is often displayed in the form of impedance loci. These are obtained by performing interharmonic penetration studies from a minimum to a maximum frequency. However, Figure 6.37, obtained at 20 Hz intervals, shows that the use of equally spaced frequency intervals gives too many points in some regions (between 50–150 and 300–800 Hz) and very poor representation in regions with rapid change of impedance with frequency.

A possible solution is to perform two cubic spline interpolations, one on resistance as a function of frequency and the second of reactance as a function of frequency, then replot resistance against reactance. Another approach is the use of an adaptive algorithm to optimally select the frequency points to be solved for, based on the rate of change of impedance with frequency [29].

In practice, a three-phase system will have a 3×3 matrix of impedance loci, as shown in Figure 6.38.

6.6.1 Post-Processing for Telephone Interference

Telephone interference occurs where the power and telephone circuits are in close proximity, which can occur anywhere along the transmission lines length. Therefore,

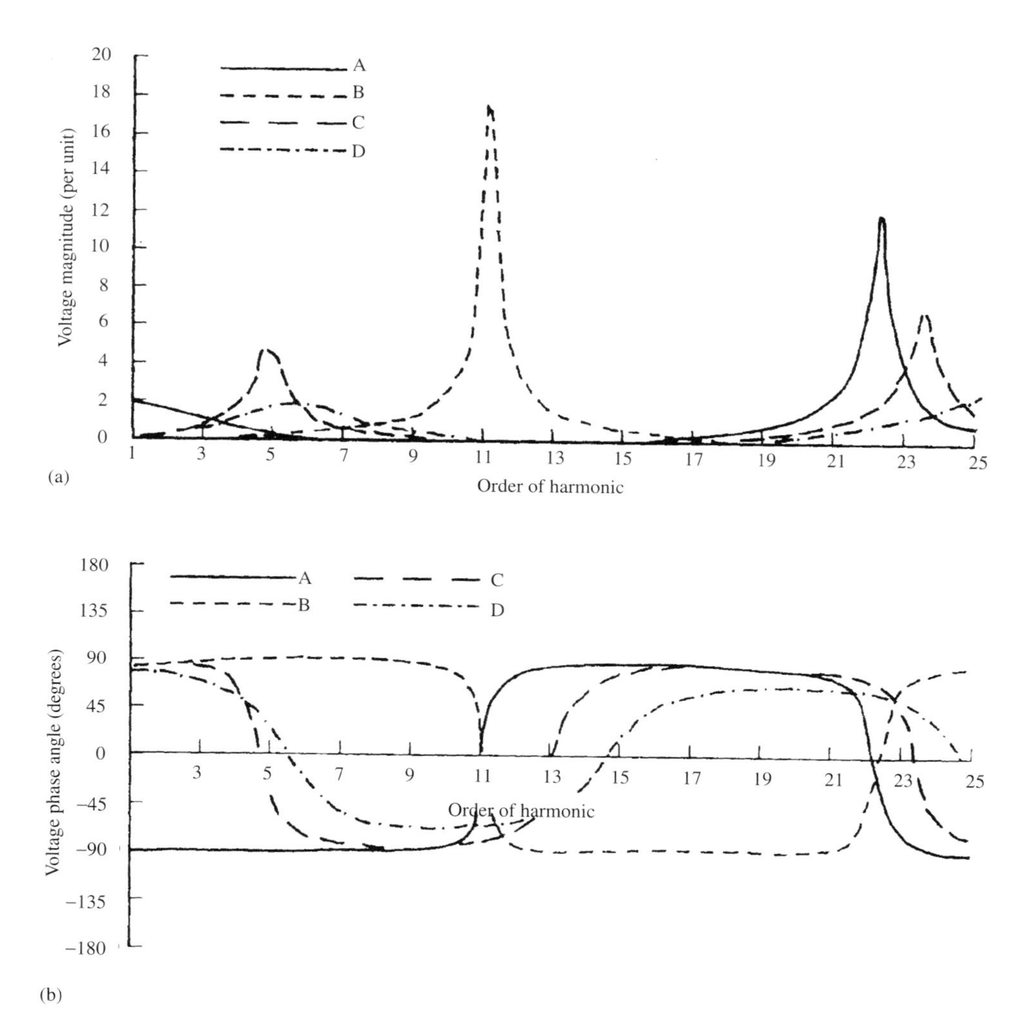

Figure 6.35 Two-dimensional plot of harmonic voltage profile

it is important to know the harmonic current distribution along a line, particularly the zero sequence current. However, the harmonic analysis described so far, based on the line equivalent pi circuit, only provides information of the sending and receiving ends currents.

For a homogeneous line, knowing the receiving end current and voltage for each harmonic frequency allows the current and voltage at any point on the line to be calculated for each frequency by using the following equations.

$$I(x) = \frac{I_R}{2Z_0}[(Z_R + Z_0)e^{\gamma x} + (Z_0 - Z_R)e^{-\gamma x}], \tag{6.94}$$

$$V(x) = \frac{I_R}{2}[(Z_R + Z_0)e^{\gamma x} + (Z_R - Z_0)e^{-\gamma x}], \tag{6.95}$$

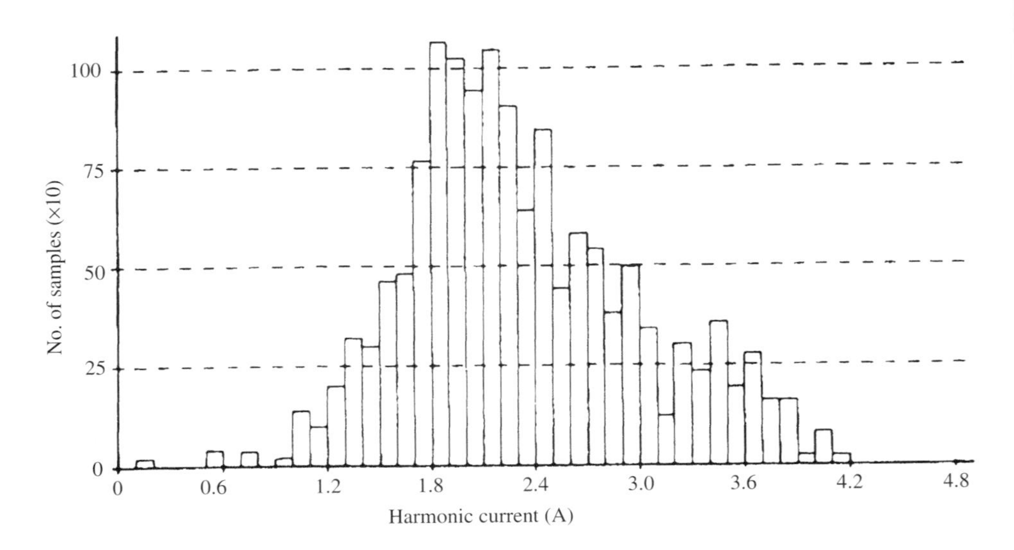

Figure 6.36 Histogram (bar) graph

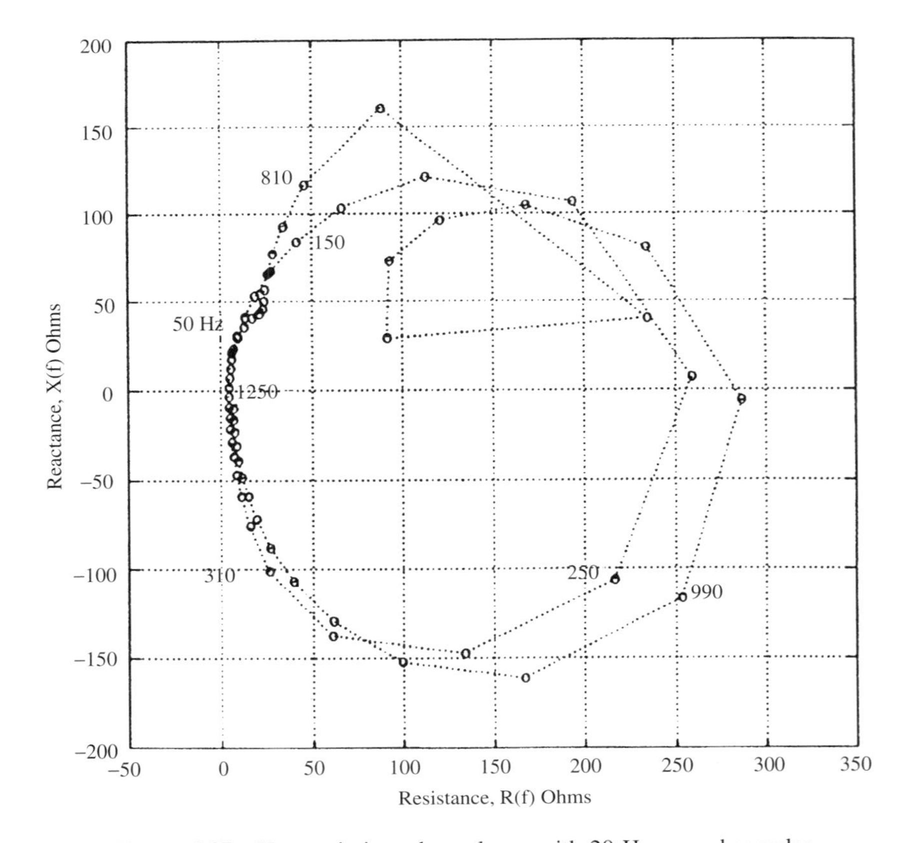

Figure 6.37 Harmonic impedance locus with 20 Hz spaced samples

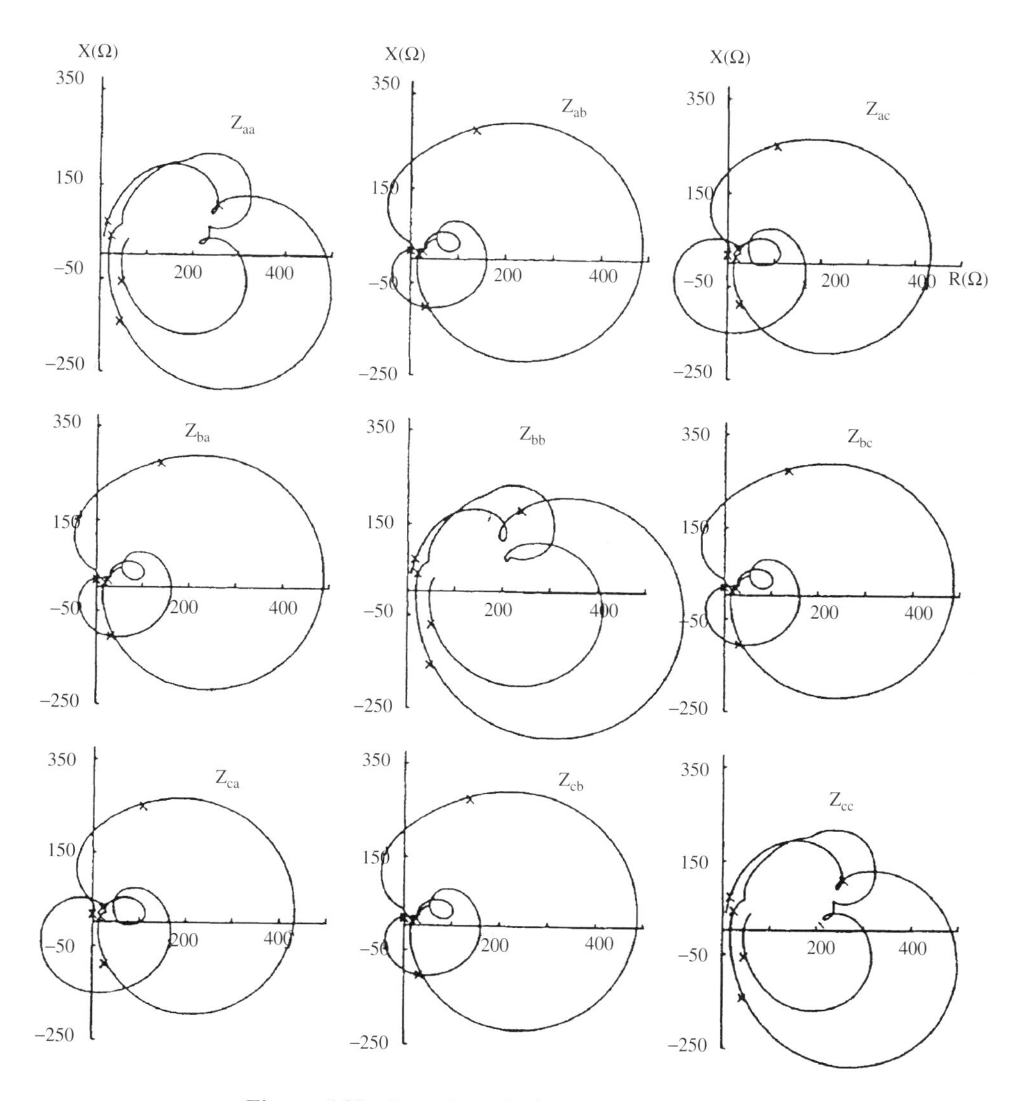

Figure 6.38 Impedance loci of a three-phase system

where x is the distance from the receiving end, I_R is the receiving end current, V_R is the receiving end voltage, and $Z_R = V_R/I_R$. The points on the transmission line at which these are maximum are obtained by considerations of the currents and voltages as forward (incident) and backward (reflected) travelling waves with respect to the receiving end.

For example, consider the current equation (6.94). The incident current at the receiving end is

$$I_R^+ = \frac{(Z_R + Z_0)}{2Z_0} I_R \tag{6.96}$$

and the reflected current at the receiving end is

$$I_R^- = \frac{(Z_0 - Z_R)}{2Z_0} I_R \tag{6.97}$$

The angles associated with these currents, at any point along the line, are given by

$$\theta^+ = \theta_R^+ + \beta x, \qquad \theta^- = \theta_R^- - \beta x, \tag{6.98}$$

where θ_R^+, θ_R^- are the angles of the current at the receiving end.

The current will be a maximum for θ^+ equal to θ^-. Thus

$$\theta_R^+ + \beta x = \theta_R^- - \beta x \tag{6.99}$$

or

$$x = \frac{\theta_R^- - \theta_R^+}{2\beta}. \tag{6.100}$$

The current will also have local maxima at intervals of one half wavelength along the line.

While the total r.m.s. voltage and current (over the fundamental and all harmonics) are of greatest importance, the location of the maximum total r.m.s. voltage and current will most likely be dominated by that harmonic which is closest to a resonant frequency of the system.

The harmonic analysis described above uses phase components, whereas for telephone interference the zero sequence component is of primary importance. The sequence components (indicated by subscripts $+-0$ for the positive, negative and zero sequences, respectively) are obtained from the phase quantities (1,2,3) using the relationship:

$$[V_{+-0}] = [T_s]^{-1}[V_{123}], \tag{6.101}$$

$$\begin{aligned} [I_{+-0}] &= [T_s]^{-1}[I_{123}] \\ &= [T_s]^{-1}[Y_{123}][T][V_{+-0}], \end{aligned} \tag{6.102}$$

where $[T_s]$ is the transformation matrix.

$$[T_s]^{-1} = \frac{1}{3}\begin{bmatrix} a & a^2 & 1 \\ a^2 & a & 1 \\ 1 & 1 & 1 \end{bmatrix} \quad [T_s] = \begin{bmatrix} a^2 & a & 1 \\ a & a^2 & 1 \\ 1 & 1 & 1 \end{bmatrix} \quad \text{and} \quad a = -\frac{1}{2} + j\frac{\sqrt{3}}{2}.$$

The sequence voltages and currents are thus related by the sequence admittance matrix,

$$[Y_{+-0}] = [T_s]^{-1}[Y_{123}][T]. \tag{6.103}$$

In practice, a transmission line will contain sections with different tower geometries. Each of these sections can then be explicitly represented by its equivalent pi circuit or these can be cascaded to obtain an overall equivalent of the complete line. The most efficient solution is to carry out the system harmonic flow analysis without explicit representation of the individual sections and then use a post-processing technique whereby the end terminals, voltages and currents are used to calculate the harmonic

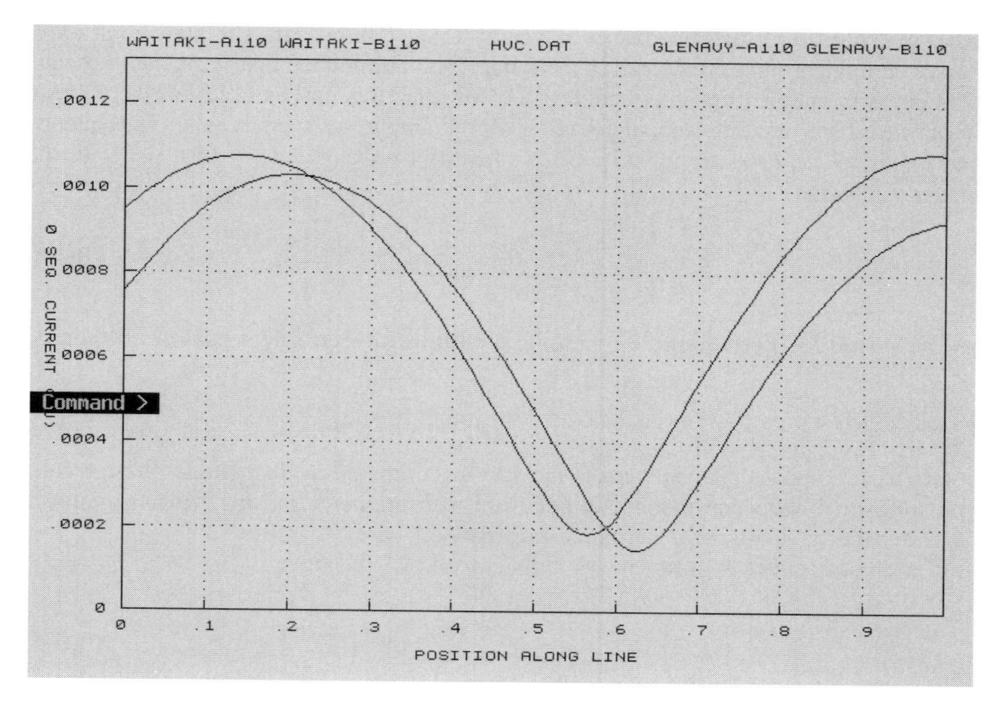

Figure 6.39 Zero sequence harmonic current profile along a double circuit transmission line

current profile along the transmission line section of interest. An example of the end result is shown in Figure 6.39 for a 110 kV double circuit line.

When performing telephone interference studies, often the harmonic sources are not well known yet need to be represented in the simulation. For example, the troublesome harmonics could be entering the 110 kV circuits from the local low voltage distribution system or from the 220 kV system. Without resorting to a system-wide Harmonic State Estimation (described in Chapter 7), it is difficult to overcome this problem for telephone interference studies. A number of simulations are performed for different injection scenarios. Usually a 1 p.u. injection is used and the post-processing program scales all the harmonic penetration results to match given measurements levels at one point in the system. Measurements are taken at a second point and these are compared with the scaled results of the other scenarios. Injection scenarios that do not result in the correct harmonic profile are then discarded. This can be taken a step further by opening some circuit breakers to produce a different harmonic flow pattern to distinguish between scenarios that are not easily distinguished from measurements at two locations.

6.6.2 Post-Processing for Test Results Comparisons

Restricted measurements on the physical network limit the ability to compare a three-phase model with test results. Also, the data obtained from live three-phase systems only includes the phase voltages and currents of the coupled phases. To compare measured and simulated impedances at a current injection busbar, it is thus necessary to derive equivalent phase impedances from the 3×3 admittance matrix. If the sequence of a given harmonic current injection is known, then this can be used to give a phase impedance incorporating mutual coupling. For example, assuming a positive sequence and making $I_1 = I\angle 0^\circ$ per unit, $I_2 = I\angle -120^\circ$ per unit, $I_3 = I\angle +120^\circ$ per unit, the matrix equation

$$\begin{bmatrix} I_1 \\ I_2 \\ I_3 \end{bmatrix} = \begin{bmatrix} Y_{11} & Y_{12} & Y_{13} \\ Y_{21} & Y_{22} & Y_{23} \\ Y_{31} & Y_{32} & Y_{33} \end{bmatrix} \cdot \begin{bmatrix} V_1 \\ V_2 \\ V_3 \end{bmatrix} \tag{6.104}$$

can be solved for V_1, V_2 and V_3, yielding the following equivalent phase impedances:

$$Z_1 = \frac{V_1}{I_1}, \quad Z_2 = \frac{V_2}{I_2}, \quad Z_3 = \frac{V_3}{I_3}. \tag{6.105}$$

Often, three-phase analysis results need to be compared against single-phase results (i.e. a positive sequence model), either from measurements or single-phase analysis. Consider the comparison of harmonic impedance.

In terms of sequence components, Equation (6.104) becomes

$$\begin{pmatrix} I_+ \\ I_- \\ I_0 \end{pmatrix} = \begin{bmatrix} Y_{++} & Y_{+1} & Y_{+0} \\ Y_{-+} & Y_{--} & Y_{-0} \\ Y_{0+} & Y_{0-} & Y_{00} \end{bmatrix} \cdot \begin{pmatrix} V_+ \\ V_- \\ V_0 \end{pmatrix}. \tag{6.106}$$

If a pure +ve sequence current is injected, all sequences of voltage are produced due to the coupling between sequences. This coupling can be derived using a Kron reduction, i.e.

$$\begin{pmatrix} I_+ \\ I_- \\ I_0 \end{pmatrix} = \begin{bmatrix} [A] & [B] \\ [C] & [D] \end{bmatrix} \begin{pmatrix} V_+ \\ V_- \\ V_0 \end{pmatrix}, \tag{6.107}$$

where

$$[A] = [Y_{++}][B] = [\,Y_{+-} \quad Y_{+0}\,],$$

$$[C] = [\,Y_{-+} \quad Y_{0+}\,]^t \quad \text{and} \quad [D] = \begin{bmatrix} Y_{--} & Y_{-0} \\ Y_{0-} & Y_{00} \end{bmatrix}.$$

Setting $I_- = 0$ and $I_0 = 0$ and rearranging gives

$$I_+ = ([A] - [B]^{-1}[D][C])V_+ = Y_+^{\text{effective}} V_+, \tag{6.108}$$

which results in the following positive sequence impedance:

$$[Z_+] = ([A] - [B]^{-1}[D][C])^{-1}. \tag{6.109}$$

6.7 Summary

An algorithm of general applicability to harmonic analysis has been described that has the following capabilities:

- models the steady-state phasor response of multi-phase networks to the presence of single or multiple current or voltage sources at harmonic, subharmonic or inter-harmonic frequencies.
- represents the individual network elements, assumed linear, by accurate frequency dependent models.
- provides graphical interfaces for the specification and display of the system to be analysed, and for the post-processing of the information obtained from the analysis.

6.8 References

1. Fortescue, C L, (1918). Method of symmetrical co-ordinates applied to the solution of polyphase networks, *AIEE Transactions*, **37**(2), pp. 1027–1140.
2. Zollenkopf, K, (1960). Bifactorisation-basic computational algorithm and programming techniques, *Conference on Large Sets of Sparse Linear Equations*, Oxford.
3. Lemoine, M, (1997). Methods of measuring harmonic impedances', *CIRED*, Publication no. 151, London, pp.5–6.
4. Brewer, G D, Chow, J H, Gentile, T J, et al, (1982). HVDC-AC harmonic interaction. I. Development of a harmonic measurement system hardware and software, *IEEE Transactions on Power Apparatus and Systems*, **PAS-101**, pp. 701–708.
5. Baker, W P, (1981). The measurement of the system impedance at harmonic frequencies, *Paper presented at the International Conference on Harmonics in Power Systems*, UMIST, Manchester.
6. Barnes, H, (1976). An automatic harmonic analyser for the supply industry, *IEE Conference. on Sources and Effects of Power System Disturbances*, London.
7. Robert, A and Deflandre, T, (1997). Guide for assessing the network harmonic impedance, 14th *International Conference and Exhibition on Electricity Distribution*, Part 1: Contributions. CIRED, IEE Conference Publication no. 438, **1**, pp. 2.3.1–1.3.10.
8. Morched, A S and Kundur, P, (1984). Identification and modeling of load characteristics at high frequencies, *IEEE Transactions on Power Apparatus and Systems*, **PAS-1-3**(3), pp. 619–630.
9. Nagpal, M, Zu, W and Sawada, J, (1998). Harmonic impedance measurement using three-phase transient, *IEEE Transcations on Power Delivery*, **13**(1).
10. Smith, J R, Hauer, J F and Trudnowski, D J, (1993). Transfer function identification in power system applications, *IEEE Transactions on Power Systems*, **8**, (3), pp. 1282–1288.
11. Chen, M S and Dillon, W E, (1974). Power system modelling, *Proceedings. IEE*, **62**, pp. 901.
12. Arrillaga, J, Arnold, C P and Harker, B J, (1983). *Computer Modelling of Electrical Power Systems*, John Wiley & Sons Ltd, London.
13. Kimbark, E W, (1950). *Electrical Transmission of Power and Signals*, John Wiley & Sons, New York.
14. Galloway, R H, Shorrocks, W N and Wedepohl, L M, (1964). Calculation of electrical parameters for short and long polyphase transmission lines, *Proceedings of the IEE*, **111**, pp. 2051–2059.
15. Bowman, W I and McNamee, J M, (1964). Development of equivalent PI and T matrix circuits for long untransposed transmission lines, *IEEE Transcations* **PAS-84**, pp. 625–632.
16. Wilkinson, J H and Reinsch, C, (1971). *Handbook for Automatic Computations, Vol. II, Linear Algebra*, Springer-Verlag, Berlin.

17. Carson, J R, (1926). Wave propagation in overhead wires with ground return. *Bell System Technical Journal*, **5**, pp. 539–556.
18. Deri, A, Tevan G, Semlyen, A and Castanheira, A, (1981). The complex ground return plane, a simplified model for homogeneous and multi-layer earth return. *IEEE Transactions on Power Apparatus and Systems*, **PAS-100**, pp. 3686–3693.
19. Semlyen, A and Deri, A, (1985). Time domain modelling of frequency dependent three-phase transmission line impedance. *IEEE Transactions on Power Apparatus and Systems*, **PAS-104**, pp. 1549–1555.
20. Acha, E, (1988). Modelling of power system transformers in the complex conjugate harmonic space, Ph.D. Thesis, University of Canterbury, New Zealand.
21. Lewis, V A and Tuttle, P D, (1958). The resistance and reactance of aluminium conductors steel-reinforced, *Transcations of the AIEE*, **PAS-77**, pp. 1189–1215.
22. Dommel, H W, (1978). Line constants of overhead lines and underground cables, Course E.E. 553 notes, University of British Columbia.
23. Bianchi, G and Luoni, G, (1976). Induced currents and losses in single-core submarine cables', *IEEE Transcations*, **PAS-95**, pp. 49–58.
24. Ribeiro, P F, (1985). Investigations of harmonic penetration in transmission systems, Ph.D. Thesis, UMIST, Manchester.
25. Pesonen, J A, (1981). Harmonics, characteristic parameters, method of study, estimates of existing values in the network, *Electra*, **77**, pp. 35–56.
26. Huddart, K W and Brewer, G L, (1996). Factors influencing the harmonic impedance of a power system, *IEE Conference on High Voltage d.c. Transmission*, (22), pp. 450–452.
27. Mahmoud, A A and Shultz, R D, (1982). A method for analyzing harmonic distribution in a.c. power systems, *IEEE Transcations*, **PAS-101**(6), pp. 1815–1826.
28. Thunberg, E and Söder, L, (1998). A harmonic Norton model of a real distribution network, *International Conference on Harmonics and Quality of Power* (*ICHQP'98*), Athens, pp. 279–286.
29. Coope, I, Watson, N R and Arrillaga, J, (1991). An adaptive scheme for the efficient calculation of impedance loci', *Proceedings of the IPENZ*, Auckland, New Zealand, pp. 117–125.

7

POWER QUALITY STATE ESTIMATION

7.1 Introduction

State estimation has been used since the late 1960s in conjunction with fundamental frequency power flow studies [1]. These assume that all current and voltage waveforms are pure sinusoids with constant frequency and magnitude; they only apply to symmetric power systems operating under balanced three-phase conditions. State estimation is now an essential part in energy management systems. Recent contributions [2–13] have extended the concept to harmonic state estimation (HSE) and identification of harmonic sources. However, full measurement of the system states, by first recording the voltage and current waveforms at nodes and lines and then deriving their frequency spectra, is prohibitive for a large system. Only partial measurement (not necessarily made at the harmonic sources) is practical and, therefore, the measurements must be complemented by system simulation.

The framework of harmonic state estimation is illustrated in Figure 7.1. It uses a three-phase system model to describe asymmetrical conditions such as circuit mutual couplings, impedance and current injection imbalance. A partial measurement set is also needed consisting of some bus voltages, injection currents and line currents, or bus injection volt-amperes and line volt-amperes. Instead of system-wide HSE, some contributions [14–17] discuss the issue with reference to the estimation of the harmonic components from the voltage or current waveform at a measurement point, i.e. point HSE.

On the basis of the network topology, a harmonic state estimator is formulated from the system admittance matrix at harmonic frequencies and the placement of measurement points. Measurements of voltage and current harmonics at selected busbars and lines are sent to a central workstation for the estimation of the bus injection current, bus voltage and line current spectra at all or selected positions in the network.

The placement of measurement points is normally assumed symmetrical (e.g. either three or no phases of injection currents of a busbar are measured). However, this requirement restricts the search for the optimal placement of measurement points in three-phase asymmetrical power systems. Moreover, the implementation of existing algorithms will, in practice, be limited by poor synchronisation of conventional instrumentation schemes, lack of continuity of measurements or lack of processing speed.

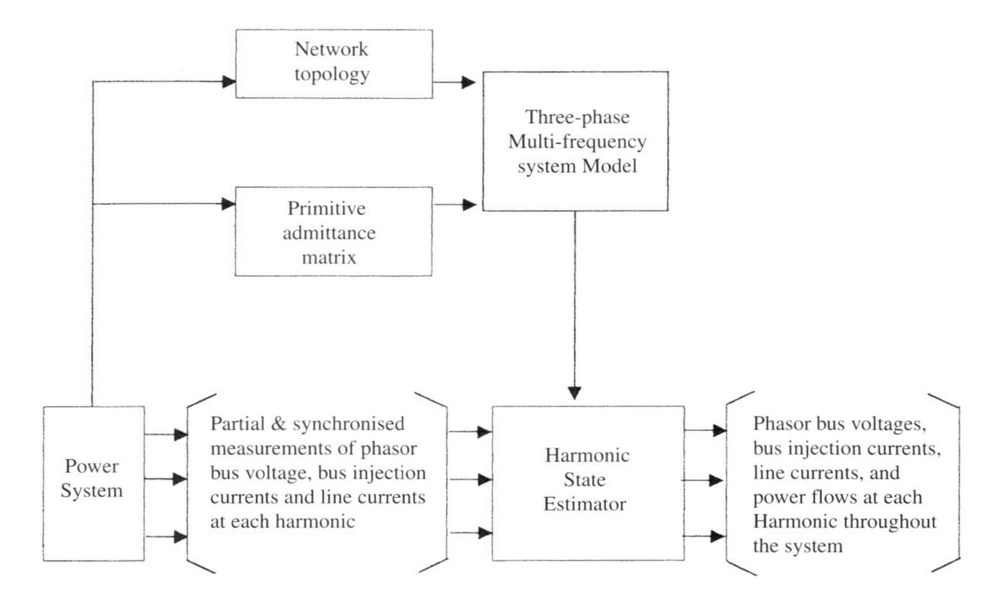

Figure 7.1 Framework for harmonic state estimation

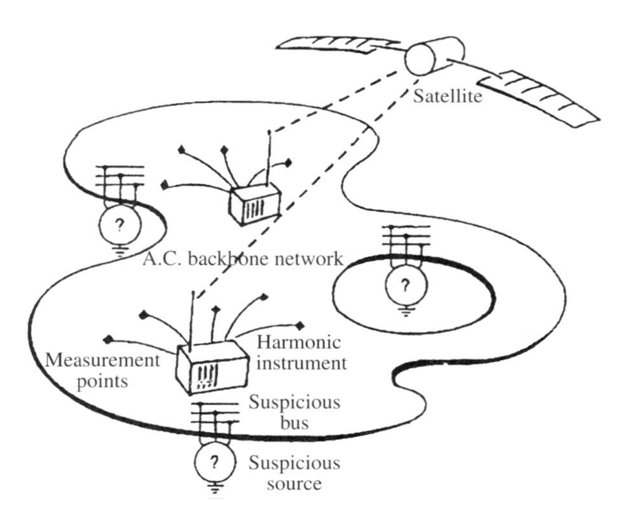

Figure 7.2 System-wide harmonic state estimation

A system-wide or partially observable HSE requires synchronised measurement of phasor voltage and current harmonics made at the different measurement points, as illustrated in Figure 7.2.

HSE turns multi-point measurement to *system-wide measurement* in a very economic way. Two important optimisation problems in HSE are the maximum observable subsystem for a given measurement placement, and the minimum number of measurement channels needed for the observability of a given system. The HSE can be implemented continuously in real time if the measurement is continuous and the processing speed fast enough. Potentially, the harmonic monitoring instrument and estimator can then be integrated into an existing supervisory control and data acquisition (SCADA) system.

State estimation is thus an alternative, or a supplement, to the direct measurement of electrical signals. Strictly speaking, a system state is a mathematically definable, although not necessarily measurable, electrical quantity such as a nodal voltage or a line current. In practice, however, the *state* concept is often extended to other variables, such as the voltage phase angle difference of a transmission line; it is also used for complex combinations of individual variables.

Many of the power quality indices discussed in Chapter 2 (e.g. Total Harmonic Distortion, Equivalent Distorting Current, etc.) are combinations (often weighted) of individual states and are not suited to direct state estimation. Instead, only the harmonic (or interharmonic) voltage and current signals are normally state estimated and the power quality indices are then calculated from these estimates.

7.2 Harmonic Measurement—State Variable Model

The task of the HSE is to generate the *best* estimate of the harmonic levels from limited measured harmonic data, corrupted with measurement noise. The three issues involved are the choice of state variables, some performance criteria and the selection of measurement points and quantities to be measured.

State variables are those variables that, if known, completely specify the system. The voltage phasors at all the busbars are usually chosen, since they allow the branch currents, shunt currents and currents injected into the busbar to be determined.

Various performance criteria are possible, the most widely used being the Weighted Least Squares (WLS), described in Appendix III. This method minimises the weighted sum of the squares of the residuals between the estimated harmonic levels and actual harmonic measurements. This can be shown to be the maximum likelihood estimate assuming that the noise distribution is Gaussian [18]. Other possible criteria are: Weighted Least Absolute Value (WLAV), Least Median of Squares (LMS) and Non-quadratic estimators.

For a given measurement set and system topology, the basic circuit laws lead to the following measurement equation:

$$\mathbf{Z} = \mathbf{h}(\mathbf{x}) + \varepsilon, \tag{7.1}$$

where $\mathbf{Z}$ and $\mathbf{x}$ are the vectors of measurements and state variables, respectively, and ε the measurement error vector which is assumed to be made of independent random variables with Gaussian distribution.

The WLS estimate is therefore the vector $\mathbf{x}$ that minimises the weighted sum of the squares of the residuals ($\mathbf{r} = \mathbf{Z} - h(\mathbf{x})$) between the actual harmonic measurements and estimated harmonic levels, i.e.

$$\text{Minimise } J(\mathbf{x}) = \tfrac{1}{2}[\mathbf{Z} - \mathbf{h}(\mathbf{x})]^{\mathrm{T}} R^{-1} [\mathbf{Z} - \mathbf{h}(\mathbf{x})], \tag{7.2}$$

where R^{-1} is the inverse of the covariance matrix.

When real and reactive power measurements are used, instead of current measurements, for branch flows and busbar injection, the measurement equation is non-linear. In such cases, the solution to Equation (7.2) must be obtained through an iterative algorithm. This method is used for fundamental frequency state estimation as power measurements are always available for revenue purposes. However, for harmonic frequencies current measurements are more readily available.

Although, in general, the measurement equation can be non-linear, by choosing the busbar harmonic voltages as state variables and measuring busbar harmonic voltages, branch harmonic currents and harmonic injection currents make Equation (7.2) linear.

In this case, the task of estimating x given z measurements in the presence of noise (ε) is expressed as

$$\mathbf{z} = [h]\mathbf{x} + \varepsilon, \tag{7.3}$$

where $[h]$ is the measurement matrix.

State estimation problems can be classified as over-determined, completely-determined or under-determined, depending on whether the number of independent measurement equations are greater, equal or less than the number of unknown state variables. A unique solution is only possible for over-determined and completely-determined systems.

Traditionally, state estimation techniques are performed on an over-determined system where the number of rows in $[h]$ exceed the number of columns. The solution to measurement equation ($Z = hx$), in the least squares sense, is obtained by solving the equation

$$\left[h^{\mathrm{T}}R^{-1}h\right]\mathbf{x} = \left[h^{\mathrm{T}}\right]\mathbf{Z}. \tag{7.4}$$

Matrix R is diagonal and contains the covariances of the measurements (if they are known). This permits applying higher weightings to measurements that are known to be more accurate. However, normally, the covariances of the measurements are unknown (and often equal since the same instrumentation is used to obtain them). Therefore, R is replaced by the identity matrix in such cases.

The measurement matrix $[h]$ has more rows than columns but $[h^{\mathrm{T}}R^{-1}h]$ (known as the gain matrix) is a square matrix and can be easily solved using standard techniques, such as LU decomposition and back-substitution, Cholesky decomposition or Gauss–Jordan elimination. A state estimator in the over-determined system allows for bad data detection and acts like a noise filter to clean up the erroneous data in the redundant measurements.

In a completely-determined system $[h]$ may or may not be square, depending on whether linear dependency exists among rows of the measurement matrix. Regardless of this, the same solution procedure can be adopted as for an over-determined system.

7.2.1 Building Up the Measurement Matrix

Each current injection measurement adds one extra row to the measurement matrix, which constitutes a row of the system admittance matrix. Each line current measurement also adds an extra row. The measurement for a line connected between busbars i and j, as shown in Figure 7.3, has two possible non-zero entries. If the sending end

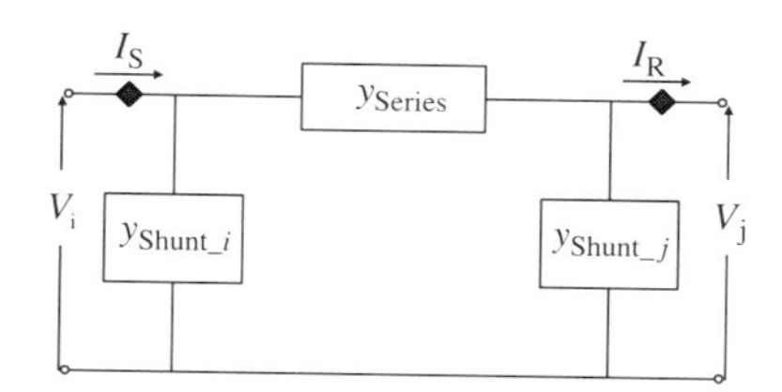

Figure 7.3 General representation of a branch

harmonic current is measured, then the entries in columns i and j are

$$\begin{aligned} h(x, i) &= y_{\text{Series}} + y_{\text{Shunt\_}i}, \\ h(x, j) &= -y_{\text{Series}}. \end{aligned} \tag{7.5}$$

If the receiving end harmonic current is measured, then the entries in columns i and j are

$$\begin{aligned} h(x, i) &= y_{\text{Series}}, \\ h(x, j) &= -y_{\text{Series}} - y_{\text{Shunt\_}j}. \end{aligned} \tag{7.6}$$

Of course, normally, $y_{\text{Shunt\_}i} = y_{\text{Shunt\_}j}$.
A more general power system requires an oriented graph including a set of all nodes (N), except the reference node, and a set of all branches (B). The appropriate matrix rows are then extracted as follows.

$$\mathbf{I}_{\text{N}} = [C_{\text{BN}}^T Y_{\text{Prim}} C_{\text{BN}}]\mathbf{V}_{\text{N}} = [Y_{\text{Nodal}}]\mathbf{V}_{\text{N}}, \quad \text{for injection measurements} \tag{7.7}$$

and

$$\mathbf{I}_{\text{M}} = [C_{\text{MB}}]\mathbf{I}_{\text{b}} = [C_{\text{MB}} Y_{\text{Prim}} C_{\text{BN}}]\mathbf{V}_{\text{b}}, \quad \text{for line measurements,} \tag{7.8}$$

where

$\mathbf{I}_{\text{N}}$ = vector of nodal injections,
$\mathbf{V}_{\text{N}}$ = vector of node voltages,
$[Y_{\text{Prim}}]$ = primitive admittance matrix,
$[C_{\text{BN}}]$ = branch-node incidence (connection) matrix,
$[C_{\text{MB}}]$ = measurement point–branch incidence (connection) matrix,
$\mathbf{I}_{\text{b}}$ = branch current vector,
$\mathbf{I}_{\text{M}}$ = line current measurements vector.

Two approaches exist for handling voltage measurements. The first is to add an extra row in the measurement matrix for each such measurement. This results in only one non-zero element per row, indicated by a one in the column corresponding to the measured busbar voltage. The second approach is to remove the measured voltages from the vector of state variables to be estimated. For instance, if the voltage at busbar m is measured, then the measurement matrix is partitioned and known terms are removed from the measurement vector to form a modified measurement vector $\mathbf{Z}'$ and a reduced set of unknowns $\mathbf{V}_{\text{u}}$, i.e.

$$\mathbf{Z} = [h_{\text{u}} \quad h_{\text{m}}] \begin{pmatrix} \mathbf{V}_{\text{u}} \\ \mathbf{V}_{\text{m}} \end{pmatrix}, \tag{7.9a}$$

$$\mathbf{Z}' = \mathbf{Z} - [h_{\mathrm{m}}]\mathbf{V}_{\mathrm{m}} = [h_{\mathrm{u}}]\mathbf{V}_{\mathrm{u}}. \tag{7.9b}$$

Both approaches achieve the same result.

7.2.2 Virtual and Pseudo-Measurements

In the under-determined case, a unique solution cannot be obtained unless extra information is supplied. Two such pieces of information are termed Pseudo-measurements and Virtual measurements. Pseudo-measurements are estimates based on historical data, but considering the lack of harmonic data it is not normally viable for HSE.

Virtual measurements provide the kind of information that does not need metering; for example, zero injection at a switching substation. The under-determined HSE problem can be transformed into an over-determined problem using this approach; this is achieved by reducing the number of unknown state variables to include only busbars that are known, or thought likely to have non-linear devices directly connected to them (provided, of course, that sufficient measurements exist in the reduced system). The measurement equation with busbars partitioned into suspicious (V_1) and non-suspicious busbars (V_2) becomes

$$\begin{bmatrix} h_{11} & h_{12} \\ h_{21} & h_{22} \end{bmatrix} \begin{pmatrix} \mathbf{V}_1 \\ \mathbf{V}_2 \end{pmatrix} = \begin{pmatrix} \mathbf{I}_1 \\ 0 \end{pmatrix}, \tag{7.10}$$

where

$\mathbf{V}_1$ = a vector of suspicious busbar voltages,
$\mathbf{V}_2$ = a vector of non-suspicious busbar voltages,
$\mathbf{I}_1$ = is a vector of nodal harmonic current injection or line currents at suspicious busbars.

Hence, from Equation (7.10) the following expression for V_2 is obtained:

$$\mathbf{V}_2 = [-h_{22}^{-1} Y_{21}]\mathbf{V}_1. \tag{7.11}$$

To reduce the number of unknowns, Equation (7.11) is used in (7.10) to eliminate the busbar voltages at non-suspicious busbars, i.e.

$$[h_{11} - h_{12}h_{22}^{-1} Y_{21}]\mathbf{V}_1 = \mathbf{I}_1. \tag{7.12}$$

Once V_1 is solved, V_2 is obtained from Equation (7.11). The problem with this technique is that it does not permit the use of voltage measurement data at non-suspicious busbars.

The same result could be obtained by adding an equation to the measurement matrix equation, specifying the nodal current injection at each non-suspicious busbar to be zero and not reducing the number of unknowns.

Initially, the number of non-zero elements is either one (for a voltage measurement), two (line current measurement) or $1 + n_i$; in the latter case, n_i is the number of busbars directly connected via a branch to a busbar ith with a current injection measurement. Even though reduction techniques, as discussed earlier, increase the number of non-zero elements, the resulting matrix is still sparse. Speed and memory requirements can therefore be reduced by the application of sparsity techniques to the solution of these equations.

7.3 Observability Analysis

Observability Analysis (OA) is required in HSE to identify its solvability. A power system is said to be observable if the set of available measurements is sufficient to calculate all the state variables of the system uniquely. Observability is dependent on the number, locations and types of available measurements, network topology, as well as the system admittance matrix. For a different network topology, or same network topology but different measurement placements, an OA is to be performed in each case.

It is important for OA not only to decide whether the system is observable and hence system-wide HSE can be performed, but also to provide information of the observable/unobservable islands as well as redundant measurement points if not completely observable. This allows the re-positioning of measurement points to maximise their usefulness.

A system is observable if a unique solution can be obtained for the given measurements. A unique solution exits if and only if the rank of $[h]$ equals the number of unknown state variables. Therefore, for observability the number of measurements must not be less than the number of state variables to be estimated. However, this condition is not sufficient, since linear dependency may exist among rows of the measurement matrix. The rank of $[h]$ does not depend on the quality of the measurements and therefore the noise vector can be assumed to be zero.

Existing OA can be divided into three groups of methods; Numerical (floating point calculations), Topological and Symbolic.

Numerical observability determination is based on assessing the rank of the gain matrix by triangular factorisation. There are several algebraically equivalent ways of expressing the state estimation equations that have good sparsity and numerical stability for large systems. However, due to ill-conditioning and finite precision arithmetic, numerical problems can occur. The factorisation method is simple and uses some of the techniques of the HSE algorithm; however, it can fail due to numerical round-off errors. These result from performing floating point calculations on large sets of poorly conditioned equations. For example, matrix elements that should be zero are not exactly zero, therefore requiring a threshold to be applied to these elements. The choice of threshold may not be obvious, since it depends both on the network and the precision of the arithmetic used.

This leads to the distinction between algebraic observability and numerical observability. A power system is algebraically observable for a given set of measurements if the rank of the gain matrix equals the number of state variables to be estimated. A power system is numerically observable if the measurement model can be solved for the state variables. If a system is numerically observable, then it must also be algebraically observable; however, the converse need not hold. It is possible for the gain matrix to have the required rank but be so ill-conditioned that it cannot be solved numerically. However, for most power systems, algebraic observability will imply numerical observability.

Floating point determination of rank is time consuming and does not give information on where the problems are.

As the name indicates, the topological approach tells whether a system is topologically observable. Although it is possible for a topologically observable system to be algebraically unobservable, this is unlikely to occur in a practical system since it only happens with a theoretical choice of network admittances. This condition is called *parametrically unobservable*.

The system is topologically observable if there exists a spanning tree of full rank. In this respect, a tree is any interconnected, loop-free collection of branches of the network and a spanning tree is a tree that is incident to every busbar. The number of possible trees for n busbars is n^{n-2}, a very large figure even for a small system [19]. To start with, the branches with flow measurements are used to build a tree. All the loops are eliminated since the flow through any branch that forms a loop can be calculated from Kirchhoff's voltage law, circuit parameters and flows in other branches. Hence, such measurements are redundant and do not contribute to the rank of the gain matrix. This leaves several connected pieces or trees, and the resulting unconnected loop-free subgraph is termed a forest. Then busbar injection information is used. However, the topological method requires procedures that are not needed to compute the state estimation and is combinatorial, thus requiring considerable computational effort.

The symbolic method seeks to overcome the numerical problems associated with floating point operations by replacing them with symbolic calculations, where each entry in the measurement matrix used for OA is either a one or a zero. While being extremely fast and simple, the basic method is not capable of finding all the observable islands and redundant measurements.

A second phase has been added to the symbolic method to try to overcome these deficiencies while retaining its simplicity and speed [20]. Figure 7.4 shows a flow chart of this symbolic observability analysis algorithm. It is based on symbolic reduction of the gain matrix, without the need for numerical computation. The search is not combinatorial and is driven by the structure of the measurement matrix. It uses a step-by-step reduction through the elimination of observable groups (subsystems). A subsystem is observable if the number of measurement equations linking the state variables is equal to the number of state variables, otherwise it is unobservable.

Unobservable subsystems can be categorised based on the number of additional measurements needed to make them observable. For example, a subsystem is referred to as *univariate conditionally observable* if the set of equations is one less than the number of state variables linked, since one more measurement will make the subsystem observable. Two observable subsystems can be combined to form the observable system even though they do not interconnect (have no state variables in common). Two univariate conditionally observable subsystems, however, can only be combined to form a bigger univariate conditionally observable subsystem if they have at least one state variable in common. There are thus two Process Phases, the first searching for the overall observable system and the second searching for univariate conditionally observable subsystems. For a subsystem to be observable, at least one state variable must be known (measured) and these are used as the reference busbars.

To understand the algorithm, some terminology is required. Non-zero entries in the columns of the measurement matrix indicate the equations in which a particular state variable participates. The non-zero entries will identify state variables that are a *direct neighbour* of the state variable being considered. In turn, direct neighbour state variables will be linked as direct neighbours to new busbars. The *all neighbours* set

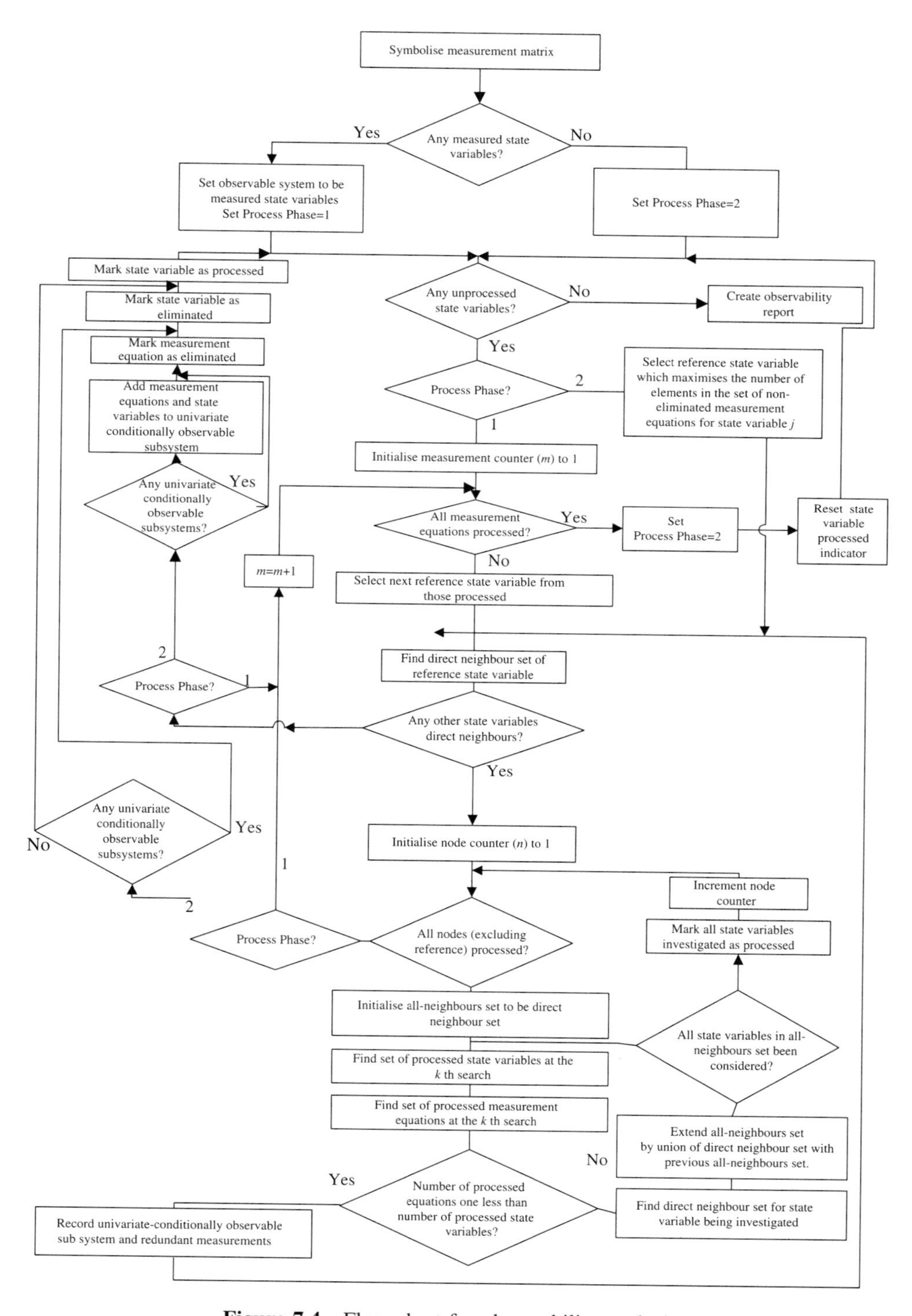

Figure 7.4 Flow-chart for observability analysis

is the set of state variables that can be reached by systematically looking at the *direct neighbours* of a state variable. The *all neighbours* set forms a subsystem, which may or may not be observable depending on the number of measurement equations linking them and the number of state variables linked. The search is formalised by sequentially searching and then marking state variables as processed. Measurement equations that relate only processed state variables are marked as processed measurement equations.

The symbolic method starts with the observable state variables and sequentially chooses one as a reference; it then searches through the symbolic measurement matrix to identify the state variables that are linked to the reference state variable as well as the number of state variables linking them. If the number of equations linking the unknown state variables is sufficient, then the unknown state variables linked to the reference are identified as observable. Initially, the observable state variables are those being measured; however, as this algorithm proceeds, more of the state variables are marked as observable; these in turn become the reference state variables and are searched for state variables linked to them and the number of equations. Although the concept of this observability analysis is simple, it requires complex bookkeeping of the processed, eliminated state variables and equations as well as the observable, univariate conditionally observable subsystem state variables and equations.

7.3.1 Example of Application

The test system shown in Figure 7.5 is used to illustrate the symbolic observability analysis described above. This system has eight busbars, all with potentially non-linear loads (i.e. injecting harmonics). The symbolised measurement matrix is

$$
\begin{array}{rr}
\text{State variable}\rightarrow & \begin{array}{cccccccc}1 & 2 & 3 & 4 & 5 & 6 & 7 & 8\end{array} \\
[h] = \begin{array}{c}1\\2\\3\\4\\5\end{array} & \begin{bmatrix}
1 & 1 & 1 & 0 & 0 & 0 & 0 & 0 \\
1 & 1 & 1 & 0 & 0 & 0 & 0 & 0 \\
0 & 0 & 0 & 1 & 1 & 1 & 1 & 1 \\
0 & 0 & 0 & 0 & 1 & 1 & 1 & 0 \\
0 & 0 & 0 & 0 & 1 & 0 & 0 & 1
\end{bmatrix} \\
\uparrow & \\
\text{Measurement equation number} &
\end{array}
$$

The measured state variables 5, 6 and 8 are marked as processed. State variable 5 is then selected as the first reference state variable and its direct neighbours are found to be state variables 5, 4, 7 (note that state variable 6 and 8 are missing from the direct neighbour set since they have already been marked as processed). Measurement equations 3 and 4 are found to link these three state variables (and are marked as processed). Since state variable 5 is measured, state variables 4 and 7 are observable using measurement equations 3 and 4.

State variables 1, 2 and 3 are not reached in the Process Phase 1 search (considering each processed state variable as a reference sequentially). Process Phase 2 needs therefore to be entered and state variable 1 searched. Two equations (1 and 2) are seen to link state variables 1, 2 and 3 and therefore an unobservable subsystem is found. It is a univariate conditionally observable subsystem, since one extra equation will

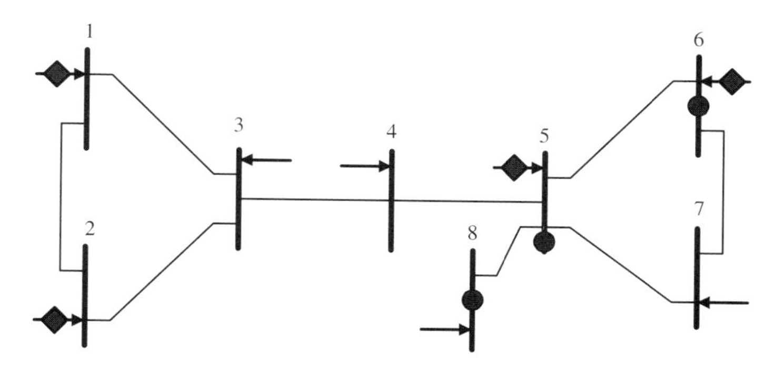

Figure 7.5 Single phase test system

make it observable. After completion of this process, measurement equation 5 is never processed, indicating that this is a redundant measurement. A detailed step-by-step application of the algorithm is made below for the test system of Figure 7.5.

Step 1: The Process Phase is set to 1 as there are voltage measurements. The vector indicating measured state variables is [0 0 0 0 1 1 0 1] and, hence, state variables 5, 6 and 8 are marked as processed; they are also marked as observable (Observable State Variables $= [SV_5, SV_6, SV_8] = [S_1, S_2, S_3]$).

Step 2: Since there are unprocessed state variables, continue to *Step 3* (Unprocessed State Variables $= [SV_1, SV_2, SV_3, SV_4, SV_7]$).

Step 4: All observable state variables have not been searched. The index (m) is set to 1 referring to the first of the three observable state variables, number 5, which is selected as the reference.

Step 5: Looking through the measurement equations, find the direct neighbours of the reference state variable (in this case 5), i.e. the equations that link SV_5 to other unprocessed state variables. The column corresponding to the state variable is traversed looking for 1s, then these rows (measurement equation) are traversed to find the 1s in the columns corresponding to unprocessed state variables; these are the direct neighbours. Equation 3 links state variable 5 to state variables 4 and 7. Equation 4 also shows 7 to be a direct neighbour. Hence, the direct neighbour set for state variable 5 is found to be state variables 5, 4, 7 ($DN_{SV5} = [SV_5, SV_4, SV_7] = [j_0, j_1, j_2]$). Note that state variables 6 and 8 are missing from the direct neighbour set since they have already been marked as processed. Since the direct-neighbour set consists of more than just the reference state variable, the counter n is initialised to 1 (which corresponds to the next state variable, i.e. 4).

Step 7: All the direct neighbours have not been searched and, therefore, the all-neighbour set (AN) is initialised to be the reference state variable (5) and the direct neighbour under consideration, i.e. (4) ($AN_{SV5} = [SV_5, SV_4]$).

Step 9: The number of processed state variables during this search is two, i.e. $[SV_5, SV_4]$. Since the number of processed equations (0 at this stage) is not one less than the number of processed state variables, proceed to *Step 11*.

Step 11: The direct neighbours of 4 are SV_4 and SV_7 and the all-neighbours set is thus extended to 5, 4 and 7. State variable 7 is in the all-neighbours set and, therefore, the direct neighbours of 7 are not searched for; so go to *Step 9*.

Step 9: The number of processed equations is now two (equations 3 and 4) and the number of processed state variables is three ($[SV_5, SV_4, SV_7]$), so a univariate conditionally observable subsystem is recorded as well as the redundant measurements. The indicators for processed state variables, eliminated state variables and eliminated equations are modified.

Step 10: Since the Process Phase is 1, then starting from measured voltage, hence the marked univariate conditionally observable subsystem is, in fact, an observable subsystem. The identified region is added to the observable system.

Step 5: Find the direct neighbours of state variable 5 again. The new direct neighbours set contains only itself, because state variables 4 and 7 have already been processed.

Step 6: As the Process Phase is 1, the index is incremented to look at the next processed state variable.

Step 4: All observable state variables have not been searched ($m = 2$) and therefore the second variable in the list is selected as the reference, i.e. state variable 6.

Step 5: Find the direct neighbours of state variable 6. In this case, the direct neighbours set contains only 6 itself; there is no other state variable.

Step 6: The Process Phase is 1 and is thus the index incremented to look at the next processed state variable.

Step 4: All observable state variables have not yet been searched ($m = 3$) and so the third state variable in the list (SV_8) is selected as the reference.

Step 5: Find the direct neighbours of state variable 7. In this case, the direct neighbours set contains only 8 itself, since there is no other state variable.

Step 6: The Process Phase is 1 and the index incremented to look at the next processed state variable.

Step 4: All observable state variables have not been processed ($m = 4$) and thus the fourth variable in the list (i.e. SV_4) is selected as the reference.

Step 5: Find the direct neighbours of state variable 4. In this case, the direct neighbours set contains only 4 itself, since there is no other state variable.

Step 6: The Process Phase is 1 and the index incremented to look at the next processed state variable.

Step 4: All observable state variables have not been processed ($m = 5$) and the fifth state variable in the list (SV_7) is selected as the reference.

Step 5: Find the direct neighbours of state variable 7. In this case, the direct neighbours set contains only 7 itself, since there is no other state variable.

Step 6: The Process Phase is 1 and the index is incremented to look at the next processed state variable.

Step 4: All observable state variables have been processed ($m = 6$ yet only 5 observable state variables) so set the Process Phase to 2 and reset the indicator of processed variables.

Step 2: There are still some unprocessed state variables (SV_1, SV_2 and SV_3).

Step 3: Since the Process Phase is 2, we need to select a reference state variable. There are three unprocessed state variables and the one to be chosen as a reference to be searched should have the maximum number of measurements. However, in this case, the three of them are the same and so state variable 1 is selected.

Step 5: The direct neighbour set of 1 is found to be SV_1, SV_2 and $SV_3(= [j_0, j_1, j_2])$. Since the direct neighbour set consists of more than just the reference state variable, the counter is initialised to 1 (hence index state variable 2).

Step 7: All the direct neighbours have not yet been searched and, therefore, the all-neighbours set is initialised to be the reference state variable (1) and the first direct neighbour counter is pointing to SV_2.

Step 8: The number of processed state variables during this search is 2 (i.e. SV_1, SV_2). Since the number of processed equations (0 at this stage) is not one less than the number of processed state variables, proceed to *Step 11*.

Step 11: The direct neighbours of 2 are SV_2, SV_1 and SV_3 and the all-neighbours set is thus extended to be state variables 1, 2 and 3. State variable 3 is in the all-neighbours set and is not searched for direct neighbours; so go to *Step 9*.

Step 9: The number of processed equations is now two and the number of processed state variables, three, so a univariate conditionally observable subsystem is recorded as well as the redundant measurements. The indicators for processed state variables, eliminated state variables and eliminated equations are modified.

Step 10: Since the Process Phase is 2, then not starting from known voltage, hence mark as univariate conditionally observable subsystem and is unobservable.

Step 5: Find the direct neighbours of state variable 1. In this case, the direct neighbours set contains only 1 itself, since state variables 2 and 3 have been processed.

Step 6: Since the Process Phase is 2, the indicators for processed state variables, eliminated state variables and eliminated equations are modified.

Step 2: Since there are no unprocessed state variables, the observability analysis report is created and the process stops.

Observable State Variables = $[SV_4, SV_5, SV_6, SV_7, SV_8]$
Measurement Equations involved with Observable Subsystem = $[3, 4]$
Measured State Variables = $[SV_5, SV_6, SV_8]$
Unobservable State Variables = $[SV_1, SV_2, SV_3]$
State Variables in univariate conditionally observable subsystem = $[SV_1, SV_2, SV_3]$
Measurement Equations in univariate conditionally observable subsystem = $[1, 2]$
Redundant measurement equations = Non-eliminated equations = $[5]$

7.3.2 Partially-Observable HSE

In harmonic state estimation, due to the cost of suitable harmonic measuring equipment it is impractical to perform complete system-wide estimation. To solve $[A]\mathbf{x} = \mathbf{b}$ for the under-determined case, where only observable islands exist, Singular Value Decomposition (SVD) needs to be applied, since standard techniques for solving such equations will fail. A description of the SVD algorithm is given in Appendix IV. For the under-determined case there is, of course, no unique solution but an infinite number of solutions. SVD, however, will provide a particular solution and a null space vector for each singularity.

The null space vectors can be multiplied by any constant and added to the particular solution to give another valid solution to the set of equations, thereby specifying the infinite number of solutions [21]. Variables corresponding to zeros in all the null space vectors will not be changed by this process and hence are completely specified by the particular solution. These variables correspond to estimates of quantities in the observable islands. The variables corresponding to non-zero elements in the null space vectors are in the unobservable regions, since they cannot be uniquely determined. Therefore, analysis of the null space vectors also provides OA as a by-product.

Figure 7.6 depicts the solution process for an under-determined system. For comparison, Figure 7.7 shows the solution process for a completely or over-determined system.

Another approach, still using SVD, suggests eliminating the state variables in the unobservable system, i.e. partitioning into unknown (denoted by subscript u), known measured values (subscript k) and *not desired* current injection estimates (subscript n):

$$\begin{bmatrix} Y_{1u} & Y_{1k} \\ Y_{2u} & Y_{2k} \\ Y_{3u} & Y_{3k} \end{bmatrix} \begin{pmatrix} \mathbf{V}_u \\ \mathbf{V}_k \end{pmatrix} = \begin{pmatrix} \mathbf{I}_{bu} \\ \mathbf{I}_{bk} \\ \mathbf{I}_{bn} \end{pmatrix}. \tag{7.13}$$

Sub-vector $\mathbf{I}_{bu}$ is then solved in the least squares sense, i.e.

$$\mathbf{I}_{bu} = Y_{1u}Y_{2u}^{-1}\mathbf{I}_{bk} + (Y_{1k} - Y_{1u}Y_{2u}^{-1}Y_{2k})\,\mathbf{V}_k, \tag{7.14}$$

where Y_{2u}^{-1} is a pseudo-inverse in the case of an under-determined system.

7.3.3 Conversion to a Real-valued Problem

Regardless of whether the HSE is achieved via the solution of the Normal equations ($\lfloor h^T R^{-1} h \rfloor \mathbf{x} = \lfloor h^T \rfloor \mathbf{Z}$) or using SVD to solve the measurement equation, the complex-valued problem is split into real and imaginary components. This is done to take

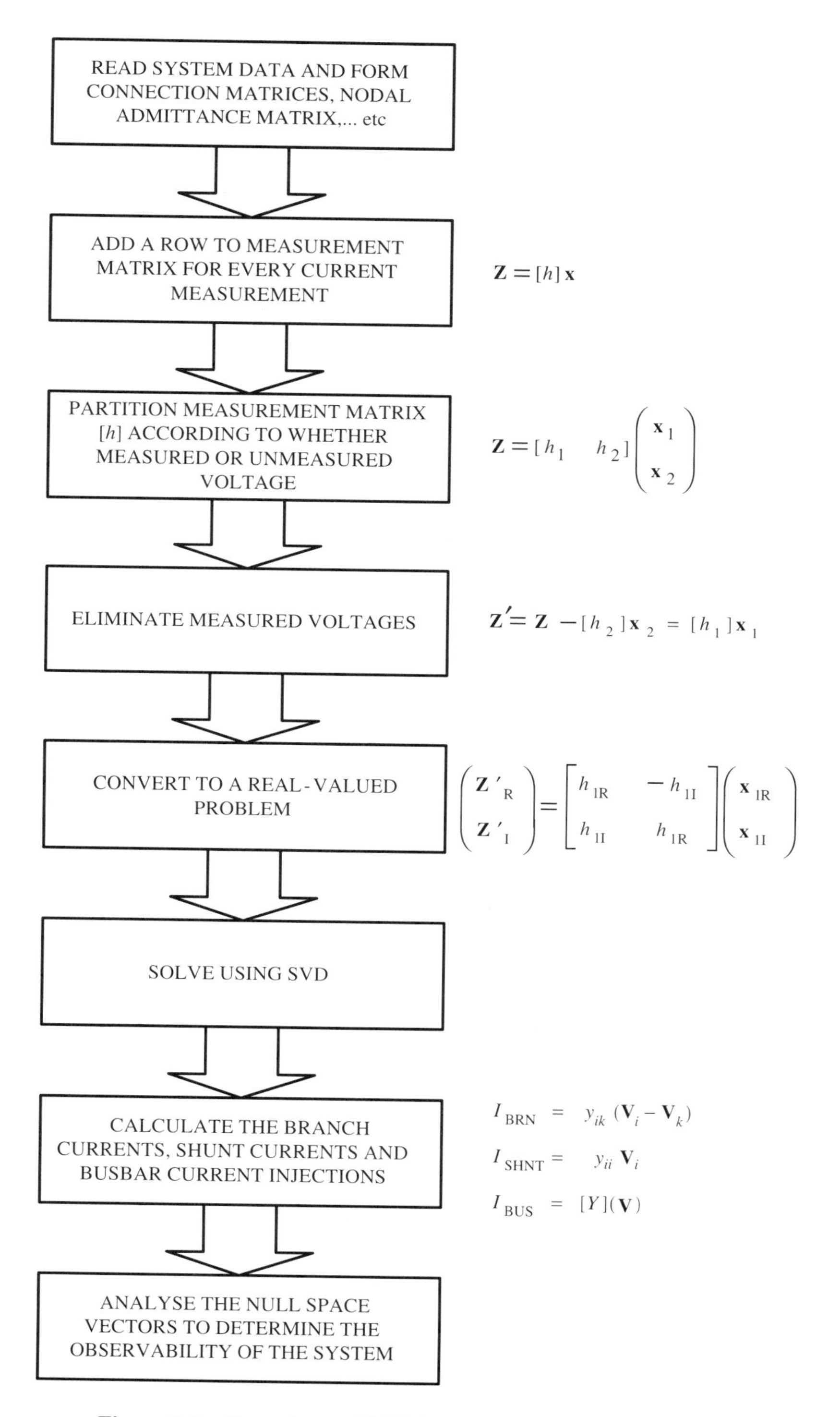

Figure 7.6 Flow-chart of HSE for an under-determined system

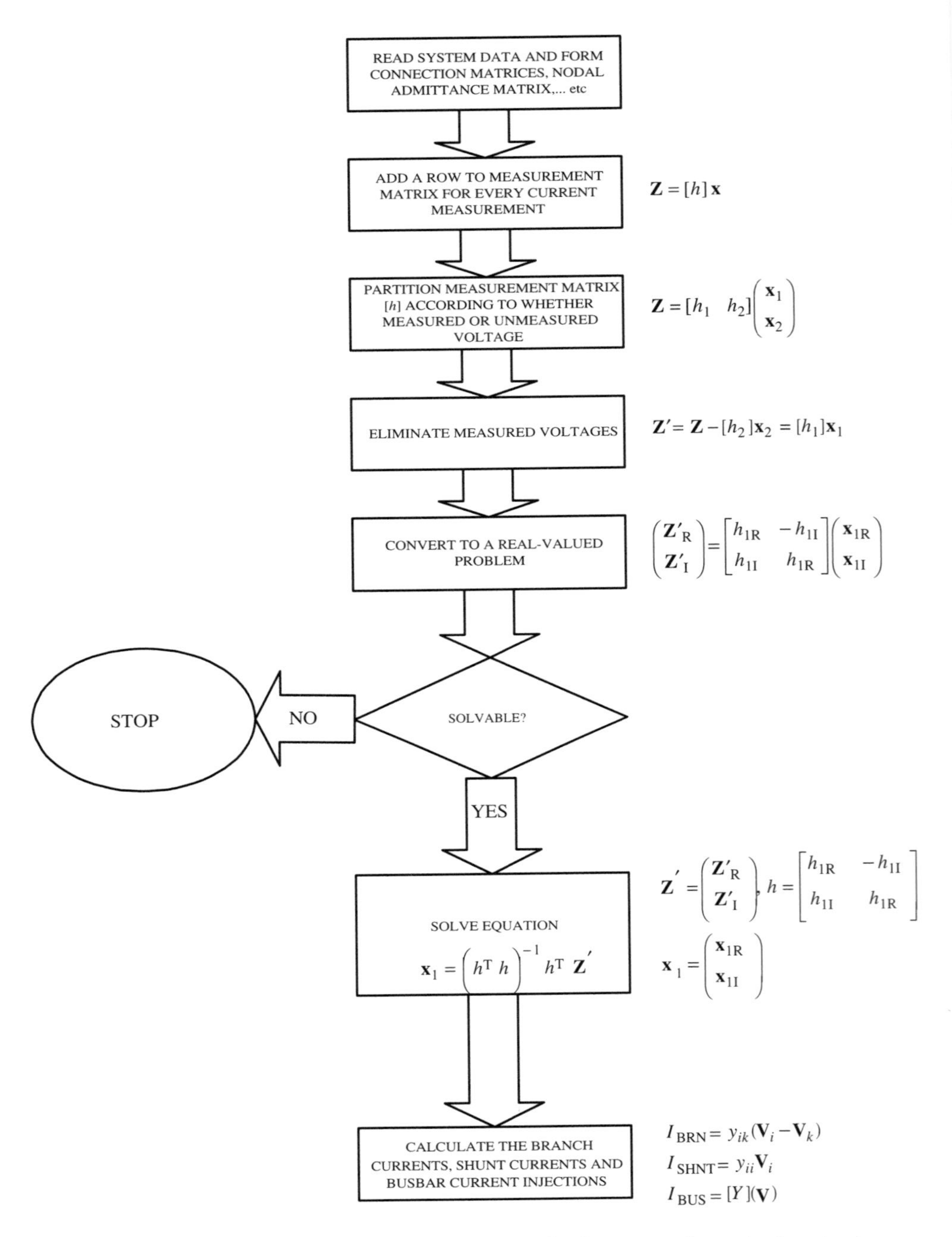

Figure 7.7 System-wide HSE algorithm (completely or over-determined system)

advantage of standard real-valued numerical procedures. The converted terms of the measurement equation $\mathbf{Z} = [H]\mathbf{X} + \mathbf{E}$ are

$$\begin{cases} x = x_R + jx_I \\ z = z_R + jz_I \\ h = h_R + jh_I \\ \varepsilon = \varepsilon_R + j\varepsilon_I \end{cases}, \tag{7.15}$$

where

$$\mathbf{Z} = \begin{bmatrix} z_R \\ z_I \end{bmatrix}, \quad [H] = \begin{bmatrix} h_R & -h_I \\ h_I & h_R \end{bmatrix}, \\ \mathbf{X} = \begin{bmatrix} x_R \\ x_I \end{bmatrix}, \quad \mathbf{E} = \begin{bmatrix} \varepsilon_R \\ \varepsilon_I \end{bmatrix}. \tag{7.16}$$

7.3.4 Ill-conditioning

When solving the linear equation $[A]\mathbf{x} = \mathbf{b}$ the relative error in A or $\mathbf{b}$ can be magnified by as much as the condition number of A. The condition number of the gain matrix $[h^T R^{-1} h]$ is the square of the measurement matrix $[h]$, therefore the normal equation formulation is not well conditioned. Due to finite precision the build-up of round-off error can be a problem. Two approaches to alleviate the numerical ill-conditioning are behind the various methods that have been proposed. The first is to avoid the formulation of the gain matrix. The second is to treat virtual measurements as equality constraints.

For an over-determined system, the use of SVD can be significantly slower than solving the normal equations and requires more storage, but is less susceptible to round-off error. Moreover, its theoretically fool-proof reliability more than makes up for the speed disadvantage.

7.4 Further Capabilities of Harmonic State Estimation

7.4.1 Optimal Placement of Measurements

The design of a measurement system to perform HSE is a very complex problem. Among the reasons for its complexity are the system size, conflicting requirements of estimator accuracy, reliability in the presence of transducer and data communication failures, adaptability to changes in the network topology and cost minimisation. In particular, the number of harmonic instruments available is always limited due to cost, and the quality of the estimates is a function of the number and location of the measurement points. Therefore, a systematic procedure is needed to design the optimal placement of measurement points.

Any of the following criteria can be used to determine the optimal placement of measurements.

- Minimise the number of measurement points for network observability,
- Minimise the sum of the state estimate variances,
- Maximise measurement system reliability,
- Minimise or limit measurement system cost, or
- Minimise the condition number of the gain matrix.

The existing optimal measurement placement algorithms include constrained non-linear optimisation, Monte Carlo simulation and perturbation, sequential elimination, and sequential addition.

A rigorous formulation of optimal measurement placement would result in the solution of a 0–1 integer-programming problem (as measurement is either present or not present); that is very difficult to solve exactly for large-scale systems. Therefore, all the approaches proposed for optimal measurement placement are based on non-rigorous formulations or heuristic solution techniques, yielding nearly optimal solutions rather than the global optimum.

Traditionally, the measurement placements have been symmetrical, i.e. either three or no phases of buses or lines were measured. This requirement restricts the search for the optimal placement of measurement points in three-phase asymmetrical power systems. Asymmetrical placement (only one or two phase measurements of voltage or current) is better in terms of the number of measurements required.

However, minimising the number of channels does not necessarily result in lower cost because the predominant cost is in the base unit, while the incremental cost for additional channels is relatively small. Therefore, it is the number of locations that should be minimised.

7.4.2 Bad-Data Analysis

Bad data can result from erroneous measurement values, incorrect system parameters or incorrect network topology. Erroneous measurement has been the main focus of bad data analysis. This can be categorised into three groups; extreme errors, gross errors and normal measurement noise. The presence of bad data degrades the accuracy of the HSE results and the problem is overcome by the detection, identification and removal of the bad data or the use of a more robust estimator replacing the weighted least squares.

Although a great deal of work has been done on bad-data analysis for fundamental frequency, in particular, detecting its presence, identifying which measurements are bad and eliminating the influence of the bad data, this usually requires the system to be over-determined and is therefore of limited applicability to HSE. However, in the presence of bad data, the residual $r_i = Z_i - h_i(\hat{x})$ should be large. A statistical hypothesis test can thus be applied on the residual values (weighted or normalised versions of the residual) to identify the presence of bad data.

7.4.3 Hierarchical HSE

The computation time involved in a centralised HSE solution increases non-linearly with the size of the power system. Distributed processing is, therefore, necessary to make the HSE feasible in a large-scale system. As an alternative to the centralised scheme, a hierarchical HSE [22] has been suggested for obtaining the system-wide HSE solution based on the decomposition co-ordination theory of large-scale systems. Its purpose is to reduce the complexity of the centralised HSE by decomposing it into lower-order subproblems; however, a careful co-ordination is needed to obtain a complete and coherent solution.

A power system network can be decomposed into a certain number of non-overlapping subsystems interconnected by tie-lines. The two ends of each tie-line, called boundary buses, belong to two distinct subsystems. For each subsystem, a first-level computer performs independently and simultaneously a local HSE, using only the measurements that are related to the local state variables. A second-level computer is devoted to dealing with synchronisation, data sharing and co-ordination tasks among the individual subsystems. The hierarchical structure of HSE uses distributed algorithms combining parallel and sequential operations in which considerable information can be exchanged via the high-speed data links between the remote computers at the first and second level. Although the information exchange means a real-time overhead, the total computation speed for the system-wide HSE will increase dramatically with hierarchisation.

It can theoretically be shown, however, that the centralised HSE, which minimises the global performance index of the estimator, provides the most accurate system-wide estimate for the given set of all measurements in the network. With respect to that, the final estimate from the hierarchical HSE is the second most optimal solution.

7.4.4 Real-Time HSE

The purpose of real-time HSE is to estimate harmonic information through a power system at a fast enough speed so that the system-wide harmonic dynamic process can be continuously monitored, and, furthermore, controlled by using this information available in time. The static HSE can be implemented continuously in real time with a very short cycle interval based on the continuous harmonic measurements, real-time databases and very fast HSE algorithms [6]. The basic bulk-processing algorithm of the static HSE can be reformulated in recursive form for efficient use in real time by using the last available estimates as the starting values of the current estimation. In most cases, static HSE is sufficient for building a database representing the current system harmonic state. Such tracking estimators, however, do not have any ability of harmonic prediction.

In cases where a predictive harmonic database is needed, dynamic HSE can be used as a harmonic state predictor, to provide the necessary harmonic information for all types of predictive harmonic analysis and control. The dynamic HSE is based on a conventional extended Kalman filter (Appendix V), consisting of alternate sequences of filtering and prediction [4–11]. At the filtering step, a measurements state variable model (the same as in the static HSE but described as functions of sampling time k) is used to produce a WLS estimate of the harmonic state variables at time k. At the prediction step, a dynamic state transition model (describing the state variables at time $k+1$ as functions of the state variables and any known system control at time k) is used to predict the state variables at time $k+1$.

Owing to the predictive ability of the dynamic HSE, the observability analysis, and bad-data detection and identification can be improved. The predicted harmonic database can be used as pseudo-measurements when the power system becomes unobservable due to insufficient measurements. By comparing the measurements with their corresponding prediction, it is possible to detect and identify the bad data in the measurement and topology errors. Compared with the static HSE, however, the dynamic HSE has two practical difficulties: the availability of the reliable dynamic state transition model, and the computational complexity and burden. Although the computational problem can be mitigated by using more efficient algorithms and possibly a hierarchical HSE technique, the availability of a simple and correct dynamic state transition model is still open to question.

7.4.5 Enhanced Implementation

The implementation of HSE could be enhanced using the program structure and database format described in Chapter 6 (Section 6.5). It can be incorporated as an extra program, sharing the same user interface and component and network models used for harmonic simulation. In that environment, the user can create one-line or three-phase diagrams, using a network editor provided by the Graphic User Interface (GUI). The data of the network topology and all the component parameters are stored into a database, and are used to generate the system admittance matrix at each harmonic frequency, which again is stored into the database. The user can specify a number of measurement points on the network diagram, and invoke an OA algorithm to show the observability of the system directly on the network diagram. The measurement placement can then be modified until the desired observability is satisfied.

With the measured harmonic voltages and currents being continuously sent to the database via data communication channels, a harmonic state estimator will be providing the system-wide harmonic estimates and corresponding harmonic performance indices, which are graphically displayed and updated in the GUI. Once the network topology,

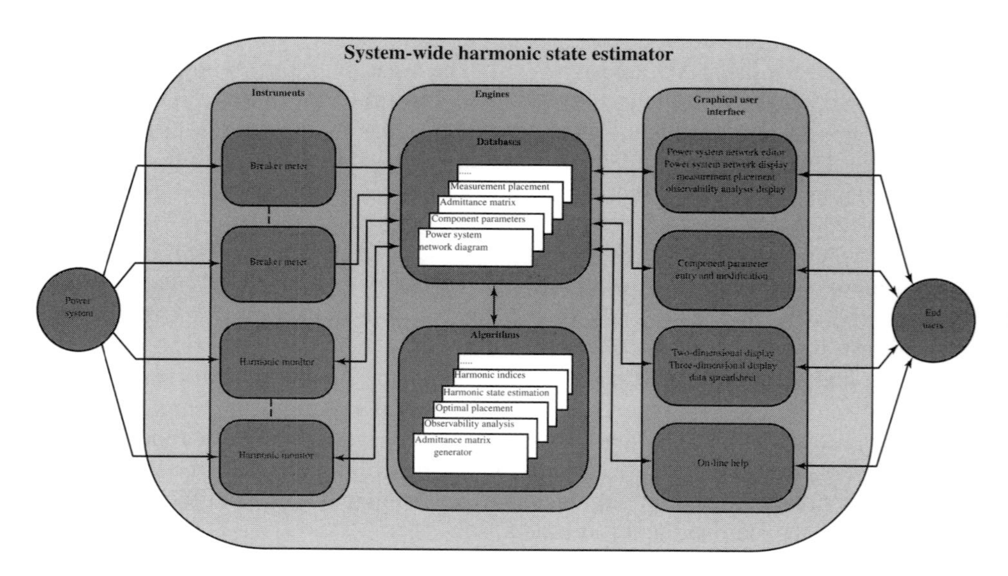

Figure 7.8 System-wide harmonic state estimator

being monitored by the breaker status meters in real time, is changed, the network diagram in the GUI is updated accordingly, and the OA and HSE algorithms are executed subsequently based on the new network topology. HSE can be implemented both on-line with the measurements continuously provided by the harmonic instruments to a real-time database, or off-line with the measurements stored in the database in advance.

The use of Object Oriented Programming (OOP) will enhance the modularity and maintainability of the HSE software and make it easily extendable and reusable according to the varying needs of the power system.

Figure 7.8 gives an overview of a possible implementation of a system-wide state estimator taking advantage of the developments in computer technology.

7.5 Test System and Results

The backbone of the test system illustrated in Figure 7.9 is taken from the 220 kV network below Roxburgh in the south island of New Zealand. The system contains nine buses, five star-delta-connected two-winding transformers, three generators at Manapouri and Roxburgh, two single transmission lines between Roxburgh and Invercargill, and three double transmission lines between Manapouri, Invercargill and

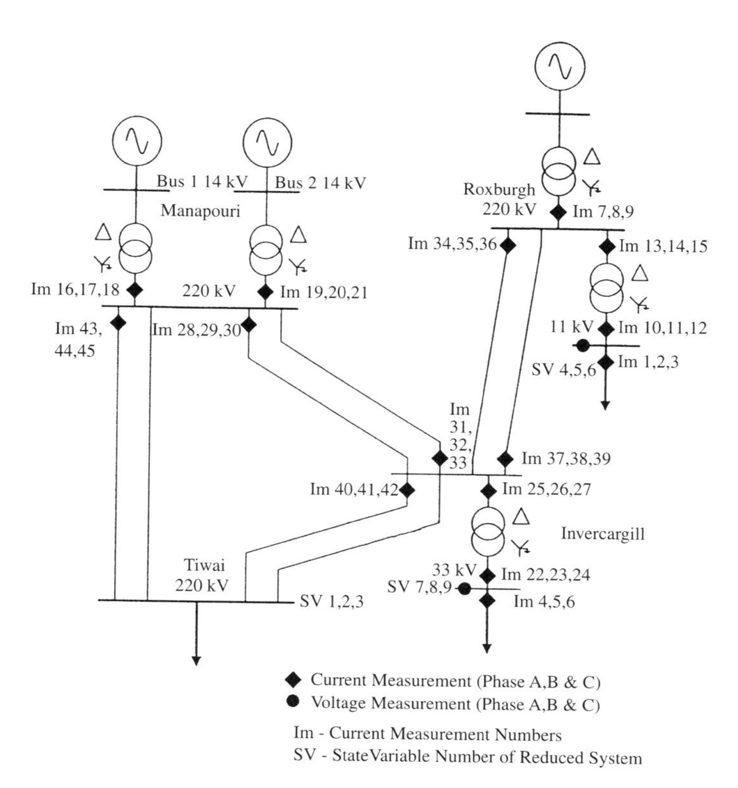

Figure 7.9 Initial measurement placement for lower south island of New Zealand test system

Tiwai. The distances are about 176 km between Manapouri and Tiwai, 132 km between Roxburgh and Invercargill, and only 24 km between Invercargill and Tiwai. The harmonic current source consists of 6-pulse rectification at the Tiwai bus.

However, for the purpose of demonstrating the ability of HSE to identify harmonic sources, no information is given about their location at the start of the HSE solution. Therefore, the harmonic sources or linear loads at Tiwai, Invercargill and Roxburgh are replaced by three suspicious sources at the corresponding locations.

The transmission lines are represented by equivalent PI circuits with corresponding couplings, and the generators are modelled simply by shunt branches connected at their terminal nodes. Inside the backbone there are 27 nodes, 18 non-suspicious nodes and 9 suspicious nodes; there are also 87 lines and 111 branches.

A node-line incidence matrix (27×87) and a line-branch incidence matrix (87×111) are generated from the network topology of the backbone. For each harmonic up to the 25th, an individual primitive admittance matrix (111×111) is also generated. Furthermore, a none-node admittance matrix (27×27) and a line-node admittance matrix (87×27) are calculated, and reduced to the suspicious nodes as two smaller admittance matrices (9×9 and 87×9), respectively.

The *actual* values were determined by performing detailed three-phase harmonic simulation of the system, as described in Chapter 6. Results from the simulation were used at the measurement locations and supplied to the HSE algorithm. The HSE estimated the unmeasured quantities to be compared with the exact solution. For the initial measurement point placements, illustrated in Figure 7.9, the results of the simulation are shown in Figures 7.10 to 7.13, with a 3-D display of current injections and nodal harmonic voltages at 27 nodes for up to the 25th harmonic.

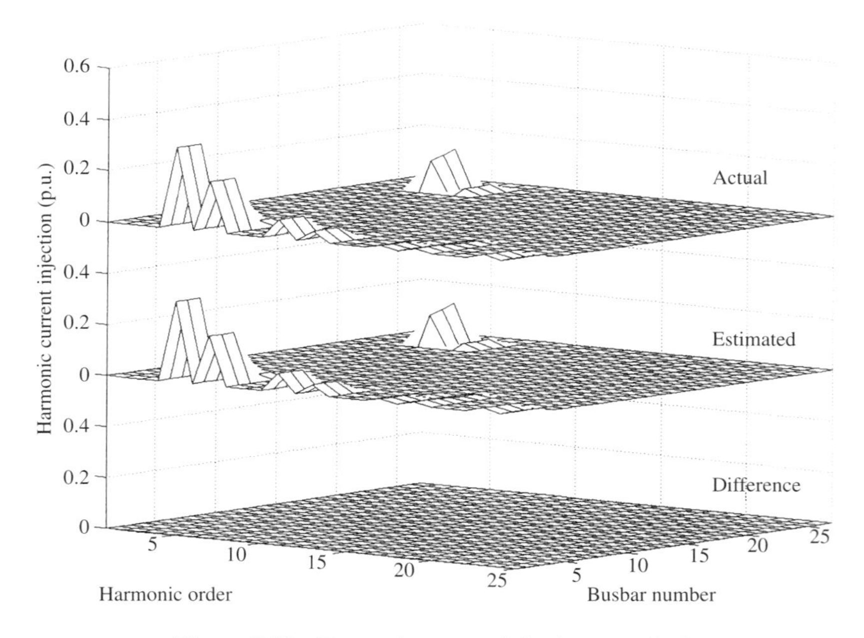

Figure 7.10 Harmonic current injection magnitude

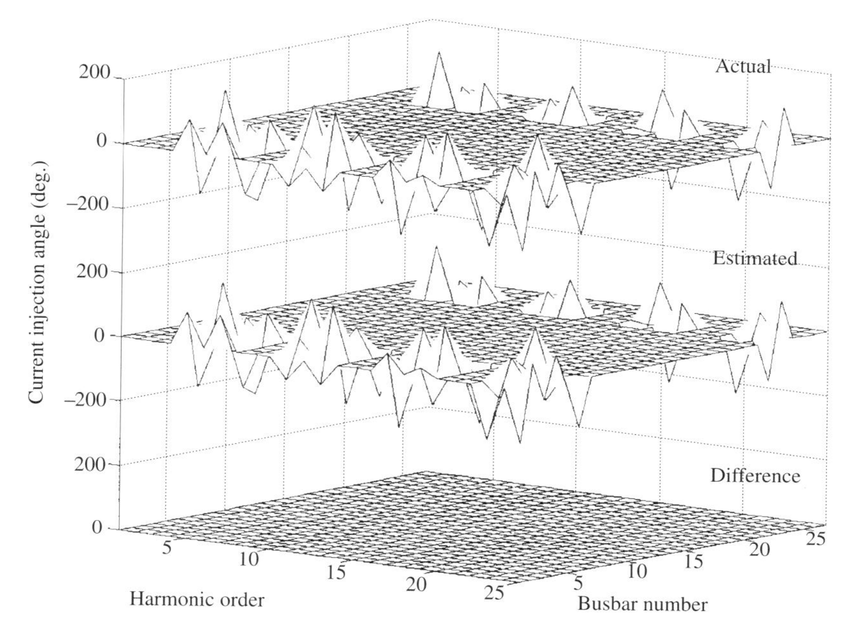

Figure 7.11 Harmonic current injection phase angle

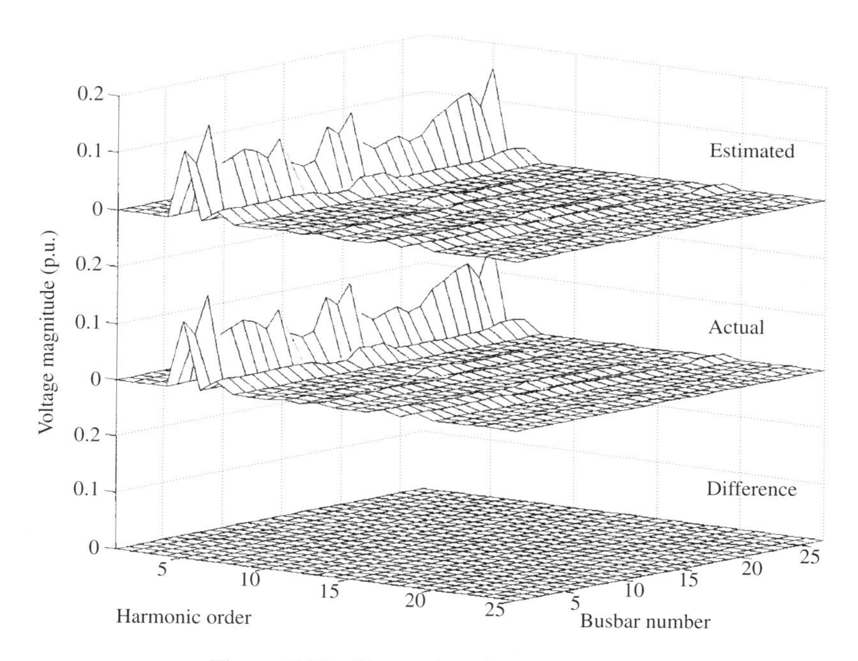

Figure 7.12 Harmonic voltage magnitude

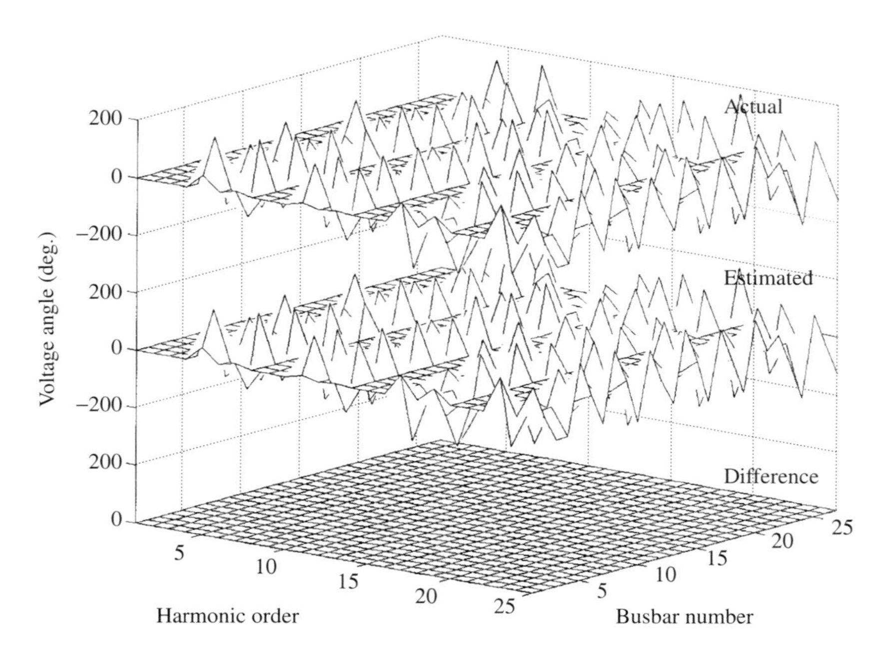

Figure 7.13 Harmonic voltage phase angle

The figures display the *estimated, actual* and *error* of the magnitude and phase for harmonic current injections and busbar harmonic voltages, respectively. Two forms of measurement noise are included in the test. One is the noise due to limiting the precision of the harmonic simulation values that are supplied to the HSE algorithm. The second source is the limited precision with which the system parameters are specified. The results show that with three monitoring locations (at Manapouri, Roxburgh and Invercargill) HSE provides accurate harmonic information throughout the test system. The effect of reducing the measurement locations is considered next; this is important because, as explained in section 7.4.1, the estimation cost is mainly determined by the number of measurement locations.

An example with only two measurement sites (Roxburgh and Invercargill) is shown in Figure 7.14, where four line current, six voltage and two current injection measurements are used. The number of virtual measurements is 18 since there are only nine nodes with loads directly connected to them and, hence, capable of having harmonic currents injected; these are Roxburgh (11 kV), Invercargill (33 kV) and Tiwai (220 kV). The comparison between the estimated harmonic levels and those of the actual solution is shown in Figures 7.15 and 7.16 for the voltage and harmonic current injections, respectively. Clearly, the HSE performance, while still reasonable, shows increased error as compared with that of the original placement.

In the previous examples, the number of measurement points has been made arbitrarily. To illustrate the use of OA to make more objective decisions, the initial measurement placement in Figure 7.9 is now subjected to symbolic observability. The corresponding symbolised measurement matrix $[h]$ involving nine state variables and 3×15 current measurements is:

Measurement equation number ⇓ / State variable ⇒	1	2	3	4	5	6	7	8	9
1	1	1	1	1	1	1	1	1	1
2	1	1	1	1	1	1	1	1	1
3	1	1	1	1	1	1	1	0	1
4	1	1	1	1	1	1	1	1	1
5	1	1	1	1	1	1	1	1	1
6	1	1	1	1	1	1	1	0	1
7	1	1	1	1	1	1	1	1	1
8	1	1	0	1	1	0	1	1	1
9	1	1	1	1	1	1	1	1	1
10	1	1	1	1	1	1	1	1	1
11	1	1	1	1	1	1	1	1	1
12	1	1	1	1	1	1	1	0	1
13	1	1	1	1	1	1	1	1	1
14	1	1	0	1	1	0	1	1	1
15	1	1	1	1	1	1	1	1	1
16	1	1	1	1	1	1	1	1	1
17	1	1	0	1	1	0	1	1	1
18	1	1	1	1	1	1	1	1	1
19	1	1	1	1	1	1	1	1	1
20	1	1	0	1	1	0	1	1	1
21	1	1	1	1	1	1	1	1	1
22	1	1	1	1	1	1	1	1	1
23	1	1	1	1	1	1	1	1	1
24	1	1	1	1	1	1	1	0	1
25	1	1	1	1	1	1	1	1	1
26	1	1	0	1	1	0	1	1	1
27	1	1	1	1	1	1	1	1	1
28	1	1	1	1	1	1	1	1	1
29	1	1	1	1	1	1	1	1	1
30	1	1	1	1	1	1	1	1	1
31	1	1	1	1	1	1	1	1	1
32	1	1	1	1	1	1	1	1	1
33	1	1	1	1	1	1	1	1	1
34	1	1	1	1	1	1	1	1	1
35	1	1	0	1	1	0	1	1	1
36	1	1	1	1	1	1	1	1	1
37	1	1	1	1	1	1	1	1	1
38	1	1	0	1	1	0	1	1	1
39	1	1	1	1	1	1	1	1	1
40	1	1	1	1	1	1	1	1	1
41	1	1	1	1	1	1	1	1	1
42	1	1	1	1	1	1	1	1	1
43	1	1	1	1	1	1	1	1	1
44	1	1	1	1	1	1	1	1	1
45	1	1	1	1	1	1	1	1	1

Each measurement equation represents either a current injection or a line current; the correspondence between equations and measurement numbers is also indicated in Figure 7.9.

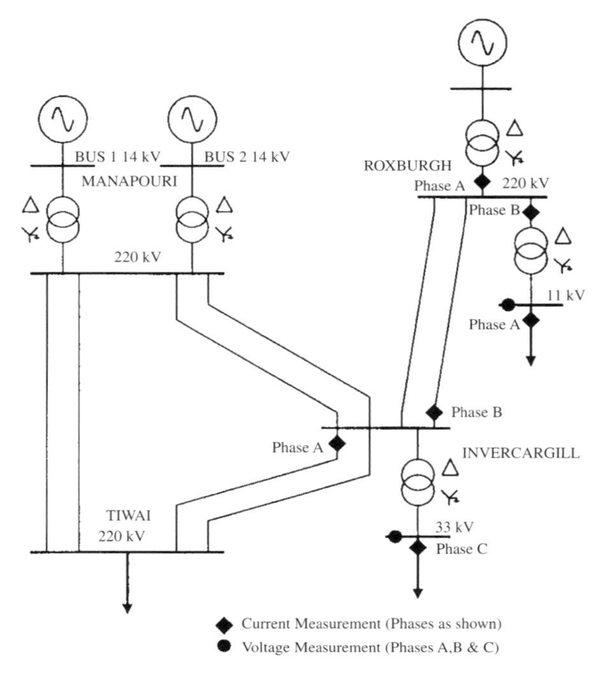

Figure 7.14 Lower south island of New Zealand with reduced measurement placement

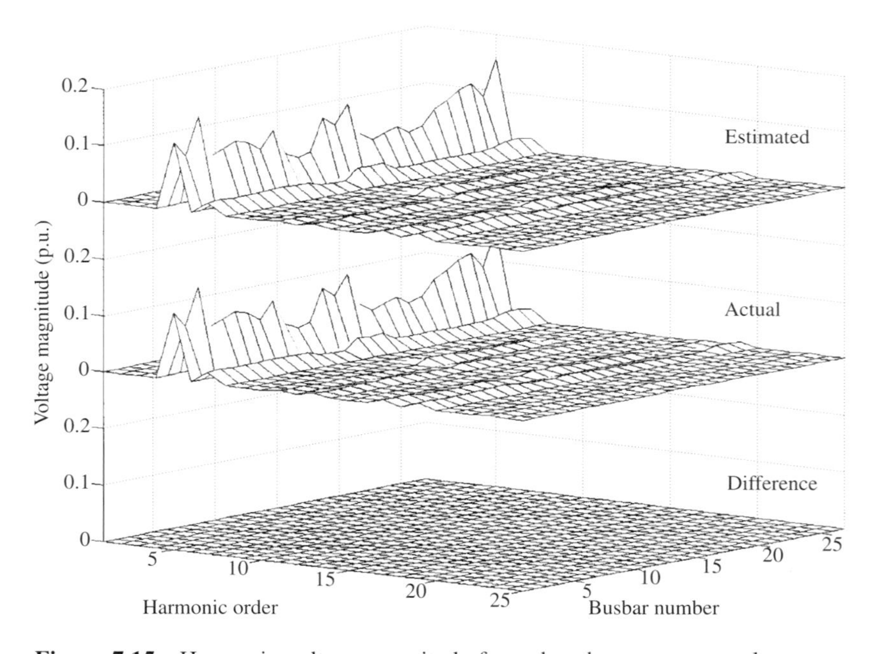

Figure 7.15 Harmonic voltage magnitude for reduced measurement placement

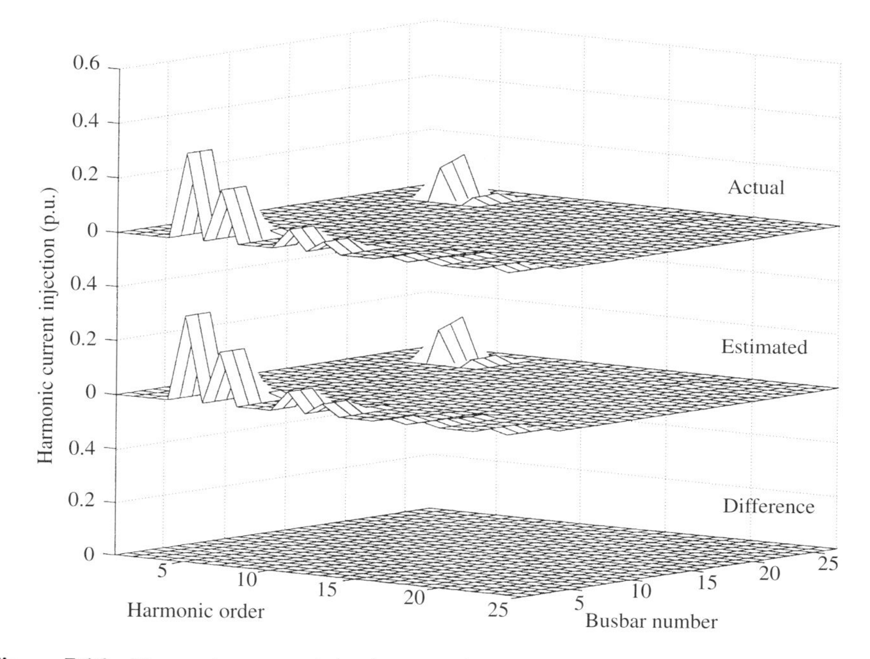

Figure 7.16 Harmonic current injection magnitude for reduced measurement placement

Applying the symbolic OA described in Section 7.3, which starts with the known (measured) state variables and then identifies the state variables linked to them, results in two groups of measurements.

The first group, shown in Figure 7.17, starts its search with state variable 4 and contains seven current measurements. Since these measurements couple two unknown state variables (i.e. 1 and 2), only two measurements are required.

The second group, displayed in Figure 7.18, contains 38 current measurements and only one unknown state variable (i.e. 3); thus, only one current measurement is required. The total number of required current measurements is the addition of the two groups, i.e. three for this test system. The OA report on the initial arrangement (Figure 7.9) suggests the measurement placement of Figure 7.19, and Figure 7.20 illustrates the difference between the actual and estimated levels of voltage distortion for this arrangement.

This method assumes that all voltage measurements contribute to the solution, and identifies how many current measurements are redundant by looking at the number of unknown state variables and the number of equations linking the state variables in each identified group. It should be noted that symbolic OA cannot detect cases where there are two dependent measurement equations (such as when current at both ends of a line are measured) because the actual values are lost (either a zero or a one).

In the OA example of section 7.3.1, the first two lines of the symbolic measurement matrix were:

$$\begin{matrix} 1 & 1 & 1 & 0 & 0 & 0 & 0 & 0 \\ 1 & 1 & 1 & 0 & 0 & 0 & 0 & 0 \end{matrix}$$

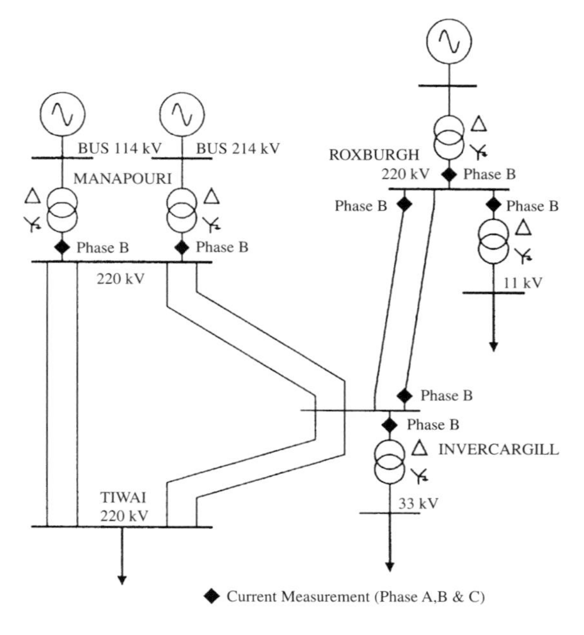

Figure 7.17 First group of seven measurements where only two required

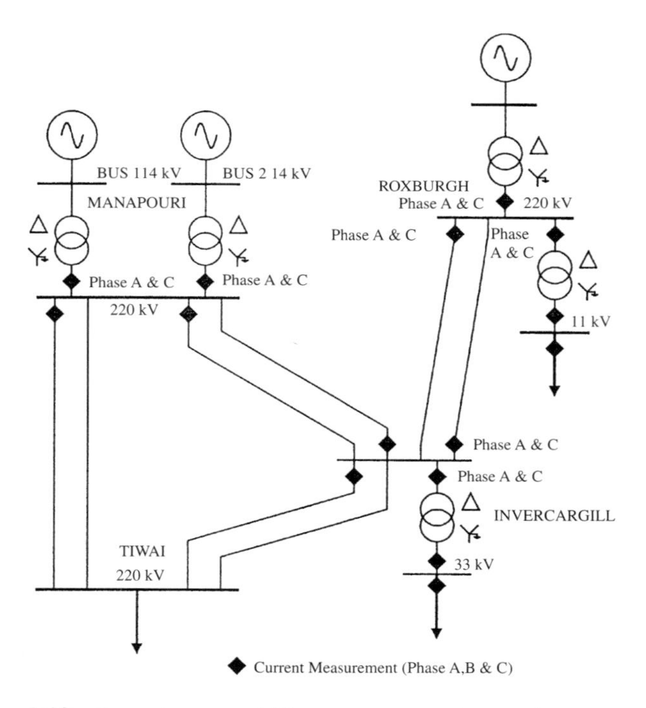

Figure 7.18 Second group of 38 measurement where only one required

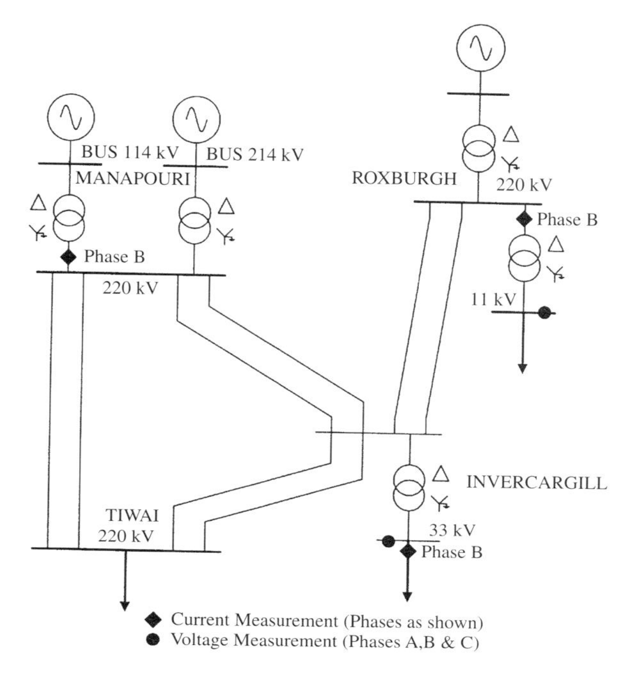

Figure 7.19 Reduced measurement points

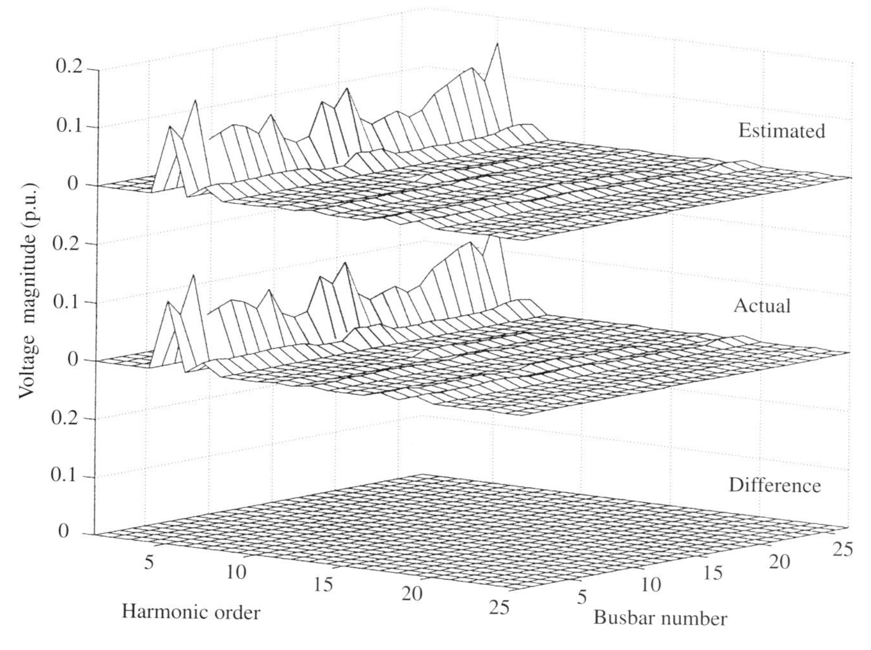

Figure 7.20 Harmonic voltage magnitude for three current measurement placement

which, although identical in structure, result from harmonic current injection at two different nodes (1 and 2) and so they are not linearly dependent. Yet, if the two harmonic line current measurements were taken (at either end of the line) between busbars 1 and 2, then the symbolic matrix would have shown

$$\begin{matrix} 1 & 1 & 0 & 0 & 0 & 0 & 0 & 0 \\ 1 & 1 & 0 & 0 & 0 & 0 & 0 & 0 \end{matrix}$$

and the OA would not treat them as dependent even though they are.

7.5.1 Load and Harmonic Source Identification

The harmonic simulation and HSE algorithms differ regarding the way loads are treated. In general, a load bus may contain linear (passive) and non-linear components. These can be modelled in detail in harmonic simulation, which represents separately the current injections and the passive components. HSE, on the other hand, may have no information on the composition of the load and is only capable of estimating the net current flow into or out of the load busbar.

Therefore, the current injection information supplied to the HSE algorithm is the sum of the harmonic current source and harmonic current flowing in the load. This can be seen in Figure 7.10 where nodes 22, 23 and 24 (Invercargill 33 kV busbar) have a passive load attached and a small amount of current can be seen to flow into it.

The harmonic voltages at the suspicious buses, and harmonic currents injected from the suspicious sources to the backbone, are provided by the estimator at the end of HSE and each suspicious source is classified as a harmonic injector or a harmonic absorber.

In general, a suspicious harmonic source can be considered as a Norton equivalent circuit at each harmonic frequency (Figure 7.21), and the following relationship applies for a harmonic of order n:

$$\hat{I}_i(n) - I_i(n) = V_i(n)Y_i(n). \tag{7.17}$$

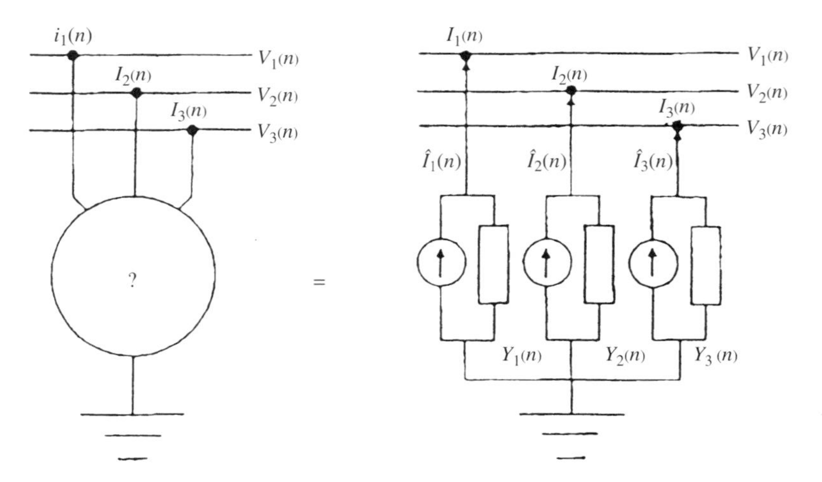

Figure 7.21 Norton equivalent for suspicious harmonic sources

In Equation (7.17), $V_i(n)$ and $I_i(n)$ are the nodal voltage and current injection, respectively, as provided by the estimator, while $\hat{I}_i(n)$ and $Y_i(n)$ are the unknown Norton harmonic current injection and admittance within the suspicious source ($i = 1, 2, 3$). In theory, it should be possible to derive some information on the nature of the load from the estimated harmonic voltages and injected currents at the bus.

When no harmonic current injection exists at a node, it is possible to identify the load impedance using the nodal harmonic voltage and current information at two different harmonics. This calculated impedance can be verified using information at other frequencies. When a harmonic source is present it is not possible to identify the components without additional information. This information may come from measurements obtained under different operating conditions (e.g. a component switched in or out) or may take the form of an assumed ratio of the harmonic current injection to the fundamental based on the converter pulse number. In the latter case, if the following two assumptions are made for the suspicious source,

$$Y_i(n) = G_i - jB_i/n, \tag{7.18}$$

$$|\hat{I}_i(n)| = \delta_i(n, n_0)|\hat{I}_i(n_0)|, \tag{7.19}$$

where G_i and B_i are unknown parameters for node i, n_0 is a chosen reference harmonic (e.g. the 11th harmonic for the cases of 6-pulse and 12-pulse converters), and $\delta_i(n, n_o)$ is a chosen ratio of $|\hat{I}_i(n)|$ to $|\hat{I}_i(n_0)|$, then for any two harmonics n_1 and n_2 which are not n_0, the set of quadratic equations (Equations (7.17)–(7.19)) is solvable to obtain the unknown Norton parameters $\hat{I}_i(n)$ and $Y_i(n)$ for each harmonic n of interest.

It can be shown by sensitivity analysis that the estimated Norton parameters using the above method are very dependent on the chosen ratio when the suspicious source contains non-zero Norton current injections, and very insensitive to the chosen ratio when the suspicious source does not contain Norton current injections. Therefore, the above method can at least be used to identify whether a suspicious source is a purely passive load and, in such a case, estimate the equivalent harmonic admittances of the passive load.

To illustrate the identification capability of HSE to loads containing static converters, the test system is now modified to include one harmonic source (a 12-pulse converter) at Tiwai, one harmonic source (a 6-pulse converter) at Invercargill, and only one linear load at Roxburgh.

As shown in Figure 7.22, only two measurement sites (at Manapouri and Roxburgh) and a total of nine channels are used to derive system-wide information and locate the harmonic sources. With the chosen measurements placement, the measurement matrix is singular and requires SVD analysis. However, as explained earlier for the underdetermined case, SVD provides multiple solutions and the particular solution selected is the one that minimises the solution's norm.

Although neither voltages nor injection currents are measured, the HSE study provides accurate information on all the current injections (Figure 7.23(a)), branch currents (Figure 7.23(b)) and nodal voltages (Figure 7.23(c)). The results of the particular solution chosen by SVD are very close to the actual values. Inspection of the null space vectors shows that the injection at Tiwai, even though far from the measurement points, can be uniquely determined. Even with this partially observable system the load at the Roxburgh 011 busbar is identified as being passive with an

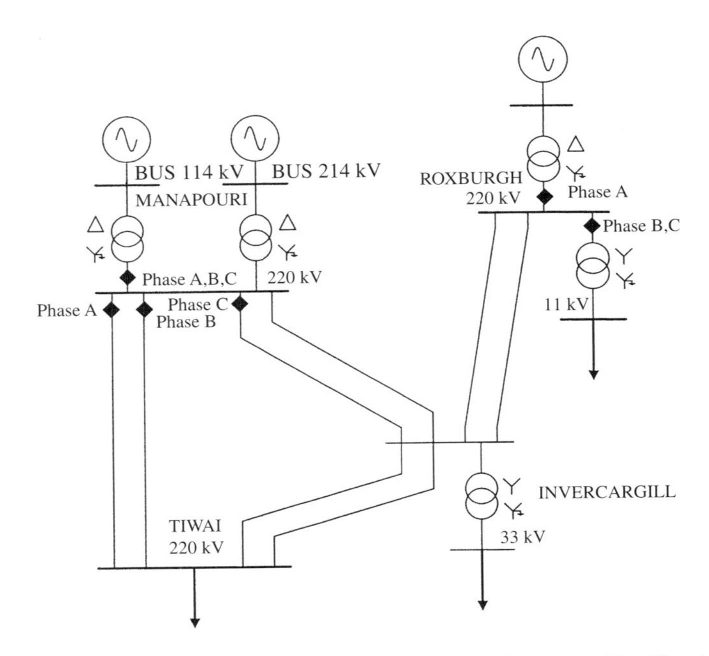

Figure 7.22 Measurement placement for harmonic source identification

Table 7.1 Identified admittance

Harmonic order	Simulated		Estimated	
	Conductance	Susceptance	Conductance	Susceptance
5	0.90001	−0.108	0.90001	−0.108
7	0.90001	−0.077144	0.90001	−0.077144
11	0.90001	−0.049091	0.90001	−0.049091
13	0.90001	−0.041539	0.90001	−0.041539
17	0.90001	−0.031765	0.90001	−0.031765
19	0.90001	−0.028421	0.90001	−0.028421
23	0.90001	−0.023478	0.90001	−0.023478
25	0.90001	−0.0216	0.90001	−0.0216

admittance of 0.900009-j0.540005 at fundamental frequency and Table 7.1 shows the comparison between actual and estimated harmonic admittances at this bus. These results illustrate that it is possible to assess the overall effect of several unknown loads connected to the same suspicious bus without the need to know their detailed configuration and parameters.

The singularity of the measurement matrix can be removed with the addition of one more measurement point (Invercargill to Tiwai), giving 10 measurements; although this allows other methods to be used to solve for the voltage, the matrix is still rank deficient and hence a multitude of solutions still exist. Some additional measurements at other locations are needed for the complete system to be observed. When the load contains

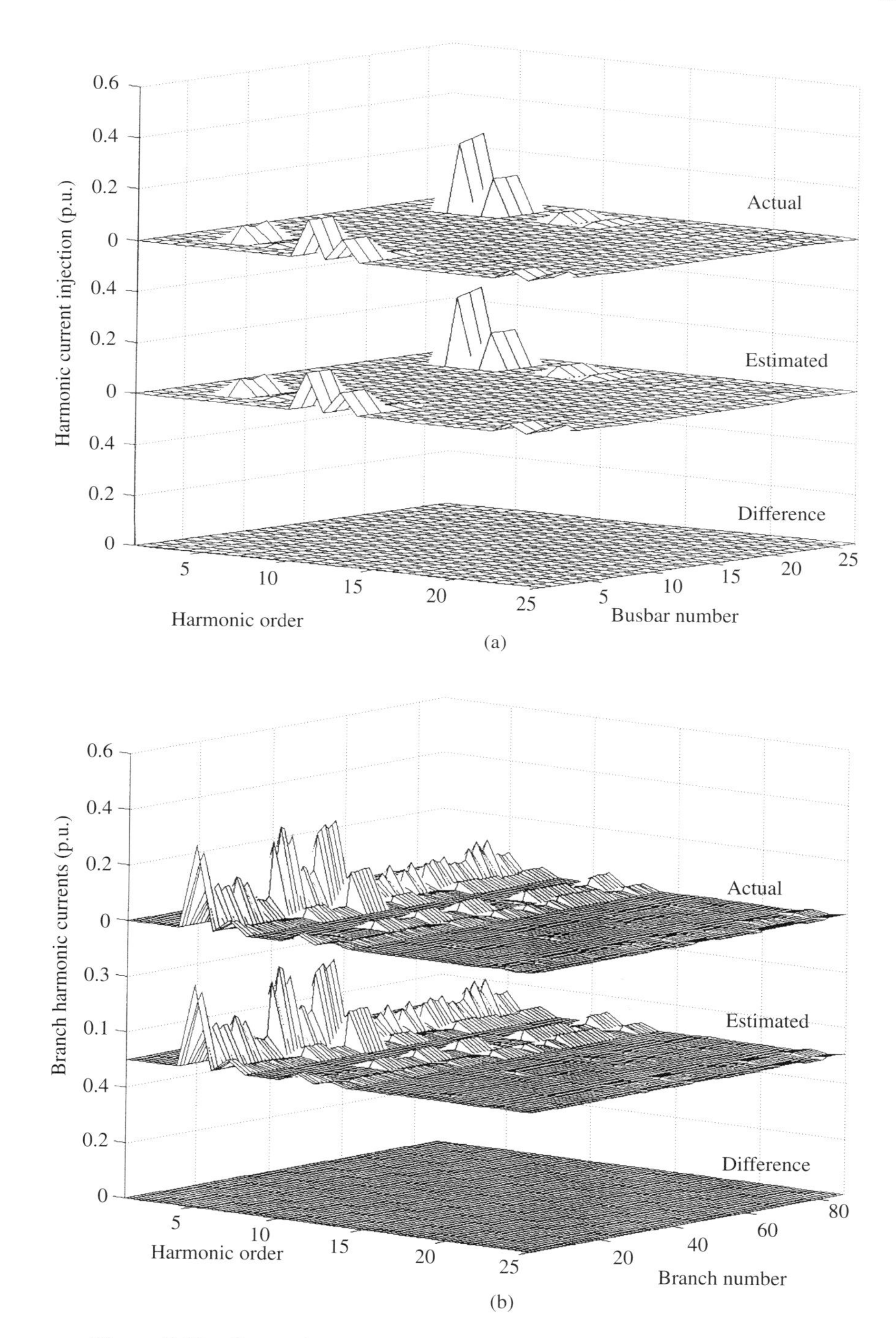

Figure 7.23 Comparison of estimated and simulated levels of harmonics

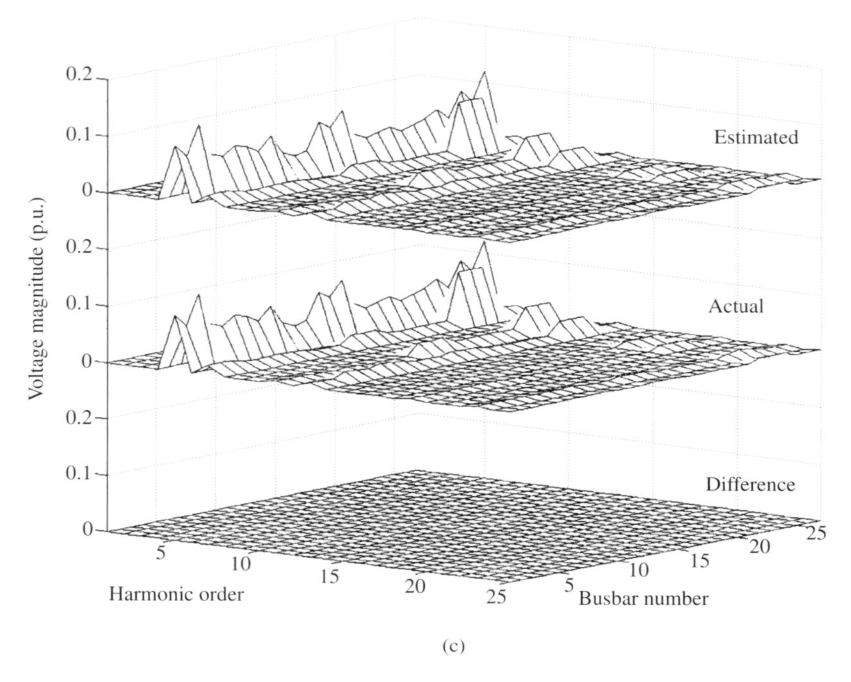

(c)

Figure 7.23 (*continued*)

harmonic sources the net injected real power depends on the relative magnitude of real power absorbed by the passive component and the real power supplied by the source.

It must be stressed, therefore, that a negative real power injection at all harmonic frequencies is not a sufficient condition for the load to be passive. An unambiguous identification of the type of load and configuration can be achieved from the readings at several frequencies and the solution of their corresponding Norton impedances.

Effect of Transformer Connections The observability of circuits with a star-*g*/delta transformer requires special consideration.

Measurement of line currents on the delta side are inadequate since there are an infinite number of possible solutions for the currents in the star-*g* winding to match the observed measurements. This is because the zero sequence currents can flow into the star-*g* winding and couple to corresponding circulating currents in the delta that are never measured.

At least one measurement on the star-*g* side is required for the system to be observable, but if sufficient measurements are made on the star-*g* side, which is normally the low voltage side and hence closer to the load, then delta side measurements can be eliminated completely.

It should be noted that complete observability can be achieved with less measurement points where star-*g*/delta transformers exist (provided, of course, the measurements are on the star-*g* side) due to the coupling between phases, i.e. one phase on the star-*g* is influenced by two phases on the delta side. With the load transformers being star-*g*/delta, the measurements shown in Figure 7.24 give complete observability.

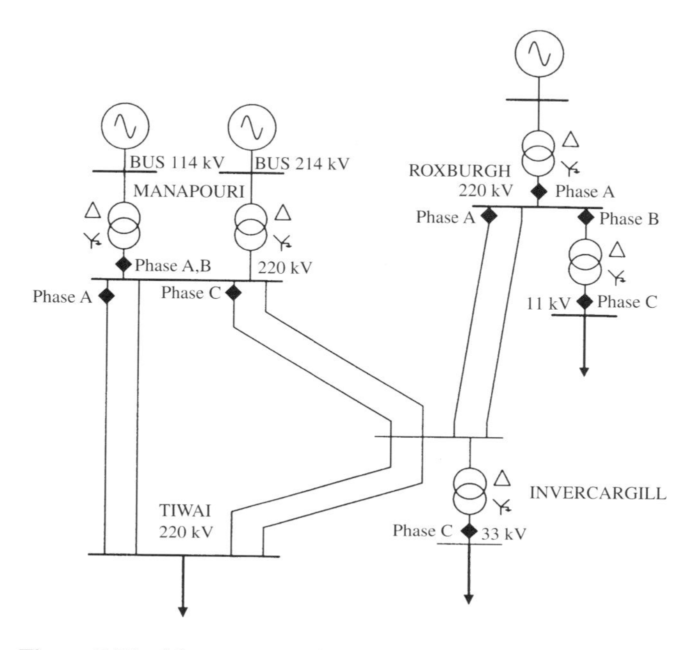

Figure 7.24 Measurement placement for complete observability

With the same measurement points and star-g/star load transformers the measurement matrix is singular, requiring SVD to obtain a solution. Adding one measurement point (e.g. Transformer MANAPOURI220/MANAPOUR1014 Phase C Sending _ End) the singularity is removed but the matrix is still rank deficient (i.e. the rank is 25 while the full rank is 27, corresponding to nine three-phase busbars).

7.6 Summary

There are fundamental differences between conventional, power-frequency-based state estimation and power-quality state estimation. In the former case, the measurements are the real and reactive powers, the voltages are close to 1 per unit and their phase angle differences are small; these conditions have led to the use of efficient decoupled non-linear solutions of the load-flow type. The signals measured in harmonic state estimation, on the other hand, are currents and voltages; their magnitudes and phase angles, unlike those at fundamental frequency, vary widely and thus the use of decoupled estimators would lead to ill-conditioned equations.

HSE is carried out with restricted measurements and relies on OA to minimise the measurement placements. Also, due to the limited number of measurements, bad-data detection is difficult and has not progressed very far. At present, power quality state estimation is only discussed with reference to harmonics (HSE) and under static conditions. However, new techniques, such as those based on Kalman filtering and artificial neural networks, are starting to be developed and tested for use in power quality estimation; these remove the static restriction by allowing tracking of the dynamic variation of the disturbing signals with time.

7.7 References

1. Schweppe, F C, Wildes, J C and Rom, D, (1970). Power system static state estimation, Part I, II and II′, *IEEE Transactions on Power Apparatus and Systems*, **PAS-89**, pp. 120–135.
2. Heydt, G T, (1989). Identification of harmonic sources by a state estimation technique, *IEEE Transactions on Power Delivery*, **4**(1), pp. 569–576.
3. Najjar, M and Heydt, G T, (1991). A hybrid nonlinear — least squares estimation of harmonic signal levels in power systems, *IEEE Transactions on Power Delivery*, **6**(1), pp. 282–288.
4. Beides, H M and Heydt, G T, (1991). Dynamic state estimation of power system harmonics using Kalman filter methodology, *IEEE Transactions on Power Delivery*, **6**(4), pp. 1663–1670.
5. Meliopoulos, A P S, Zang, F and Zellingher, S, (1994). Power system harmonic state estimation, *IEEE Transactions on Power Delivery*, **9**(3), pp. 1701–1709.
6. Du, Z P, Arrillaga, J and Watson, N R, (1996). Continuous harmonic state estimation of power systems, *Proceedings of the IEE*, **143 Pt.C**(4), pp. 329–336.
7. Du, Z P, Arrillaga, J, Watson, N R and Chen, S, (1998). Implementation of harmonic state estimation, 8th *International Conference on Harmonics and Quality of Power (ICHQP'98)*, Athens, Greece, pp. 273–278.
8. Du, Z P, Arrillaga, J and Watson, N R, (1996). A new symbolic method of observability analysis for harmonic state estimation of power systems, *Proceedings of the IEE*, **1**, China, pp. 431–435.
9. Farach, J E, Grady, W M and Arapostathis, A, (1993). An optimal procedure for placing sensors and estimating the location of harmonic sources in power systems, *IEEE Transactions on Power Delivery*, **8**(3), pp. 1303–1310
10. Farach, J E, Grady, W M and Arapostathis, A, (1996). Optimal harmonic sensor placement in fundamental network topologies, *Proceedings of the IEE, General Transmission and Distribution*, **143**(6), pp. 608–612.
11. Ma, H and Girgis, A A, (1996). Identification and tracking of harmonic sources in a power system using a Kalman filter, *IEEE Transactions on Power Delivery*, **11**(3), pp. 1659–1665.
12. Hartana, R K and Richards, G G, (1990). Constrained neural network based identification of harmonic sources, *Proceedings of Industry Applic. Society Annual Meeting*, pp. 1743–1748.
13. Hartana, R K and Richards, G G, (1990). Harmonic source monitoring and identification using neural network, *IEEE Transactions on Power Systems*, **5**(4), pp. 1098–1104.
14. Soliman, S A, Christensen, G S, Kelly, D H and El-Naggar, K M, (1990). A state estimation algorithm for identification and measurement of power system harmonics, *Electric Power Research*, (19), pp. 195–206.
15. Al-Kandari, A and Soliman, S A, (1975). Digital dynamic identification of power system sub-harmonics based on the least absolute value, *Electric Power Systems Research* (28), pp. 99–104.
16. Osowski, S, (1994). SVD technique for estimation of harmonic components in a power system: A statistical approach, *Proceedings IEE on Generation, Transmission and Distribution*, **141**(5), pp. 473–479.
17. Moo, C S and Chang, Y N, (1995). Group-harmonic identification in power systems with non-stationary waveforms, *Proceedings of the IEE on Generation, Transmission and Distribution*, **142**(5), pp. 517–522.
18. Wu, F F, (1990). Power system state estimation — A survey, *Electrical Power and Energy Systems*, **12**(2), pp. 79–87.
19. Krumphotz, G R, Clements, K A and Davis, P W, (1980). Power system observability: A practical algorithm using network topology, *IEEE Transactions on Power Apparatus and Systems*, **PAS-99** (4), pp. 1534–1542.

20. Du, Z P, Arrillaga, J and Watson, N R, (1996). A new symbolic method of observability analysis for harmonic state estimation of power system, *Proceedings International Conference on Electrical Engineering*, **1**, Beijin, pp. 431–435.
21. Press, W H, Flannery, B P, Teukolsky, S A and Vetterling, W T, (1990). *Numerical Recipes: The Art of Scientific Computing (FORTRAN version)*, Cambridge University Press.
22. Lin, S Y, (1992). A distributed state estimator for electric power systems, *IEEE Transactions on Power Systems*, **7**(2), pp. 551–557.

Appendix I

SIGNAL PROCESSING FOR DIGITAL INSTRUMENTS

I.1 Data Sampling

Using digital signal processing to compute harmonic levels requires that voltage and current waveforms be represented by a set of digital samples of magnitudes, taken at discrete instants in time. The *sampling* process is accomplished by multiplying the voltage or current waveform, represented by $h(t)$ in Figure I.1(a), by the sampling function $\delta_0(t)$, illustrated in Figure I.1(b). This results in a signal

$$\hat{h}(t) = \sum_{k=-\infty}^{\infty} h(kT)\delta(t - kT), \tag{I.1}$$

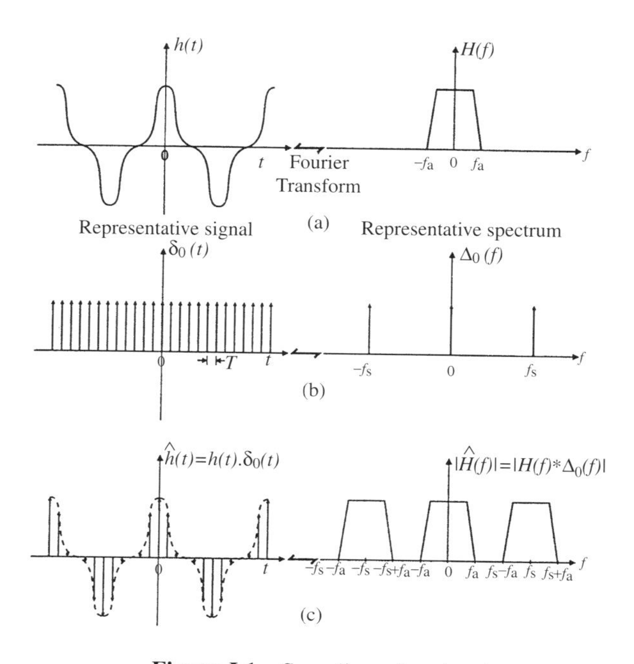

Figure I.1 Sampling of a signal

which is a set of uniformly spaced samples, T seconds apart, illustrated in Figure I.1(c). The sampling period is defined as T and the sample rate as

$$f_s = \frac{1}{T}. \tag{I.2}$$

For a unique correspondence between the continuous function $h(t)$ and its samples $\hat{h}(t)$, the sampling period T must be chosen to satisfy the requirements of the Nyquist sampling theorem [1], which essentially states that the signal $h(t)$ must be band-limited to f_a and that

$$f_s > 2f_a. \tag{I.3}$$

This is illustrated in Figure I.1(c) which depicts the Fourier Transform of the sampled signal. If this condition is not met, the spectrum centred at f_s will interfere with that at 0, distorting the sampled signal by the aliasing effect.

In practice, it is impossible to completely band-limit a signal, a problem circumvented by low pass filtering the signal before sampling, and sampling at such a rate that aliasing is negligible. The level below which aliasing is negligible is the signal noise floor, determined predominantly by quantisation noise, discussed in the following section.

I.2 Signal Quantisation

The sampled signal must be quantised to a number of discrete magnitudes for it to be represented by a finite word length machine. Quantisation is usually performed during analogue-to-digital conversion, and creates a quantisation noise voltage. An approximate theoretical r.m.s. signal-to-quantisation noise ratio of an N-bit ADC for a full-scale sinewave input is given in [2]:

$$SNR(\text{dB}) = 6.02N + 1.76. \tag{I.4}$$

This provides a noise level below which aliasing is negligible, since the aliased signal will not be resolved by the converter, giving an indication of the filter response required to avoid noticeable aliasing. Although this equation is only an approximation, it provides a pessimistic indication of the attenuation required to avoid aliasing.

This filter response is depicted in Figure I.2, with the aliased response shown dotted about f_s. The filter has a roll-off from the cut-off frequency f_c to f_a which is sufficiently steep to ensure an attenuation of A dB at $f_s = 2f_a$. This assumes that the signal

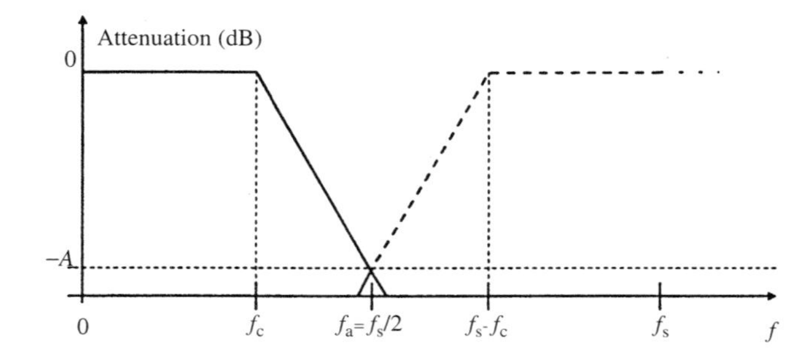

Figure I.2 Anti-aliasing filter response

being sampled may contain frequency components of full power beyond the cut-off frequency. Such a stringent filter may not be required for real signals.

I.3 Anti-Aliasing FIR Filter Design

The sampled signal $\hat{h}(t)$ must be re-sampled at a lower rate for FFT computation. Because it may contain frequency components beyond the now lower Nyquist limit when re-sampling (or down converting), it must be band-limited by an anti-aliasing filter. This is performed on the digital samples using a FIR filter.

On the basis of the frequency response of the filter required for anti-aliasing, a direct form of the time-invariant FIR filter, as depicted in Figure I.3 is implemented.

The filter coefficients, $h(k)$, are found from this response by essentially taking its inverse Fourier Transform and modifying the resulting truncated impulse response by a Hamming window to reduce the Gibbs phenomenon [3]. The filter length is determined by two main factors; namely, the amount of processing time available on the DSP (the longer the filter the more processing is required), and how true the actual response is to be to the ideal response (the longer the filter, the better the response). A length of 128 taps ($N = 128$) is a good compromise between processing overhead and good response characteristics. Figure I.4 shows the actual frequency magnitude response of the anti-aliasing FIR filter.

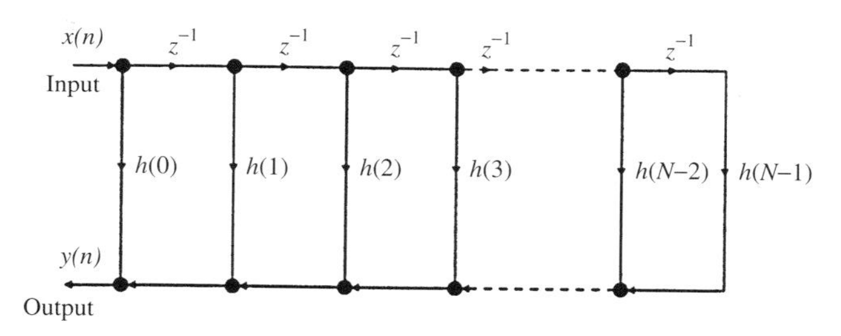

Figure I.3 Direct form structure for a FIR digital filter

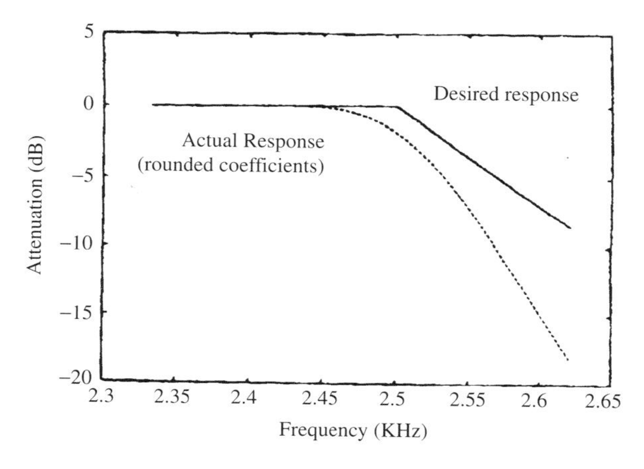

Figure I.4 128-tap anti-aliasing FIR filter response for a 50 Hz fundamental

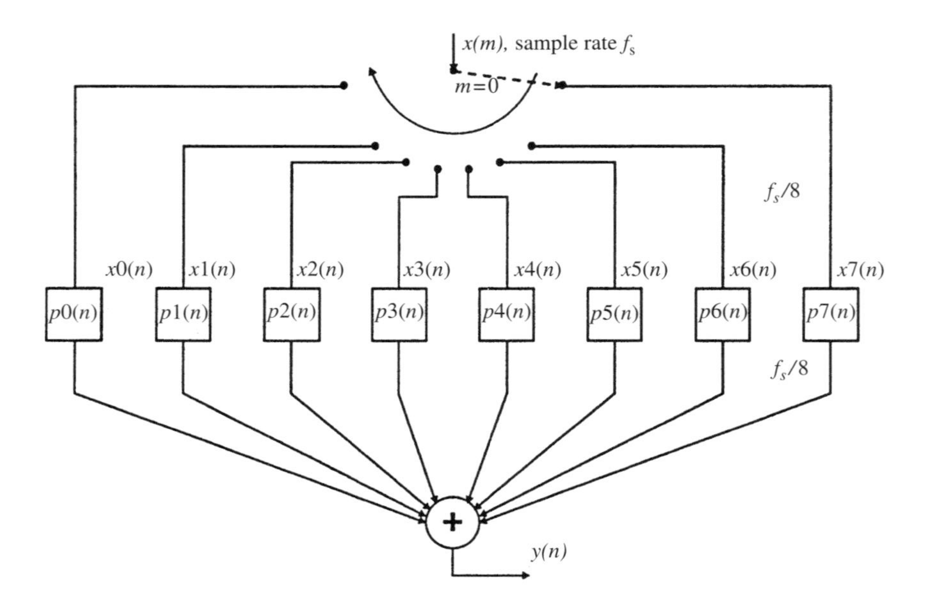

Figure I.5 The commutator model for an 8-to-1 polyphase decimator. Each of the polyphase filters is a decimated version of the full filter impulse response $h(k)$

To compute one output from the FIR filter, a total of N operations is required. For a 50 Hz fundamental frequency with an oversampling rate of $M = 8$, the actual sampling frequency is 8×6.4 kHz $= 51.2$ kHz. Hence, operating this 128-taps FIR filter at this high frequency will be very inefficient. However, the FIR filter output is followed by an 8-to-1 decimator, meaning that the filter output need only be computed every eighth sample.

A more subtle approach can be used to implement the filter efficiently. This approach involves breaking the filter up into a polyphase network, consisting of M smaller filters which contribute to the filter output for different time slots, and can be understood in terms of the *commutator model*, illustrated in Figure I.5 [3].

The coefficients of the M-to-1 polyphase decimator are

$$P_\rho(n) = h(nM + \rho), \tag{I.5}$$

for $\rho = 0, 1, 2, \ldots, M - 1$, and all n, where ρ denotes the ρth polyphase filter. The commutator effectively takes M input samples of the signal $x(m)$ and distributes them to the polyphase branches in the reverse sequence $\rho = 7, 6, 5, \ldots, 0$. When each of the polyphase filters has received a new input, the polyphase filters are computed and their outputs summed to give a single output sample $y(n)$. This means that for each input sample (at a rate of f_s), only a $128/8 = 16$ tap filter needs to be computed.

I.4 References

1. Linder, D A, (1959). A discussion of sampling theorems, *Proceedings of the IRE*, **47**(7) pp. 1219–1226.
2. Bennett, W R, (1948). Spectra of quantized signals, *B.S.T.J.* **27,** pp. 446–472.
3. Crochiere, R E, and Rabiner, L R, (1983). *Multirate Digital Signal Processing*, Prentice Hall, Inc.

Appendix II

CONTINUOUS HARMONIC ANALYSIS IN REAL TIME (CHART)

The CHART monitoring system is based on the Intel multi-processor bus standard, known as Multibus II, with the capability to compute voltage and current harmonics for as many as 36 channels up to the 50th harmonic in real time. Harmonics are computed for every fundamental cycle continuously and can be monitored on-line, leading to a significant reduction in required storage capacity and the ability to capture and store significant events.

Traditionally, synchronised simultaneous measurements could only be achieved with relatively elaborate direct communication links between the various monitors in use. This design was regularly plagued by the delays in the communication channels and the great distances between the measurement sites. However, the arrival of the Global Positioning System (GPS) technology eliminated the need for the direct communication mediums between units. CHART made use of this technology. In CHART, the acquisitions of samples on all units are synchronised to the GPS signals. Since the GPS is designed to have an accuracy of 1 ms or better, the measurements taken by different CHART units can also be synchronised to that accuracy.

Figure II.1 shows an overview of the CHART system, which is made up of three major parts

- Multiple front end interfaces to transducers (Remote Data Conversion Module or RDCM).
- A parallel processing system (Parallel Processing Unit or PPU).
- A network of user control and display workstations (Control And Display Unit or CADU).

Each analogue signal from a transducer is converted into a digital format and transmitted to the PPU via fibre optic cable. In the PPU, the digital samples are processed and can be stored in the PPU hard disk. At the same time, the processed data can also be transferred to CADU for display purposes. The CADU also provides interfaces for the user to control the operation of the PPU and hence control the functioning of the RDCM. The PPU obtains the accurate timing signals from the GPS satellite receiver and redirects them to all RDCMs simultaneously. This ensures that the sampling processes in all RDCMs are synchronised.

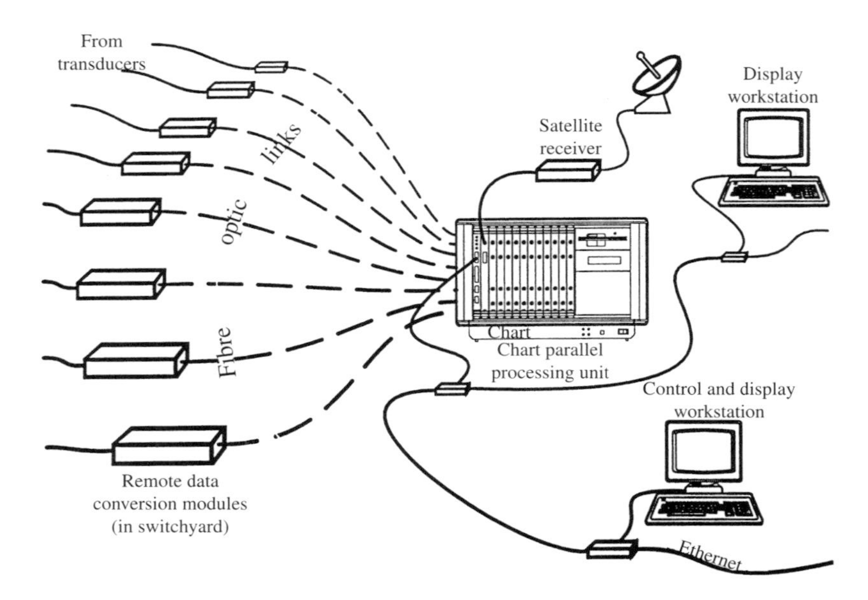

Figure II.1 CHART system overview

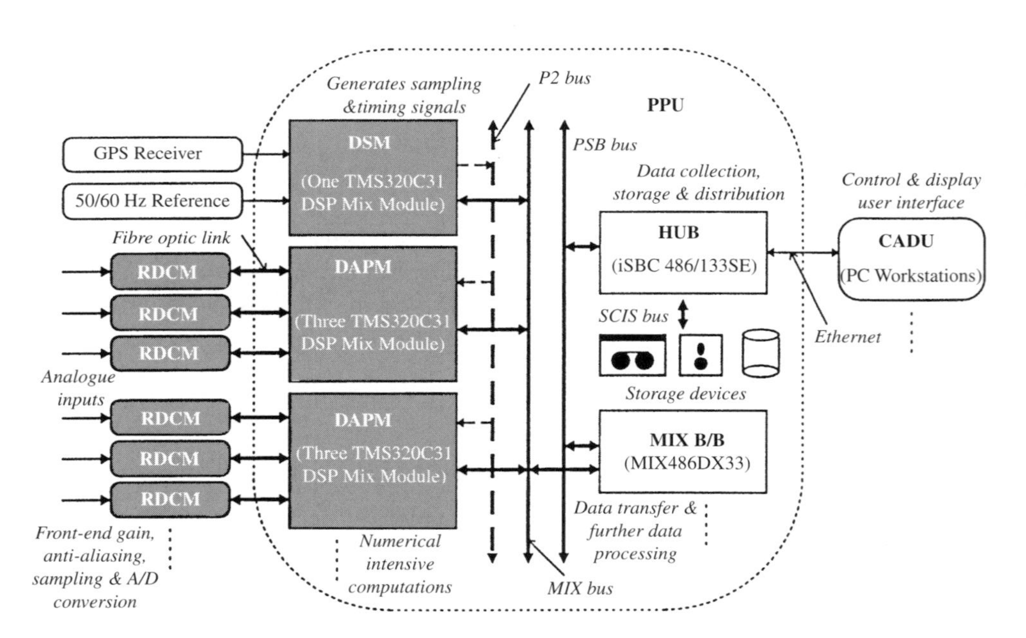

Figure II.2 CHART system details

The RDCM is a custom-designed module housed in a weather-proof case, complete with battery supply pack and charger. The PPU is an Intel MultibusII parallel computer with specialised hardware for the acquisition and processing of the data samples. The CADU is an IBM laptop computer running Microsoft Windows^{TM} 3.11.

Figure II.2 shows the system details including the internal contents of the PPU. The custom hardware components (highlighted in Figure II.2) are

- Remote Data Conversion Module or RDCM,
- Digital Services Module or DSM, and
- Data Acquisition and Processing Module or DAPM.

The function of the RDCM is to convert analogue signals into digital format. The current version is optimised for harmonic analysis and has been equipped with the following features.

- software controlled amplification or attenuation of the analogue signal from transducer-autoranging,
- 16-bit A/D converter at 100 kHz maximum sampling frequency,
- self-calibration and compensation for gain drifting and d.c. offset.

Stand-alone Remote Data Conversion Modules (RDCM) are used to convert the current transformer and voltage transformer outputs to 16-bit digital signals, and to transmit them to the parallel processing system. The RDCMs are typically located next to busgear (either interior or exterior), near the current and voltage transducers to which they are connected. They receive their analogue to digital conversion command from the parallel processing system (via a fibre optical cable), thereby ensuring that conversion occurs simultaneously on each channel. Sampling is synchronous with the power system fundamental frequency.

The DSM generates the sampling signals to be used by all RDCMs. This generation process is synchronised to the accurate timing signals obtained from a GPS receiver. The sampling frequency can be adjusted according to the network fundamental frequency in order to ensure the coherent sampling which is needed for harmonic analysis. Each DAPM possesses three Digital Signal Processor (DSPs) which perform numerical processing such as FFT on the acquired samples. A total of three channels of data samples can be handled by a DAPM simultaneously. The other two processor boards; namely, the MIX baseboard and Hub processor are Intel made single board computers operating *servers* for other boards.

In parallel with the hardware development, the software development is classified into several categories.

- The DAPM software for controlling the operation of RDCM, and for acquiring and processing the data samples (harmonic analysis).
- The DSM software for generation of sampling signals, synchronisation with GPS timing signals and for broadcasting the date and time information to the rest of the system.
- The MIX baseboard software for loading software onto DAPM and DSM, for transferring processed samples from DAPM, and date and time information from DSM to the Hub processor. It also allows further processing of the data, particularly for combining samples from different channels to derive power-related or three-phase information.

- The Hub processor software to collect data, to store data in the hard disk and to distribute the data to the CADU.
- The CADU software providing user interfaces for displaying the captured samples as well as for controlling the operation of the entire unit.

II.1 Bibliography

1. Miller, A J V, Lake, C B and Dewe, M B, (1990). Multichannel real-time harmonic analysis using the Intel Multibus II bus architecture, *Proceedings of the 7th International Intel Real-Time Users Group Conference*, St Louis, Missouri, USA, pp. 11–24.
2. Miller, A J V and Dewe, M B, (1992). Multichannel continuous harmonic analysis in real time, *IEEE Transactions on Power Delivery*, **7**(4), pp. 1813–1819.
3. Miller, A J V and Dewe, M B, (1993). The application of multi-rate digital signal processing techniques to the measurement of power system harmonic levels, *IEEE Transactions on Power Delivery*, **8**(2), pp. 531–539.
4. Chen, S and Dewe, M B, (1996). System and hardware considerations of a flexible multi-channel real time data acquisition system for power quality, *IEEE Conference on Harmonics and Quality of Power* (*ICHQP*). Las Vegas, pp. 189–195.
5. Chen, S and Dewe, M B, (1996). Virtual operating system design *via* flexible multi-channel real time data acquisition system for power quality assessment, *IEEE Conference on Harmonics and Quality of Power (ICHQP)*. Las Vegas, pp. 182–188.

Appendix III

LEAST SQUARES FITTING

III.1 Basic Method[1]

The application of FFT techniques to measure the harmonic content requires special care due to possible problems with aliasing, spectral leakage and picket-fence. The use of windowing techniques to alleviate these problems has been discussed in Chapter 4. Least squares fitting, or its variant least absolute value, can be applied to extract harmonic information without these drawbacks.

An outline of least squares fitting, the basis of most state estimation methods, is given here. Let the vector $\mathbf{Z}$ represent the measurements taken, $\mathbf{x}$ the state vector, and $\mathbf{w}$ the measurement error.

$$\mathbf{Z} = \mathbf{h}(\mathbf{x}) + \mathbf{w}. \qquad \text{(III.1)}$$

If a linear measurement model is used, the equation becomes

$$\mathbf{Z} = [H]\mathbf{x} + \mathbf{w}. \qquad \text{(III.2)}$$

The matrix H has the same or more rows than columns since the solution requires that there must be more data points than variables. The aim is to minimise the squared errors J,

$$J = \mathbf{w}^{\mathrm{T}}\mathbf{w}. \qquad \text{(III.3)}$$

This is calculated by making its derivative equal to zero, i.e.

$$\frac{\mathrm{d}J}{\mathrm{d}\mathbf{x}} = 0$$

and making use of Equation (III.2),

$$\begin{aligned}\frac{\mathrm{d}J}{\mathrm{d}\mathbf{x}} &= \mathrm{d}[H]\mathbf{x} - Z)^{\mathrm{T}}([H]\mathbf{x} - \mathbf{Z})/\mathrm{d}\mathbf{x} \\ &= 2H^{\mathrm{T}}([H]\mathbf{x} - \mathbf{Z}),\end{aligned} \qquad \text{(III.4)}$$

which gives

$$\mathbf{x} = [H^{\mathrm{T}}H]^{-1}H^{\mathrm{T}}\mathbf{Z}. \qquad \text{(III.5)}$$

III.2 Weighted Least Squares

An alternative to the basic method is the weighted least squares (WLS), in which weights are given to the various samples and thus J becomes

$$J = \mathbf{w}^{\mathrm{T}}[R]\mathbf{w}, \tag{III.6}$$

where the weights, based on the standard deviation of the measurement error, are arranged in a diagonal matrix $[R]$.

Non-linearities in the measurement process can be accommodated by a Taylor series expansion of the non-linearity around the base point. This, then, becomes an iterative process in which the estimate state vector depends on the base point of linearisation and, therefore, on the previous estimate of the state vector.

The least squares estimate now results from the minimisation of

$$J(\mathbf{x}) = \tfrac{1}{2}[\mathbf{Z} - \mathbf{h}(\mathbf{x})]^{\mathrm{T}} R^{-1} [\mathbf{Z} - \mathbf{h}(\mathbf{x})] \tag{III.7}$$

and thus making $\dfrac{\partial J(\mathbf{x})}{\partial \mathbf{x}} = 0$

$$\Rightarrow g(\mathbf{x}) = H^{\mathrm{T}}(x) R^{-1} [\mathbf{Z} - \mathbf{h}(\mathbf{x})] = 0, \tag{III.8}$$

where the measurement Jacobian matrix $H(x) = \partial h(\mathbf{x})/\partial \mathbf{x}$.

The iterative solution of this equation using the Newton method is

$$\frac{\partial g(\mathbf{x}^k)}{\partial \mathbf{x}} \Delta \mathbf{x} = -g(\mathbf{x}^k), \tag{III.9}$$

where superscript k represents the iteration number. Evaluating $\partial g(\mathbf{x})/\partial \mathbf{x}$ gives:

$$\frac{\partial g(\mathbf{x})}{\partial \mathbf{x}} = \frac{\partial^2 \mathbf{h}(\mathbf{x})}{\partial \mathbf{x}^2} R^{-1} [\mathbf{Z} - h(\mathbf{x})] - H^{\mathrm{T}}(x)[R]^{-1} H(x). \tag{III.10}$$

Ignoring the second derivative term gives

$$\frac{\partial g(\mathbf{x})}{\partial \mathbf{x}} \approx -H^{\mathrm{T}}(x)[R]^{-1} H(x). \tag{III.11}$$

The normal equation is thus

$$[G(x^k)]\Delta \mathbf{x} = H^{\mathrm{T}}(x^k)[R]^{-1}[\mathbf{Z} - \mathbf{h}(\mathbf{x}^k)], \tag{III.12}$$

where the gain matrix is

$$[G(x^k)] = [H^{\mathrm{T}}(x^k)[R]^{-1} H(x^k)]. \tag{III.13}$$

For the special case of a linear measurement equation $h(\mathbf{x}) = [h]\mathbf{x}$ and $g(\mathbf{x}) = \partial h(\mathbf{x})/\partial \mathbf{x} = [h]$, therefore, the condition $\dfrac{\partial J(\mathbf{x})}{\partial \mathbf{x}} = 0$ results in

$$g(\mathbf{x}) = [h]^{\mathrm{T}} R^{-1} [\mathbf{Z} - [h]\mathbf{x}] = 0. \tag{III.14}$$

Hence, the normal equation to be solved is

$$\left[[h]^{\mathrm{T}}[R]^{-1}[h]\right] \mathbf{x} = [h]^{\mathrm{T}}[R]^{-1}\mathbf{Z}. \tag{III.15}$$

It should be noted that the diagonal elements of the matrix $[H^{\mathrm{T}}RH]^{-1}$ are the variances and the off-diagonal elements the co-variances between the elements of vector x.

III.3 Bibliography

1. Heydt, G T, (1991). *Electric Power Quality*, Stars in a Circle Publications, West Lafayette, USA.

Appendix IV

SINGULAR VALUE DECOMPOSITION (SVD)

IV.1 General Considerations

Singular value decomposition is a highly reliable and computationally stable technique to solve matrices that are either singular or close to singular (ill-conditioned). In such cases, SVD will indicate the problem and return a useful numerical answer. Some terminology is required before discussing the SVD method. Consider the linear equation

$$[A]\mathbf{x} = \mathbf{b}, \tag{IV.1}$$

where $[A]$ has m rows and n columns (i.e. $m \times n$), $\mathbf{x}$ an unknown vector with n elements and $\mathbf{b}$ a known vector containing m elements. The geometrical interpretation of vectors (equations) results in the idea of spaces. Column-space of A is the space spanned by the columns of A and the row-space the space spanned by the rows of A. The range of A is the subspace which can be reached by the mapping that A represents. The dimension of the Range is the rank of A. The null-space is the space spanned by all vectors that satisfy the equation $[A]\mathbf{x} = \mathbf{0}$.

Obviously if $\mathbf{x}_\mathrm{p}$ is a particular solution of Equation (IV.1) and $\mathbf{x}_\mathrm{NV}$ is a null-space vector, then $\mathbf{x} = \mathbf{x}_\mathrm{p} + k\mathbf{x}_\mathrm{NV}$ is also a valid solution of Equation (IV.1), i.e.

$$[A](\mathbf{x}_\mathrm{p} + k\mathbf{x}_\mathrm{NV}) = [A]\mathbf{x}_\mathrm{p} + k[A]\mathbf{x}_\mathrm{NV} = \mathbf{b} + k\mathbf{0} = \mathbf{b}. \tag{IV.2}$$

Two problems can occur when solving Equation (IV.1). Either the rows of A or its columns may be dependent.

Consider the case when m is larger than n (i.e. there are more measurements than unknowns); then, probably there will not be a $\mathbf{x}$ vector that perfectly fits the data $\mathbf{b}$. In other words, vector $\mathbf{b}$ will not be a combination of the columns of $[A]$. The solution of $[A]\mathbf{x} = \mathbf{p}$ where $\mathbf{p}$ is the closest point in the column space to $\mathbf{b}$, can be found from the normal equations, as described in Appendix II. However, in the presence of dependent columns, that solution will not be unique. The particular solution chosen is one that minimises the length (norm) of $\mathbf{x}$. Therefore, from the preceding discussion, it can be said that regarding *existence of a solution*, at least one solution exists if, and only if, the column vectors span the m-dimensional vector space (i.e. the rank of $[A]$ is m). This can only occur if $m \leq n$. Regarding *uniqueness of a solution*, it can be said that only one solution exists if, and only if, the columns are linearly independent (i.e. the rank of $[A]$ is n). This can only occur if $m \geq n$.

IV.2 SVD Factorisation

The SVD method represents the $m \times n$ matrix A as the product of three matrices, i.e.

$$[A] = [U][W][V]^{\mathrm{T}}. \qquad \text{(IV.3)}$$

In Equation (IV.3) $[W]$ is a diagonal matrix $(n \times n)$ with positive or zero elements, which are the singular values of A. Matrices $[U]$ and $[V]^{\mathrm{T}}$ are orthogonal matrices, $[U]$ being a column orthogonal $(m \times n)$ matrix and $[V]^{\mathrm{T}}$ the transpose of a $(n \times n)$ orthogonal matrix, i.e.

$$[U]^{\mathrm{T}}[U] = [V]^{\mathrm{T}}[V] = [1].$$

Moreover $[V]$ is square and hence also row-orthogonal ($[V][V]^{\mathrm{T}} = [1]$).

SVD constructs special orthonormal bases for the null-space and Range of a matrix. Not only are they orthonormal but if $[A]$ multiplies a column of $[V]$, a multiple of a column of $[U]$ is obtained. It can be shown that $[U]$ is the eigenvector matrix of $[A][A]^{\mathrm{T}}$ and $[V]$ is the eigenvector matrix of $[A]^{\mathrm{T}}[A]$. Moreover $[W][W]^{\mathrm{T}}$ is a diagonal matrix of eigenvalues. The columns of $[U]$ corresponding to the non-zero singular values are an orthonormal set of basis vectors that span the range of A. The columns of $[V]$ corresponding to the zero singular values are an orthonormal set of basis vectors that span the null space.

The solution can be shown to be

$$\mathbf{x} = [V][W]^{-1}[U][A]\mathbf{b}. \qquad \text{(IV.4)}$$

If some of the singular values are zero or near zero, then a zero is placed in the diagonal element of $[W]^{-1}$ (instead of $[1/w]$). This is equivalent to throwing away one linear combination of the set of equations. The condition number of a matrix is the ratio of the largest to smallest singular value. A singularity is considered near zero when its value approaches or is below the largest singular value times the machine's precision (e.g. 10^{-6} for single precision and 10^{-12} for double precision).

In Equation (IV.4)

$$[A]^{+} = [V][W]^{-1}[U] \qquad \text{(IV.5)}$$

is the pseudo-inverse of $[A]$.

IV.3 Numerical Example

Consider the set of equations

$$\begin{bmatrix} 1 & 1 & 1 \\ 3 & -2 & 1 \\ 4 & -1 & 2 \end{bmatrix} \begin{pmatrix} x_1 \\ x_2 \\ x_3 \end{pmatrix} = \begin{pmatrix} 3 \\ 2 \\ 5 \end{pmatrix}.$$

Matrix A is of rank 2 since the third equation is a linear combination of the previous two (simply the sum). The SVD decomposition gives the singular values as 5.892048 0.000000 1.812117. The particular solution supplied by SVD is always the one of minimal length, which is 1.000000 1.000000 1.000000.

The matrices obtained by decomposition are

$$[U] = \begin{bmatrix} 0.160610 & 0.577350 & -0.800544 \\ -0.612986 & 0.577350 & 0.539365 \\ -0.773597 & -0.773597 & -0.2611791 \end{bmatrix}$$

$$[V]^{\mathrm{T}} = \begin{bmatrix} -0.864548 & 0.312109 & -0.393885 \\ 0.486664 & 0.324443 & -0.811107 \\ -0.125360 & -0.892931 & -0.432388 \end{bmatrix}$$

Hence, the null space vector is (0.486664 0.324443 −0.811107). Therefore, the infinite number of solutions are given by

$$\begin{pmatrix} x_1 \\ x_2 \\ x_3 \end{pmatrix} = \begin{pmatrix} 1.0 \\ 1.0 \\ 1.0 \end{pmatrix} + k \begin{pmatrix} 0.486664 \\ 0.32444 \\ -0.811107 \end{pmatrix}$$

where k can be any real number.

For example, k = 1.2328835 gives the solution $\begin{pmatrix} x_1 \\ x_2 \\ x_3 \end{pmatrix} = \begin{pmatrix} 1.6 \\ 1.4 \\ 0 \end{pmatrix}$.

These values can be substituted into the matrix equation and shown also to provide a valid solution.

Bibliography

1. Press, W H, Flannery, B P, Teukolsky, S A and Vettering, W T, (1989). *Numerical Recipes, The Art of Scientific Computing, Fortran Version*. Cambridge University Press.
2. Golub, G H and van Loan, C F, (1989). *Matrix Computation*, 2nd edition, The Johns Hopkins University Press.
3. Strang, G, (1988). *Linear Algebra and its Applications*, 3rd edition, Harcourt Brace Jovanovich, Inc.

Appendix V

KALMAN FILTERING

In the basic least squares fitting, the estimated parameters are assumed constant during the observation period and only the measurement is corrupted by noise. The Kalman filter is a least squares estimate in which a state equation is added to allow its application to a dynamic system where the estimated parameters are varying. Its principal feature is the recursive processing of the noise measurement risk. This makes it ideally suited for on-line estimation of varying parameters.

In the Kalman algorithm, the discretised state and equations are

$$\mathbf{x}_{k+1} = \boldsymbol{\phi}_k + \nu_k, \tag{V.1}$$

$$\mathbf{Z}_k = [H_k]\mathbf{x}_k + \mathbf{w}_k. \tag{V.2}$$

The system covariance matrices for w_k and ν_k are

$$\mathrm{E}[\mathbf{w}_k \quad \mathbf{w}_k^{\mathrm{T}}] = [\mathbf{Q}_k] \quad \text{and} \quad \mathrm{E}[\nu_k \quad \nu_k^{\mathrm{T}}] = [R_k].$$

Again, the objective is to minimise the sum of the squares of the error. Since the vectors involved are stochastic, the expectation (E) is used

$$J = \mathrm{E}((\mathbf{x}_x - \hat{\mathbf{x}}_x)^{\mathrm{T}}(\mathbf{x}_x - \hat{\mathbf{x}}_x)). \tag{V.3}$$

The initial variable is assumed zero, i.e. $\hat{x}_{(0)} = 0$ and therefore the initial covariance matrix is

$$[\hat{P}_{(0)}] = \mathrm{E}[(\hat{\mathbf{x}} - \hat{\mathbf{x}}_{(0)}) \quad (\hat{\mathbf{x}} - \hat{\mathbf{x}}_{(0)})^{\mathrm{T}}] = \mathrm{E}[(\hat{\mathbf{x}}) \quad (\hat{\mathbf{x}})^{\mathrm{T}}]. \tag{V.4}$$

In general, the initial covariance matrix to be used will depend on prior knowledge of the harmonic sources and load levels at some buses. As an approximation, the harmonic injections at different buses are assumed to be uncorrelated and the above covariance matrix is diagonal; this assumption has little effect because the Kalman filter is not sensitive to moderate changes in the initial covariance.

Assuming, for illustration purposes, that the system and measurement equations are linearised, the sequential recursive computation steps for the harmonic injection estimate by the Kalman filter are

Step 1. Compute Kalman filter gain:

$$[G_k] = \hat{P}_k H_k^{\mathrm{T}}([H_k \hat{P}_k H_k^{\mathrm{T}}] + R_k)^{-1}. \tag{V.5}$$

Step 2. Update the estimate of the state vector:

$$\mathbf{x}_k^{\mathrm{new}} = \mathbf{x}_k + [G_k](\mathbf{Z}_k - [H_k]\mathbf{x}). \tag{V.6}$$

Step 3. Calculate the error covariance of the new estimate:

$$[\mathbf{P}_k] = [(I - G_k H_k)][\hat{\mathbf{P}}_k]. \tag{V.7}$$

Step 4. Project ahead:

$$[\hat{P}_{k+1}] = [\phi_k P_k \phi_k^{\mathrm{T}}] + [Q_k] \quad \text{and} \quad \mathbf{x}_{k+1}^{\mathrm{new}} = [\phi_k]\mathbf{x}_k. \tag{V.8}$$

Several alternatives to the original Kalman filter have been proposed. Amongst them, the Extended Kalman Filter, which maintains the non-linear structure of the state equations and the Parallel Kalman Filter, which, as the name indicates, is written for parallel processing.

V.1 Bibliography

1. Kalman, R E, (1960). A new approach to linear filtering and prediction problems, *Journal of Basic Engineering*, pp. 35–45.
2. Heydt, G T, (1991). *Electric Power Quality*, Stars in a Circle Publications, West Lafayette, USA.
3. Ma, H and Girgis, A A, (1996). Identification and tracking of harmonic sources in a power system using a Kalman filter, *IEEE Transactions on Power Delivery*, **11** (3), pp. 1659–1665.

Appendix VI

HSE DEMONSTRATION

The simple single-phase test system of Figure VI.1 is used to illustrate the numerical input and output information of the harmonic penetration algorithm. The results of this program, at the measurement points indicated in the figure, are then used as pseudo-measurements in the Harmonic State Estimation demonstration.

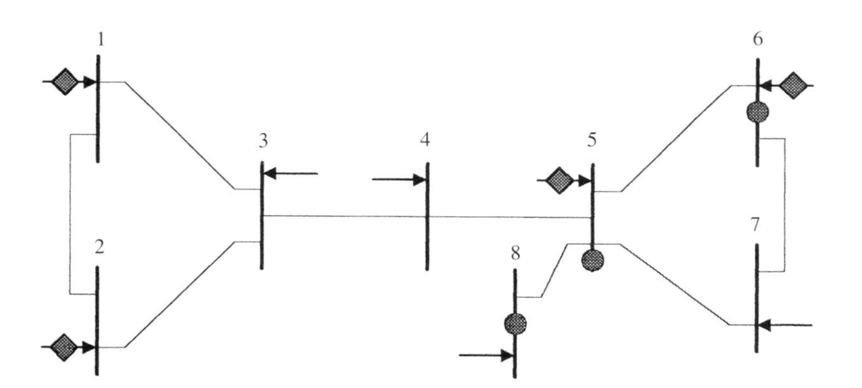

Figure VI.1 Single-phase test system

```
                    Harmonic State Estimation Demonstration Program
                    ===============================================


 Harmonic Data Read in for System Components

 Busbar Data
----Busbar----  Nominal_Voltage
 No Name              (kV)
  1 ONE              220.0
  2 TWO              220.0
  3 THREE            220.0
  4 FOUR             220.0
  5 FIVE             220.0
  6 SIX              220.0
  7 SEVEN            220.0
  8 EIGHT            220.0

 Branch Data
From To       R (p.u.)       X (p.u.)       B (p.u.)
  1  2        0.00099        0.00820        0.02600
  1  3        0.00020        0.00800        0.00000
  2  3        0.00010        0.01200        0.00000
  3  4        0.03000        0.18700        0.30000
  4  5        0.00080        0.01200        0.05000
  5  6        0.00100        0.02000        0.03000
  6  7        0.00150        0.05200        0.04000
  5  7        0.00956        0.05498        0.09092
  5  8        0.01000        0.06000        0.10000

 Shunt Data
 Busbar      P (MW)       Q (MVAr)
  1         100.0          50.0
  2          50.0          80.0
  3          45.0          45.0
  4          10.0          10.0
  6          20.0           5.0
  7         100.0          25.0
  8         100.0          25.0

 Current Injection Data
Busbar        Magnitude        Angle
  No.           (p.u.)         (Deg.)
  1           0.005000         25.00
  2           0.005000         10.00
  5           0.010000         45.00
  6           0.015000        115.00
  7           0.020000        -60.00
  8           0.010000       -145.00

 Measurement Points
No Meas.  Quant.  Send
   Type   Bus/Brn  /Rec
  1 VOLT       5
  2 VOLT       6
  3 VOLT       8
  4 IINJ       1
  5 IINJ       2
  7 IINJ       5
  8 IINJ       6


               Harmonic Penetration
               --------------------

 Y Nodal
  1  1           18.63490 + j       -245.60810
  1  2          -14.51185 + j        120.19918
  1  3           -3.12305 + j        124.92192
  2  1          -14.51185 + j        120.19918
  2  2           15.70625 + j       -204.31373
  2  3           -0.69440 + j         83.32755
  3  1           -3.12305 + j        124.92192
  3  2           -0.69440 + j         83.32755
```

```
3  3        5.10382 + j      -213.76289
3  4       -0.83638 + j         5.21342
4  3       -0.83638 + j         5.21342
4  4        6.46735 + j       -88.10302
4  5       -5.53097 + j        82.96460
5  4       -5.53097 + j        82.96460
5  5       13.79726 + j      -166.57532
5  6       -2.49377 + j        49.87531
5  7       -3.06982 + j        17.65465
5  8       -2.70270 + j        16.21622
6  5       -2.49377 + j        49.87531
6  6        3.24804 + j       -69.10509
6  7       -0.55427 + j        19.21478
7  5       -3.06982 + j        17.65465
7  6       -0.55427 + j        19.21478
7  7        4.62409 + j       -37.05397
8  5       -2.70270 + j        16.21622
8  8        3.70270 + j       -16.41622

BUSBAR VOLTAGES
No. ------Voltage------
    Magnitude      Angle
      (pu)        (Deg.)
 1  0.23903E-02    39.67
 2  0.23925E-02    39.79
 3  0.23875E-02    39.08
 4  0.24198E-02    18.56
 5  0.24348E-02    17.41
 6  0.23477E-02    16.55
 7  0.29016E-02    16.90
 8  0.25319E-02     0.73

-----------------Branch Currents------------
From To  ------Sending------    ------Receiving----
         Mag.         Angle     Mag.          Angle
         (pu)         (Deg.)    (pu)          (Deg.)
  1  2   0.68696E-03 -158.93    0.66965E-03 -153.89
  1  3   0.31113E-02   34.32    0.31113E-02   34.32
  2  3   0.25260E-02   30.41    0.25260E-02   30.41
  3  4   0.45443E-02   44.57    0.44040E-02   35.66
  4  5   0.42546E-02   39.73    0.42111E-02   38.20
  5  6   0.46678E-02  -47.69    0.47327E-02  -48.06
  6  7   0.10698E-01  110.01    0.10593E-01  110.04
  5  7   0.84848E-02  114.03    0.82440E-02  114.23
  5  8   0.11969E-01   26.70    0.11896E-01   25.57

----------------- Shunt Currents------------
Brn      ------Sending------
No.      Mag.          Angle
No.      (pu)         (Deg.)
 1  1    0.26725E-02    66.24
 2  2    0.22571E-02    97.79
 3  3    0.15194E-02    84.08
 4  4    0.34221E-03    63.56
 5  6    0.48399E-03    30.59
 6  7    0.29909E-02    30.94
 7  8    0.26098E-02    14.76


         Harmonic State Estimation
         -------------------------

*****************************
**** Measurement Equation ****
*****************************

Number of Measurment Equations (Complex) =  4

 Complex Measurement Matrix (H)
H( 1, 1)=     18.6349+j*   -245.6081;
H( 1, 2)=    -14.5119+j*    120.1992;
H( 1, 3)=     -3.1230+j*    124.9219;
H( 2, 1)=    -14.5119+j*    120.1992;
H( 2, 2)=     15.7062+j*   -204.3137;
H( 2, 3)=     -0.6944+j*     83.3275;
H( 3, 4)=     -5.5310+j*     82.9646;
```

```
H( 3, 5)=     13.7973+j*   -166.5753;
H( 3, 6)=     -2.4938+j*     49.8753;
H( 3, 7)=     -3.0698+j*     17.6546;
H( 3, 8)=     -2.7027+j*     16.2162;
H( 4, 5)=     -2.4938+j*     49.8753;
H( 4, 6)=      3.2480+j*    -69.1051;
H( 4, 7)=     -0.5543+j*     19.2148;

  Complex Measurement Vector (Z)
Z( 1)=     0.0045+j*     0.0021;
Z( 2)=     0.0049+j*     0.0009;
Z( 3)=     0.0071+j*     0.0071;
Z( 4)=    -0.0063+j*     0.0136;

 **** Partition Measurement Equation based of Measured Voltages ****
  |Z1| |     | |V1|
  |--|=|h1|h2|x|--|
  |Z2| |     | |V2|

  Where V2=Measured voltages

 Busbar No=  1 Not Measured. Index in h1 (NoUnMeasBusV) =  1
 Busbar No=  2 Not Measured. Index in h1 (NoUnMeasBusV) =  2
 Busbar No=  3 Not Measured. Index in h1 (NoUnMeasBusV) =  3
 Busbar No=  4 Not Measured. Index in h1 (NoUnMeasBusV) =  4
 Busbar No=  5 Measured. Index in h2 (NoMeasBusV) =  1
 Busbar No=  6 Measured. Index in h2 (NoMeasBusV) =  2
 Busbar No=  7 Not Measured. Index in h1 (NoUnMeasBusV) =  5
 Busbar No=  8 Measured. Index in h2 (NoMeasBusV) =  3

 == h1 Matrix ==
H( 1, 1)=     18.6349+j*   -245.6081;
H( 1, 2)=    -14.5119+j*    120.1992;
H( 1, 3)=     -3.1230+j*    124.9219;
H( 2, 1)=    -14.5119+j*    120.1992;
H( 2, 2)=     15.7062+j*   -204.3137;
H( 2, 3)=     -0.6944+j*     83.3275;
H( 3, 4)=     -5.5310+j*     82.9646;
H( 3, 5)=     -3.0698+j*     17.6546;
H( 4, 5)=     -0.5543+j*     19.2148;

  === h2 Matrix ==
H( 3, 1)=     13.7973+j*   -166.5753;
H( 3, 2)=     -2.4938+j*     49.8753;
H( 3, 3)=     -2.7027+j*     16.2162;
H( 4, 1)=     -2.4938+j*     49.8753;
H( 4, 2)=      3.2480+j*    -69.1051;

  Calculate Measurement Vector Z_dash by eliminating Measured State Variables
  |Z_dash| = |Z1| - |h2|x|V2|

  == Z_dash Vector ==
MeasEqn | Zdash
  No.   |
Zdash( 1)=       0.0045+j*      0.0021;
Zdash( 2)=       0.0049+j*      0.0009;
Zdash( 3)=      -0.1000+j*      0.2325;
Zdash( 4)=      -0.0177+j*      0.0529;

 ****************************************
 *** Convert into Real valued problem ***
 ****************************************
  |Zr| | h1r |-h1i |    |V1r|
  |--|=| --------- | x |---|
  |Zi| | h1i | h1r |    |V1i|

  Number of Equations (real) =  8
  Number of Unknowns = 10

 == h1 Real-valued Matrix ==
H1real( 1, 1)=   18.634900;
H1real( 1, 2)=  -14.511852;
H1real( 1, 3)=   -3.123048;
H1real( 1, 6)=  245.608103;
H1real( 1, 7)= -120.199179;
H1real( 1, 8)= -124.921924;
```

```
H1real( 2, 1)=  -14.511852;
H1real( 2, 2)=   15.706248;
H1real( 2, 3)=   -0.694396;
H1real( 2, 6)= -120.199179;
H1real( 2, 7)=  204.313726;
H1real( 2, 8)=  -83.327547;
H1real( 3, 4)=   -5.530973;
H1real( 3, 5)=   -3.069815;
H1real( 3, 9)=  -82.964602;
H1real( 3,10)=  -17.654648;
H1real( 4, 5)=   -0.554273;
H1real( 4,10)=  -19.214781;
H1real( 5, 1)= -245.608103;
H1real( 5, 2)=  120.199179;
H1real( 5, 3)=  124.921924;
H1real( 5, 6)=   18.634900;
H1real( 5, 7)=  -14.511852;
H1real( 5, 8)=   -3.123048;
H1real( 6, 1)=  120.199179;
H1real( 6, 2)= -204.313726;
H1real( 6, 3)=   83.327547;
H1real( 6, 6)=  -14.511852;
H1real( 6, 7)=   15.706248;
H1real( 6, 8)=   -0.694396;
H1real( 7, 4)=   82.964602;
H1real( 7, 5)=   17.654648;
H1real( 7, 9)=   -5.530973;
H1real( 7,10)=   -3.069815;
H1real( 8, 5)=   19.214781;
H1real( 8,10)=   -0.554273;

 == Z_dash Real-valued Vector ==
Z_dash (Real)  ( 1) =   0.004532;
Z_dash (Real)  ( 2) =   0.004924;
Z_dash (Real)  ( 3) =  -0.099994;
Z_dash (Real)  ( 4) =  -0.017747;
Z_dash (Real)  ( 5) =   0.002113;
Z_dash (Real)  ( 6) =   0.000868;
Z_dash (Real)  ( 7) =   0.232482;
Z_dash (Real)  ( 8) =   0.052877;
 =================================
 === SVD Decomposition Matrices ===
 =================================

 Matrix U
   0.000000  0.093446  0.799469 -0.058019  0.590551  0.000000  0.000000  0.000000
0.000000  0.000000
   0.000000 -0.063273 -0.590011 -0.086327  0.800269  0.000000  0.000000  0.000000
0.000000  0.000000
   0.000000  0.000000  0.000000  0.000000  0.000000 -0.992913  0.107820 -0.036711
0.033925  0.000000
   0.000000  0.000000  0.000000  0.000000  0.000000 -0.048414  0.012439  0.629168 -
0.775660  0.000000
   0.000000  0.799469 -0.093446  0.590551  0.058019  0.000000  0.000000  0.000000
0.000000  0.000000
   0.000000 -0.590011  0.063273  0.800269  0.086327  0.000000  0.000000  0.000000
0.000000  0.000000
   0.000000  0.000000  0.000000  0.000000  0.000000  0.107820  0.992913  0.033925
0.036711  0.000000
   0.000000  0.000000  0.000000  0.000000  0.000000  0.012439  0.048414 -0.775660 -
0.629168  0.000000
   1.000000  0.000000  0.000000  0.000000  0.000000  0.000000  0.000000  0.000000
0.000000  0.000000
   0.000000  0.000000  0.000000  0.000000  0.000000  0.000000  0.000000  0.000000
0.000000  1.000000
 Diagonal of Matrix W
   0.000000351.434739351.434739176.156420176.156420 85.159070 85.159070 18.768991
18.768991  0.000000
 Matrix V-Transpose
   0.576553  0.575673  0.579801  0.000000  0.000000  0.000000  0.000401  0.004250
0.000000  0.000000
  -0.752957  0.609766  0.143580  0.000000  0.000000  0.153703 -0.128127 -0.024153
0.000000  0.000000
   0.153703 -0.128127 -0.024153  0.000000  0.000000  0.752957 -0.609766 -0.143580
0.000000  0.000000
```

```
  -0.276350 -0.528145  0.798713  0.000000  0.000000 -0.025443 -0.037834  0.068355
0.000000  0.000000
  -0.025443 -0.037834  0.068355  0.000000  0.000000  0.276350  0.528145 -0.798713
0.000000  0.000000
   0.000000  0.000000  0.000000  0.169530  0.061267  0.000000  0.000000  0.000000
0.960324  0.212801
   0.000000  0.000000  0.000000  0.960324  0.212801  0.000000  0.000000  0.000000 -
0.169530 -0.061267
   0.000000  0.000000  0.000000  0.160778 -0.774748  0.000000  0.000000  0.000000
0.152277 -0.592223
   0.000000  0.000000  0.000000  0.152277 -0.592223  0.000000  0.000000  0.000000 -
0.160778  0.774748
   0.000000 -0.000401 -0.004250  0.000000  0.000000  0.576553  0.575673  0.579801
0.000000  0.000000

 Singular Values Set to Zero
I=  1
I= 10

 Null-Space Vector (Column i of Matrix V)
1 Null-Space Vector (i= 1)
   0.576553  0.575673  0.579801  0.000000  0.000000  0.000000  0.000401  0.004250
0.000000  0.000000
2 Null-Space Vector (i=10)
   0.000000 -0.000401 -0.004250  0.000000  0.000000  0.576553  0.575673  0.579801
0.000000  0.000000
==========================
    Solution vector is
  -0.000005 -0.000003  0.000009  0.002294  0.002776  0.000012  0.000019 -0.000031
0.000770  0.000844

BUSBAR VOLTAGES Solved for
Busbar   Voltage
No.  Real, Imag.  ---- Magnitude  Angle (Deg.)
 1  -0.00001   0.00001   0.00001 113.93938
 2   0.00000   0.00002   0.00002  99.70991
 3   0.00001  -0.00003   0.00003 -73.98664
 4   0.00229   0.00077   0.00242  18.55759
 7   0.00278   0.00084   0.00290  16.90072

          ********************************
          * Comparison of Busbar Voltages *
          ********************************

      ---- Actual----       --- Estimated--
 I      Real       Imag      Real       Imag
 1     0.00184   0.00153  -0.00001   0.00001
 2     0.00184   0.00153   0.00000   0.00002
 3     0.00185   0.00150   0.00001  -0.00003
 4     0.00229   0.00077   0.00229   0.00077
 5     0.00232   0.00073   0.00232   0.00073
 6     0.00225   0.00067   0.00225   0.00067
 7     0.00278   0.00084   0.00278   0.00084
 8     0.00253   0.00003   0.00253   0.00003

      ---- Actual----       --- Estimated--
 I    Mag.       Angle      Mag.      Angle     Error BusVMeas(I)
      (pu)      (Deg.)      (pu)      (Deg.)     (%)
 1    0.00239    39.670   0.00001   113.939    99.85    0
 2    0.00239    39.794   0.00002    99.710    99.61    0
 3    0.00239    39.077   0.00003   -73.987   100.53    0
 4    0.00242    18.558   0.00242    18.558     0.00    0
 5    0.00243    17.409   0.00243    17.409     0.00    1
 6    0.00235    16.554   0.00235    16.554     0.00    1
 7    0.00290    16.901   0.00290    16.901     0.00    0
 8    0.00253     0.726   0.00253     0.726     0.00    1
 ----------------------------------------
 BusVMeas(I) = 0 : Unmeasured Voltage
               1 : Measured Voltage

          ********************************
          * Comparison of Branch Currents *
          ********************************


          ----------------- Branch Currents -------------
```

```
Brn From To  --Sending (Actual)--        ----Sending (HSE)----
No.          -Receiving (Actual)-        ----Receiving (HSE)--      Error
             Mag.       Angle(Deg.)        Mag.    Angle(Deg.)        (%)
1    1   2 S  0.68696E-03  -158.93      0.82073E-03   167.38          19.5
           R  0.66965E-03  -153.89      0.82036E-03   167.36          22.5
2    1   3 S  0.31113E-02   34.317      0.56665E-02   19.792          82.1
           R  0.31113E-02   34.317      0.56665E-02   19.792          82.1
3    2   3 S  0.25260E-02   30.407      0.42387E-02   14.145          67.8
           R  0.25260E-02   30.407      0.42387E-02   14.145          67.8
4    3   4 S  0.45443E-02   44.575      0.12784E-01   118.41         181.3
           R  0.44040E-02   35.658      0.12428E-01   118.71         182.2
5    4   5 S  0.42546E-02   39.734      0.42546E-02   39.734           0.0
           R  0.42111E-02   38.200      0.42111E-02   38.200           0.0
6    5   6 S  0.46678E-02  -47.693      0.46678E-02  -47.693           0.0
           R  0.47327E-02  -48.064      0.47327E-02  -48.064           0.0
7    6   7 S  0.10698E-01   110.01      0.10698E-01   110.01           0.0
           R  0.10593E-01   110.04      0.10593E-01   110.04           0.0
8    5   7 S  0.84848E-02   114.03      0.84848E-02   114.03           0.0
           R  0.82440E-02   114.23      0.82440E-02   114.23           0.0
9    5   8 S  0.11969E-01   26.698      0.11969E-01   26.698           0.0
           R  0.11896E-01   25.571      0.11896E-01   25.571           0.0
```

```
          *******************************
          * Comparison of Shunt Currents *
          *******************************
     ------------------ Shunt Currents -------------
Bus  -------Actual-------          ------Estimated------         Error
No.      Mag.    Angle(Deg.)        Mag.        Angle(Deg.)        (%)
 1   0.26725E-02   66.235         0.15052E-04    140.50          -99.4
 2   0.22571E-02   97.788         0.17891E-04    157.70          -99.2
 3   0.15194E-02   84.077         0.20348E-04   -28.987          -98.7
 4   0.34221E-03   63.558         0.34221E-03    63.558            0.0
 6   0.48399E-03   30.591         0.48399E-03    30.591            0.0
 7   0.29909E-02   30.937         0.29909E-02    30.937            0.0
 8   0.26098E-02   14.763         0.26098E-02    14.763            0.0
```

```
          ****************************************
          * Comparison of Nodal Current Injection *
          ****************************************

          ------ Actual ------          ----- Estimated ----          ------ Error ------
 I          Real           Imag             Real          Imag            Real             Imag
 1        0.00453 + j    0.00211        0.00453 + j    0.00211        0.00000 + j
0.00000
 2        0.00492 + j    0.00087        0.00492 + j    0.00087        0.00000 + j
0.00000
 3        0.00000 + j    0.00000       -0.01553 + j    0.00827       -0.01553 + j
0.00827
 4        0.00000 + j    0.00000        0.00955 + j   -0.00833        0.00955 + j    -
0.00833
 5        0.00707 + j    0.00707        0.00707 + j    0.00707        0.00000 + j
0.00000
 6       -0.00634 + j    0.01359       -0.00634 + j    0.01359        0.00000 + j
0.00000
 7        0.01000 + j   -0.01732        0.01000 + j   -0.01732        0.00000 + j
0.00000
 8       -0.00819 + j   -0.00574       -0.00819 + j   -0.00574        0.00000 + j
0.00000
```

```
       ---- Actual----         --- Estimated---
 I     Mag.       Angle        Mag.        Angle     Error
                  (Deg.)                   (Deg.)     (%)
 1     0.0050   25.0000      0.0050    25.0000      0.00
 2     0.0050   10.0000      0.0050    10.0000      0.00
 3     0.0000    0.0000      0.0176   151.9620   9999.99
 4     0.0000    0.0000      0.0127   -41.1066   9999.99
 5     0.0100   45.0000      0.0100    45.0000      0.00
 6     0.0150  115.0000      0.0150   115.0000      0.00
 7     0.0200  -60.0000      0.0200   -60.0000      0.00
 8     0.0100 -145.0000      0.0100  -145.0000      0.00
```

```
*******************************************
** Add Null Space to Particular Solution **
*******************************************
```

```
Using Voltage at Busbar 1 to Calculate
Null-space multipliers

 Particular Soln Real(V1)   =        -0.000005
 Actual Real(V1) =          0.001840
 V(1,FirstSingul)=          0.576553
 Multiplier K1=          0.003201

Particular Soln AIMAG(V1)   =         0.000012
Actual AIMAG(V1) =          0.000669
V(1,SecondSingul)=          0.576553
Multiplier K2=          0.002625

*** New Solution vector is ***
  0.001840  0.001838  0.001853  0.002294  0.002776  0.001526  0.001531  0.001505
0.000770  0.000844

BUSBAR VOLTAGES Solved for
Busbar   Voltage
No.  Real, Imag.  ---- Magnitude  Angle (Deg.)
 1   0.00184   0.00153   0.00239  39.67028
 2   0.00184   0.00153   0.00239  39.79367
 3   0.00185   0.00150   0.00239  39.07694
 4   0.00229   0.00077   0.00242  18.55759
 7   0.00278   0.00084   0.00290  16.90072

          *********************************
          * Comparison of Busbar Voltages *
          *********************************

      ---- Actual----        --- Estimated--
 I      Real        Imag      Real        Imag
 1    0.00184   0.00153   0.00184   0.00153
 2    0.00184   0.00153   0.00184   0.00153
 3    0.00185   0.00150   0.00185   0.00150
 4    0.00229   0.00077   0.00229   0.00077
 5    0.00232   0.00073   0.00232   0.00073
 6    0.00225   0.00067   0.00225   0.00067
 7    0.00278   0.00084   0.00278   0.00084
 8    0.00253   0.00003   0.00253   0.00003

      ---- Actual----        --- Estimated--
 I    Mag.       Angle      Mag.       Angle      Error BusVMeas(I)
      (pu)      (Deg.)      (pu)      (Deg.)      (%)
 1    0.00239    39.670   0.00239    39.670     0.00    0
 2    0.00239    39.794   0.00239    39.794     0.00    0
 3    0.00239    39.077   0.00239    39.077     0.00    0
 4    0.00242    18.558   0.00242    18.558     0.00    0
 5    0.00243    17.409   0.00243    17.409     0.00    1
 6    0.00235    16.554   0.00235    16.554     0.00    1
 7    0.00290    16.901   0.00290    16.901     0.00    0
 8    0.00253     0.726   0.00253     0.726     0.00    1
-----------------------------------
BusVMeas(I) = 0 : Unmeasured Voltage
              1 : Measured Voltage

          *********************************
          * Comparison of Branch Currents *
          *********************************

                   ------------------ Branch Currents -------------
Brn From To   --Sending (Actual)--          ----Sending (HSE)----
No.           -Receiving (Actual)-          ----Receiving (HSE)--        Error
              Mag.       Angle(Deg.)          Mag.     Angle(Deg.)        (%)
1   1   2 S   0.68696E-03  -158.93         0.68696E-03  -158.93           0.0
          R   0.66965E-03  -153.89         0.66965E-03  -153.89           0.0
2   1   3 S   0.31113E-02   34.317         0.31113E-02   34.317           0.0
          R   0.31113E-02   34.317         0.31113E-02   34.317           0.0
3   2   3 S   0.25260E-02   30.407         0.25260E-02   30.407           0.0
          R   0.25260E-02   30.407         0.25260E-02   30.407           0.0
4   3   4 S   0.45443E-02   44.575         0.45443E-02   44.575           0.0
          R   0.44040E-02   35.658         0.44040E-02   35.658           0.0
5   4   5 S   0.42546E-02   39.734         0.42546E-02   39.734           0.0
          R   0.42111E-02   38.200         0.42111E-02   38.200           0.0
6   5   6 S   0.46678E-02  -47.693         0.46678E-02  -47.693           0.0
```

			R	0.47327E-02	-48.064	0.47327E-02	-48.064	0.0
7	6	7	S	0.10698E-01	110.01	0.10698E-01	110.01	0.0
			R	0.10593E-01	110.04	0.10593E-01	110.04	0.0
8	5	7	S	0.84848E-02	114.03	0.84848E-02	114.03	0.0
			R	0.82440E-02	114.23	0.82440E-02	114.23	0.0
9	5	8	S	0.11969E-01	26.698	0.11969E-01	26.698	0.0
			R	0.11896E-01	25.571	0.11896E-01	25.571	0.0

****
* Comparison of Shunt Currents *
****

Bus No.	Shunt Currents: Actual Mag.	Actual Angle(Deg.)	Estimated Mag.	Estimated Angle(Deg.)	Error (%)
1	0.26725E-02	66.235	0.26725E-02	66.235	0.0
2	0.22571E-02	97.788	0.22571E-02	97.788	0.0
3	0.15194E-02	84.077	0.15194E-02	84.077	0.0
4	0.34221E-03	63.558	0.34221E-03	63.558	0.0
6	0.48399E-03	30.591	0.48399E-03	30.591	0.0
7	0.29909E-02	30.937	0.29909E-02	30.937	0.0
8	0.26098E-02	14.763	0.26098E-02	14.763	0.0

****
* Comparison of Nodal Current Injection *
****

I	Actual Real		Actual Imag	Estimated Real		Estimated Imag	Error Real		Error Imag
1	0.00453	+ j	0.00211	0.00453	+ j	0.00211	0.00000	+ j	0.00000
2	0.00492	+ j	0.00087	0.00492	+ j	0.00087	0.00000	+ j	0.00000
3	0.00000	+ j	0.00000	0.00000	+ j	0.00000	0.00000	+ j	0.00000
4	0.00000	+ j	0.00000	0.00000	+ j	0.00000	0.00000	+ j	0.00000
5	0.00707	+ j	0.00707	0.00707	+ j	0.00707	0.00000	+ j	0.00000
6	-0.00634	+ j	0.01359	-0.00634	+ j	0.01359	0.00000	+ j	0.00000
7	0.01000	+ j	-0.01732	0.01000	+ j	-0.01732	0.00000	+ j	0.00000
8	-0.00819	+ j	-0.00574	-0.00819	+ j	-0.00574	0.00000	+ j	0.00000

I	Actual Mag.	Actual Angle (Deg.)	Estimated Mag.	Estimated Angle (Deg.)	Error (%)
1	0.0050	25.0000	0.0050	25.0000	0.00
2	0.0050	10.0000	0.0050	10.0000	0.00
3	0.0000	0.0000	0.0000	0.0000	0.00
4	0.0000	0.0000	0.0000	0.0000	0.00
5	0.0100	45.0000	0.0100	45.0000	0.00
6	0.0150	115.0000	0.0150	115.0000	0.00
7	0.0200	-60.0000	0.0200	-60.0000	0.00
8	0.0100	-145.0000	0.0100	-145.0000	0.00

INDEX